ADVANCES IN CHILD DEVELOPMENT AND BEHAVIOR

Series Editor

JANETTE B. BENSON

Department of Psychology,
Morgridge College of Education,
University of Denver,
Denver, CO, United States

VOLUME FIFTY THREE

ADVANCES IN
CHILD DEVELOPI
AND BEHAVIOR

The Development of Ear
Mathematics Education

VOLUME FIFTY THREE

ADVANCES IN CHILD DEVELOPMENT AND BEHAVIOR

The Development of Early Childhood Mathematics Education

Edited by

JULIE SARAMA
University of Denver, Denver, CO, United States

DOUGLAS H. CLEMENTS
University of Denver, Denver, CO, United States

CARRIE GERMEROTH
University of Denver, Denver, CO, United States

CRYSTAL DAY-HESS
University of Denver, Denver, CO, United States

ELSEVIER

ACADEMIC PRESS
An imprint of Elsevier

Academic Press is an imprint of Elsevier
50 Hampshire Street, 5th Floor, Cambridge, MA 02139, United States
525 B Street, Suite 1800, San Diego, CA 92101-4495, United States
The Boulevard, Langford Lane, Kidlington, Oxford OX5 1GB, United Kingdom
125 London Wall, London EC2Y 5AS, United Kingdom

First edition 2017

Notices
Knowledge and best practice in this field are constantly changing. As new research and experience broaden our understanding, changes in research methods, professional practices, or medical treatment may become necessary.

Practitioners and researchers must always rely on their own experience and knowledge in evaluating and using any information, methods, compounds, or experiments described herein. In using such information or methods they should be mindful of their own safety and the safety of others, including parties for whom they have a professional responsibility.

To the fullest extent of the law, neither the Publisher nor the authors, contributors, or editors, assume any liability for any injury and/or damage to persons or property as a matter of products liability, negligence or otherwise, or from any use or operation of any methods, products, instructions, or ideas contained in the material herein.

ISBN: 978-0-12-813448-1
ISSN: 0065-2407 (Series)

For information on all Academic Press publications
visit our website at https://www.elsevier.com/books-and-journals

Publisher: Zoe Kruze
Acquisition Editor: Kirsten Shankland
Editorial Project Manager: Andrea Gallego Ortiz
Production Project Manager: Magesh Kumar Mahalingam
Cover Designer: Mark Rogers

Typeset by SPi Global, India

CONTENTS

CONTRIBUTORS

Arthur J. Baroody
University of Illinois at Urbana-Champaign, Champaign, IL; University of Denver, Denver, CO, United States

Allison M. Bock
Center for Early Education and Development, University of Minnesota, Minneapolis, MN, United States

Jenny Yun-Chen Chan
Institute of Child Development, University of Minnesota, Minneapolis, MN, United States

Douglas H. Clements
University of Denver, Denver, CO, United States

Crystal Day-Hess
University of Denver, Denver, CO, United States

Ann Dowker
Oxford University, Oxford, United Kingdom

Carrie Germeroth
University of Denver, Denver, CO, United States

Ann Gervasoni
Monash University, Melbourne, Vic; Charles Sturt University, Albury, NSW, Australia

Robin Jacob
University of Michigan, Ann Arbor, MI, United States

Michelle Maier
MDRC, New York, NY, United States

Shira K. Mattera
MDRC, New York, NY, United States

Michèle M.M. Mazzocco
Institute of Child Development, University of Minnesota, Minneapolis, MN, United States

Pamela A. Morris
MDRC; New York University, New York, NY, United States

Bob Perry
Monash University, Melbourne, Vic; Charles Sturt University, Albury, NSW, Australia

Natalia Rojas
New York University, New York, NY, United States

Julie Sarama
University of Denver, Denver, CO, United States

PREFACE

Increased recognition of the importance of early mathematics had resulted in increased attention to interventions hoping to build the foundation for development and learning. Fortunately, research on young children's thinking and learning of mathematics and the interventions that support them are increasingly available and many of the most promising are described in this volume.

That is, this volume is a compilation of recent research and development in this critical area conducted by scholars doing some of the most important work in early mathematics interventions. Interventions in early mathematics are important to *all* children, and specially designed interventions are needed for certain populations—this is a critical equity issue. In the remainder of this preface, we introduce each chapter.

Crystal Day-Hess and Douglas H. Clements describe a new network conceptualized and funded by the Heising–Simons Foundation: The Development and Research in Early Mathematics Education (DREME) Network, whose purpose is to advance the field of early mathematics research and improve the opportunities to develop math competencies offered to children birth through age 8 years, with an emphasis on the preschool years. The work of the four main network projects has implications for interventions. After introducing the DREME Network, its goals, and the four projects, the authors provide an important early result of one of the projects, which is directly developing an intervention for children, based on selecting high-quality instructional activities culled from the growing number of math curriculum resources. The team reviewed 457 activities from six research-based math curricula and found a wide divergence across curricula in all these categories and provide comprehensive reports for those interested in selecting, using, or developing early mathematics curricula.

Arthur J. Baroody addresses four key issues regarding the important and often misunderstood issue of *concrete instruction*: What is concrete? What is a worthwhile concrete experience? How can concrete experiences be used effectively in early childhood mathematics instruction? Is there evidence such experiences work? Dr. Baroody argues that concrete experiences are those that build what is familiar to a child and can involve objects, verbal analogies, or virtual images, and thus, we must consider children's internal factors in using "concrete" instruction.

Julie Sarama and Douglas H. Clements discuss the development, implementation, and dissemination/scale up of interventions in early mathematics across the challenging diversity of populations and contexts in the early childhood system in the United States. Emphasizing their work with underrepresented populations, they analyze a theoretical model, its implementation, and its evaluation that appeared to avoid the pollution and dilution that often plagues efforts to achieve broad success.

In a related work, Carrie Germeroth and Julie Sarama document that teachers make the difference in such large-scale efforts, but even providing stand-alone professional development is usually not effective in changing practice. What is needed is site-embedded ongoing support in the form of coaching or mentoring. Their theoretical framework and extension work with two large projects allow them to summarize what is known and describe implications and challenges for the field.

Ann Gervasoni and Bob Perry describe a similarly successful and influential project from Australia. *Let's Count* is a preschool mathematics intervention implemented with "disadvantaged" communities across Australia. Let's Count has shown impressive positive effects on mathematics learning and dispositions of young children, early childhood educators, and families. Results suggest that when adults notice children's mathematics, then children's learning thrives, and the positive dispositions and confidence of parents and educators increase.

Shira Mattera, Pamela Morris, Robin Jacob, Michelle Maier, and Natalia Rojas discuss a very large-scale project in New York City, *Making Pre-K Count.* Their careful design allowed them to not only causally test the effects of early math skills but also to examine a number of additional questions about scale-up, the influence of contextual factors and the counterfactual environment, the mechanism of long-term fade-out, and the role of measurement in early childhood intervention findings.

The final two chapters address children with identified special needs. Building on her extensive work in this area, Ann Dowker reviews research on the diverse interventions for primary school children with difficulties in mathematics. She considers intensity, context, instructional roles, games, and formative assessment, among other factors. She also discusses the important issue of interventions whose goal is to help children learn mathematics, but via training children in domain-general skills, such as Piagetian operations, metacognition, and executive functions.

Michèle Mazzocco, Jenny Yun-Chen Chan, and Allison Bock address the relations between early mathematics and a domain-general skill:

executive function (EF). Although they document that EF skills seem to be closely related to mathematics, they take a unique view that individual differences in the nature of this relation should be considered when studying the role of EF skills in early mathematics. Are all children high in one domain high in the other and vice versa? If not, what does this imply for interventions?

These authors and their chapters represent the most current and best thinking in early childhood education, mathematics education, special education, and cognitive and developmental psychology. The broad topics provide a comprehensive view of what is currently available. In straightforward language that does not simplify, but directly illustrates and explains, the authors create a roadmap for other researchers, practitioners, curriculum developers, trainers, and policy makers to inform the development of early math curricula, professional development, and interventions.

Julie Sarama,
Douglas H. Clements,
Carrie Germeroth, and Crystal Day-Hess

CHAPTER ONE

The DREME Network: Research and Interventions in Early Childhood Mathematics

Crystal Day-Hess[1], Douglas H. Clements
University of Denver, Denver, CO, United States
[1]Corresponding author: e-mail address: crystal.day-hess@du.edu

Contents

Abstract

The DREME Network was created to advance the field of early mathematics research and improves the opportunities to develop math competencies offered to children birth through age 8 years, with an emphasis on the preschool years. All four main Network projects will have implications for interventions. Section 1 introduces the Network and its four projects. The remainder of the chapter focuses on one of these four projects, Making More of Math (MMM), in depth. MMM is directly developing an intervention for children, based on selecting high-quality instructional activities culled from the burgeoning curriculum resources. We first report a review of 457 activities from 6 research-based curricula, which describes the number of activities by content focus, type (nature), and setting of each activity. Given the interest in higher-order thinking skills and self-regulation, we then identified activities that had the potential to, develop both mathematics and executive function (EF) proficiencies. We rated these, selecting the top 10 for extensive coding by mathematics content and EF processes addressed. We find a wide divergence across curricula in all these categories and provide comprehensive reports for those interested in selecting, using, or developing early mathematics curricula.

Advances in Child Development and Behavior, Volume 53
ISSN 0065-2407
http://dx.doi.org/10.1016/bs.acdb.2017.03.002

Responding to increased interest in early mathematics education, the DREME Network was created to advance the field through research and development. All four main Network projects will have implications for interventions. This chapter introduces the Network and its projects and describes one project in detail: Making More of Math (MMM), which is directly developing an intervention for children.

1. THE DREME NETWORK

The Development and Research in Early Mathematics Education (DREME) Network was created to advance the field of early mathematics research and improve young children's opportunities to develop math skills—providing important information and resources to inform mathematics-focused interventions at multiple levels. The Network focuses on math from birth through age 8 years, with an emphasis on the preschool years. Network members and colleagues collaborate to conduct basic and applied research and develop innovative tools that address high-priority early math topics and inform and motivate other researchers, educators, policymakers, and the public (https://dreme.stanford.edu/).

1.1 A Need for More Opportunity and Access

There is a critical need for new knowledge and resources to guide and facilitate effort to promote young children's math learning and increase equity and excellence in math achievement (National Research Council, 2009). Although important research has been conducted in recent years, it does not adequately address the need for effective math instruction as an intervention, especially for preschool-aged children. It also does not speak to the organizational conditions that enable teachers to deliver this kind of instruction and intervention. Recent developments in the fields of education and developmental science make this an opportune time to invest in building the field of research and development on early math to support teaching and learning interventions.

1.2 New Evidence on the Importance of Early Math

Several empirical studies have found that compared to literacy and social–emotional development at kindergarten entry, early math concepts were the most powerful predictors of later learning (Duncan, Claessens, & Engel, 2004; Duncan et al., 2007; Watts, Duncan, Siegler, & Davis-Kean, 2014), perhaps making them a prime candidate for interventions to support multiple areas of learning and growth. Across the nation, most children who have low math scores in their kindergarten year continue to lag behind their better-prepared peers throughout their school years (Clements & Sarama, 2014; National Research Council, 2009). The least prepared and the lowest performing are disproportionately children from low-income families, who are disproportionately children of color (Clements & Sarama, 2009, 2014; Denton & West, 2002; National Research Council, 2009). Additionally, children with developmental delays and disabilities often struggle in math throughout their school years due to the lack of appropriate resources (e.g., Geary, 2013; Kovas, Haworth, Dale, & Plomin, 2007; Olkun, Altun, Göçer Şahin, & Akkurt Denizli, 2015). Clearly, any serious intervention effort to close the achievement gap needs to include children before school entry and in the early elementary grades.

1.3 Increasing Acceptance of Academic Instruction in Preschool

In the past, there has been resistance to discipline-based instruction in preschool, in favor of learning through play and a focus on social–emotional development. A review of the National Association for the Education of Young Children (NAEYC) Guidelines over the last few decades, however, reveals a shift toward productively integrating academic instruction with playful learning and efforts to develop social skills (Clements, Fuson, & Sarama, 2017; Fuson, Clements, & Sarama, 2015; Sarama & Clements, 2009a). This shift toward embracing academic instruction is clearly evident in reading, less so in math. But the general acceptance that young children can and should learn basic academic skills provides fertile ground for promoting effective and developmentally appropriate strategies for teaching math to preschool-aged children. This is consistent with research showing that, given opportunities to learn, young children can develop an informal knowledge of mathematics that is surprisingly broad, complex, and sophisticated (National Research Council, 2009; Sarama & Clements, 2009b; Thomson, Rowe, Underwood, & Peck, 2005). Young children are also interested in mathematics; for example, preschoolers engage it in spontaneously in their free play (Seo & Ginsburg, 2004). However, high-quality, playful mathematics instruction, rather than free-play time, promotes children's learning (Geary & Liu, 1996; Klein, Starkey, Clements, Sarama, & Iyer, 2008). Such instruction is effective (Case, Griffin, & Kelly, 1999; Clements & Sarama, 2007, 2011; Klein et al., 2008; Lewis Presser, Clements, Ginsburg, & Ertle, 2015), scalable (Clements, Sarama, Spitler, Lange, & Wolfe, 2011), and capable of addressing the severe equity problem in US early mathematics education (Morgan, Farkas, Hillemeier, & Maczuga, 2014; National Research Council, 2009; Rouse, Brooks-Gunn, & McLanahan, 2005). Thus, enhancing learning is not an "imposition" on children nor a "schoolification" of preschool, but a positive contribution, especially for those children with the fewest opportunities to learn outside of school.

1.4 DREME's Central Goals

With these current developments in mind, one of the main goals of the DREME Network is to convene a national network of scholars and researchers to conduct rigorous basic and applied research and development projects that address high-priority early math topics needed to inform and motivate other researchers, educators, policymakers, and the public. The DREME Network consists of four projects with the potential to inform mathematics intervention at multiple levels—child, parent, teacher, school, and district.

1.4.1 Increasing Capacity: Creating Resources for Early Childhood Teacher Educators

Improving early math learning requires teachers to know the content, understand children's thinking, engage in pedagogical practices that support learning, and see themselves as capable math teachers. Currently, most preservice teachers do not have access to robust course offerings in early childhood math and there are few faculty members at the college level who have sufficient expertise in early math teaching and learning to teach the needed courses (ATME, 2017; Sarama & DiBiase, 2004). In-service professional development opportunities are also few and far between in part because not enough experts in early math teaching are available to provide them.

To support the training of prospective and practicing early childhood teachers, the Network is creating a system of resources that will be available on the web. The resources are organized into flexible professional development modules that can be used in a variety of settings (e.g., live or online college courses, continuing education institutes, and ongoing in-service workshops). Each module will offer a variety of materials designed to help students (prospective or practicing preschool teachers) understand a specific mathematical topic. Ultimately, there will be six modules on the website: Overview, Counting, Spatial Relations, Operations, Patterns/Algebra, and Measurement/Data. The website will include be a variety of materials which provide activities, worksheets, handouts, and other resources (both video and published) developed for the Teacher Educator to use. The resources are designed to help Teacher Educators support prospective and practicing teachers of young children develop learning opportunities that are intentional, appropriate, engaging, and playful.

1.4.2 MMM Instruction: Using Math Activities to Support Math and Executive Function Skills in Early Childhood

Empirical research has shown a link between mathematical thinking in early childhood and executive function (EF) skills (Clements, Sarama, & Germeroth, 2016). EF skills share a focus on deliberate and effortful cognitive processes, including the ability to maintain attention during a task (sustained attention), the ability to resist making an automatic or desired response (inhibitory control), the flexibility to consider multiple perspectives (cognitive flexibility), and the ability to simultaneously think about and manipulate information in our "mental workspace" (working memory) (Zelazo, Craik, & Booth, 2004). In other words, EF skills enable children to pay attention, wait their turn, consider different approaches to solving problems, and remember and execute many steps when following directions. Thus, EF skills play a critical role in math as well as other domains. Moreover, both math and EF skills are linked to a wide range of important long-term academic outcomes.

Preschool EF skills predict later math achievement, and students who excel at math tend to also have strong EF skills (Clements et al., 2016). It is clear that an association between math and EF exists, but it is not fully understood. Do these skills need to be taught separately? Does one cause the other? Are they mutually reinforcing? Does support for EF skills matter more for children who are at risk for low math achievement? How can competencies in math and EF be enhanced through high-quality early childhood education?

Based on what is known about EF and its development, the Network is designing a variety of mathematical activities to test how these activities shape children's mathematical thinking and EF behaviors. Our goal is to develop instructional materials that help teachers prioritize teaching practices that simultaneously promote the development of children's mathematical thinking and EF. The knowledge gained from this work will also be useful in guiding parents' interactions with children. Section 2 reports on one major accomplishment of this project.

1.4.3 Parents' and Early Caregivers' Engagement in Math Activities With Young Children

At home, children begin their initial explorations into everyday mathematics, progressively developing and refining their skills as well as their learning expectations, motivation, and beliefs about math and their own math ability

(National Research Council, 2009). Yet, there is wide variation—linked in part to socioeconomic status and culture—in the kinds of math learning experiences children enjoy at home and the ways parents stimulate their children's development and academic learning (Clements & Sarama, 2014). Further, two-thirds of young children today spend significant time in nonparental care, which includes family childcare as well as organized preschool. Quantity and quality of math learning stimulation in these early childhood care and education (ECCE) settings vary enormously and have lasting consequences for achievement not only for math but also for reading. Relative to the large body of research on language and literacy, less attention has been given to young children's learning at home and in ECCE settings.

The DREME Network seeks to address these limitations. The Network's goal is to advance knowledge on effective ways to increase the quantity and quality of parents' and caregivers' engagement in their young children's math learning. Network collaborators are focusing specifically on caregivers of children growing up in poverty, many of whom are not native English speakers. The initial phases of this project blend research and development as a first step toward the experimental testing of books, games, and activities designed to maximize high-quality math interactions with preschool children. Collaborators are conducting observations of parent and informal childcare teachers engaging in math activities with young children, using materials that are widely available at present or have been developed by Network members. The goal is to determine which activities hold the most promise for increasing children's math learning. Additionally, collaborators are piloting, and will ultimately test, teaching modules designed to help parents and caregivers engage in productive "math talk" with young children and to understand the development of children's mathematical thinking.

Early work on this project has been focused on observing parents and childcare teachers engaged in play and reading activities with children. In doing so, we have observed caregiver–child interacting with a variety of learning materials, storybooks (paperback and electronic), and counting books. In addition, the team has been analyzing the mathematical affordances of learning activities, storybooks, and counting books. From these studies, it is evident that there is considerable variability across caregivers in the amount of math talk and scaffolding of numerical and spatial skills that they provide. Moreover, there is considerable variability in the quality of learning affordances provided by toys and books that were, seemingly, designed to support early numerical or spatial skill development. In the illustrations of many storybooks that explicitly deal with counting, for example, there are often mismatches between number words and numerical representations of objects or highly cluttered scenes in which the number of objects is difficult for even advanced learners to identify. While the study is in its earliest stages, we are seeing that caregivers could often benefit from increased knowledge and support, and perhaps reduced anxiety, when it comes to engaging with their young children in ways that could promote early math learning. In addition, we are finding considerable room for improving the affordances of commonly available learning materials and books, as well as a need to help caregivers select the highest quality learning materials for early math learning.

1.4.4 Preschool–Elementary Continuity and Coherence

Policy makers, educators, and researchers complained that math learning in preschool is often disconnected from math learning in the early elementary grades. This disconnect can lead to students experiencing uneven instructional practices which can compromise their learning. In response, some state and district policy makers across the country are working toward creating greater policy alignment in elements of policy affecting pre-K and elementary schools. Others focus on creating greater curricular coherence by coordinating learning goals and curriculum within and across grades. But there is limited research on the impact of these policies and practices on students' experiences and learning outcomes as they move from preschool through the early elementary grades (Perry, MacDonald, & Gervasoni, 2015).

To address these limitations, the Network is investigating the relationship between efforts to create policy alignment and curricular coherence on students' learning opportunities, experiences, and achievement in mathematics from pre-K through grade 2. The Network is developing and piloting measures of children's mathematical knowledge, experiences of mathematics instruction, their opportunities to learn (curriculum and teachers' instructional practices), and their understanding of the nature of math. Collaborators have conducted cross-sectional analyses of children pre-K through grade 2 in a district that is working toward policy alignment and curricular coherence. In longitudinal studies of two districts endeavoring to improve coherence, they are using our newly developed and refined measures to assess the degree to which students in these districts have coherent learning opportunities and experiences across grades in a given school. In particular, they are studying how district-level policies are mediated by school-level policies and practices to affect children's math learning experiences and outcomes.

The four projects vary on a number of dimensions as well as in their emphasis on research vs development. Perhaps most important, there is synergy among the projects; the work that is undertaken in one will be used to inform others to, ultimately improve and support children's mathematical learning and knowledge.

2. MMM INSTRUCTION: AN IN-DEPTH LOOK AT A DREME NETWORK PROJECT

This section provides a detailed look at one DREME Network project, *MMM Instruction: Using Math Activities to Support Math and EF Skills in Early Childhood*. We present an overview of the project and then describe one core research activity that we have completed.

2.1 Early Math, EF, and Curricula

There is presently a burgeoning interest in early mathematics. However, there is a wide divergence in both general perspectives toward children's development and learning of mathematical proficiencies and in the nature and content of available mathematics curricula. As part of a larger study, we analyzed and described the activities constituting six research-based preschool curricula. We then rated these for their potential to develop not only mathematical concepts and procedures, but also EF processes, given the comutual benefits documented between the development of mathematical and EF proficiencies (and the importance of the latter for school success).

2.1.1 Early Mathematics Curricula

Although several new research-based preschool curricula have been developed and most evaluated (see Table 1), there are no analyses of the similarities and differences among these efficacious programs and the instructional activities of which they are constituted. Such analyses should include the content focus (e.g., What topics and standards are addressed?), setting (e.g., whole or small group, learning center, educational technology), and type and nature (e.g., didactic, game, construction) of the instructional activities. Before describing our approach to such analyses, we discuss a unique focus of our approach, the interactions between early mathematics and EF.

Table 1 Evaluations of Early Mathematics Programs

Program/ Curriculum	Content Assessed	Age/ Grade	SES	No. of Children	Design	Results	Comments	References
Building Blocks	Math	PK	Low (Head Start and state funded)	68	No random assignment; quasiexperimental comparison	Moderate effect size (ES =0.85) for number, large (1.47) for geometry	Caveats regeneralizability: a superrealization (Cronbach et al., 1980, p. 109); small scale; teachers had access to research assistants	Clements and Sarama (2007)
Building Blocks	Math	PK	Low (Head Start and state funded)	276	Random assignment to group groups: Building Blocks, comparison (different curriculum intervention with professional development), and control	Large effect size (1.07) compared to business-as-usual control and medium effect size (0.47) compared to the different mathematics curriculum (pre-K mathematics curriculum)	Benefitted both low- and middle-SES children equally	Clements and Sarama (2008)
TRIAD/ Building Blocks	Math Language	PK	Low (two cities in different states, about 50% African-American, 25% Latino)	1375	Random assignment of schools to three treatments, TRIAD with Follow Through (TRIAD-FT), TRIAD with no follow through (TRIAD-NFT), and business-as-usual control. At pre-K the two TRIAD groups were identical	Moderate effect (0.72) on math, with stronger positive benefits African-American children. On language, positive effects, statistically significant on four of six subscores (0.16 to 0.36)	Research focused on the TRIAD model of scale-up rather than the curriculum. Classroom culture specific to mathematics and the use of the Building Blocks software were mediators of the effects	Clements et al. (2011) and Sarama, Lange, Clements, and Wolfe (2012)

Continued

Table 1 Evaluations of Early Mathematics Programs—cont'd

Program/Curriculum	Content Assessed	Age/Grade	SES	No. of Children	Design	Results	Comments	References
Follow Through (to Building Blocks)	Math	K	As above	1305	As above	Both TRIAD groups higher than control (ES = 0.38 TRIAD-FT and 0.30 TRIAD-NFT)	As predicted, effect sizes decreased after pre-K, but less so in the TRIAD-FT than the TRIAD-NFT group. However, TRIAD-FT was not statistically different from TRIAD-NFT (ES = 0.14). The FT treatment may have been too little and teachers may have resisted changing their just-adopted (undemanding) K math curriculum	Sarama, Clements, Wolfe, and Spitler (2012)
TRIAD Follow Through (to Building Blocks)	Math	1	As above	1305	As above	Both TRIAD groups higher than control (ES = 0.51 TRIAD-FT and 0.28 TRIAD-NFT) and FT higher than NFT (ES = 0.24)	The effects of TRIAD-FT increased in first grade (compared to Kindergarten), possibly because the early conceptual work formed a foundation for new mathematical topics (e.g., arithmetic)	Clements, Sarama, Wolfe, and Spitler (2013)

Building Blocks + OWL	Math Literacy Executive Function Emotions	PK	Low	2018	Comparison of children who "just made" and "just missed" the age cutoff (quasiexperimental regression-discontinuity design)	Medium-to-moderate impacts on children's language, literacy, numeracy and mathematics skills, and small impacts on children's executive functioning and a measure of emotion recognition	Cannot attribute results to any single aspect of the program, including a coaching system and consistent literacy (OWL), language, and mathematics (Building Blocks) curricula	Weiland and Yoshikawa (2013)
Big Math for Little Kids (BMLK)	Math Language	PK and K	Low	762	Randomly assigned childcare centers to BMLK or control	Medium effect size (0.32) on math; informal language evidence was positive	Two years of the curriculum throughout pre-K and kindergarten	Lewis Presser et al. (2015)
Rightstart	Math	K-1	Low, mixed language	~100 in all Rightstart groups; ~100 comparison in some studies	No random assignment; matched controls in some studies	In separate studies, higher number of children passed a number knowledge test than in comparison groups. In one, Rightstart children surpassed both a second low-SES group and a mixed-SES group	Caveats regarding generalizability: a superrealization; most teachers were research assistants; teachers received substantial help from the program developers; small scale; and expert teachers certain components of the curriculum are difficult to implement (Gersten et al., 2008)	Griffin and Case (1997) and Griffin, Case, and Siegler (1994)

Continued

Table 1 Evaluations of Early Mathematics Programs—cont'd

Program/ Curriculum	Content Assessed	Age/ Grade	SES	No. of Children	Design	Results	Comments	References
Pre-K Mathematics Curriculum + Building Blocks (software)	Math	PK	Low	276	Classrooms randomly assigned to intervention or control	Math intervention higher than control with moderate effect (0.55)	First year of implementation. Caveats: teachers were volunteers, they received substantial support, and this was not taken to scale. Differences were not detectable a year later	Klein et al. (2008) and Preschool Curriculum Evaluation Research Consortium (2008)
Pre-K Mathematics Curriculum	Math	PK	Low and middle		No random assignment	Math intervention higher than control with moderate effects	Affected both low- and middle-SES children. Caveats: Unknown internal validity, as comparison group was from a different year and was never pretested. Lower effects than Building Blocks in a comparison (Clements, 2008)	Starkey, Klein, and Wakeley (2004)

2.2 EF and Mathematics

Although there has been much recent attention to young children's development of EF and early mathematics, few studies have integrated the two. We first briefly describe EF, then its relationship to mathematics learning in the early years. Researchers and other educators have used the term EF to refer to the processes involved in intentionally controlling ones' impulses, attention, thinking, and behavior. Although the field lacks a common set of processes and definitions, the broad concept of EF can be viewed as a unit functionally, but has also been analyzed into several processes (e.g., Best, Miller, & Naglieri, 2011; Huizinga, Dolan, & van der Molen, 2006; Lehto, Juujarvi, Kooistra, & Pulkkinen, 2003; Miyake, Friedman, Emerson, Witzki, & Howerter, 2000; Raver, 2013; Schoemaker, Bunte, Espy, Deković, & Matthys, 2014). Three processes are frequently distinguished: (a) attention shifting and cognitive flexibility, or switching a "mental set" from one aspect of a situation to another as the situation requires; (b) inhibitory control, involving suppressing unproductive responses or strategies, such as controlling a proponent response to think about better strategies or ideas; and (c) working memory, a system that is responsible for the short-term holding and processing of information. Several studies have shown positive correlations between EF and achievement in young children (e.g., Alloway & Alloway, 2010; Alloway, Alloway, & Wootan, 2014; Bierman, Nix, Greenberg, Blair, & Domitrovich, 2008; Blair, Protzko, & Ursache, 2011; Blair & Razza, 2007; Bull, Espy, Wiebe, Sheffield, & Nelson, 2011; Bull, Johnston, & Roy, 1999; Cameron et al., 2012; Cerda, Im, & Hughes, 2014; Cortina et al., 2013; Gathercole & Pickering, 2000; Gawrilow et al., 2014; McClelland et al., 2014; Ng, Tamis-LeMonda, Yoshikawa, & Sze, 2014; Roebers, Cimeli, Röthlisberger, & Neuenschwander, 2012; Welsh, Nix, Blair, Bierman, & Nelson, 2010), although there are exceptions (e.g., Edens & Potter, 2013). This has led some to posit that EF is foundational to learning and must be developed before academic instruction (Best et al., 2011; e.g., Clark, Pritchard, & Woodward, 2010; for a review, see Clements et al., 2016; LeFevre et al., 2013). For this to be practical, of course, EF must be malleable.

EF seems to be amenable to learning through instruction, such as with computer games (Otero, Barker, & Naglieri, 2014; Razza & Raymond, 2015; Rueda, Rothbart, McCandliss, Saccomanno, & Posner, 2008), direct training of specific EF tasks (Dowsett & Livesey, 2000; Espinet, Anderson, & Zelazo, 2012; Röthlisberger, Neuenschwander, Cimelia, Michel, &

Roebers, 2012), or particular curricula (e.g., Bierman, Domitrovich, et al., 2008; Bierman, Nix, et al., 2008; Diamond, Barnett, Thomas, & Munro, 2007; Diamond & Lee, 2011; Klingberg, 2009; Lillard & Else-Quest, 2007; Lyons & Zelazo, 2011; Otero et al., 2014; Perels, Merget-Kullmann, Wende, Schmitz, & Buchbinder, 2009; Raver et al., 2011; Riggs, Greenberg, Kusché, & Pentz, 2006; Weiland, Ulvestad, Sachs, & Yoshikawa, 2013; Weiland & Yoshikawa, 2013). However, effects and any transfer to other tasks are often small and sometimes nonexistent even when the curriculum is implemented with good fidelity (Clements, Sarama, Layzer, Unlu, & Germeroth, 2012; Farran, Lipsey, & Wilson, 2011; Lonigan & Phillips, 2012; Morris et al., 2014). Thus, there appears to be some successful strategies for developing EF in young children, although the training and the results may need to be narrowly focused and persistence of these effects has not been evaluated (Clements et al., 2016). There is also surprisingly little evidence for a causal relationship from EF to achievement (Clements et al., 2016; Jacob & Parkinson, 2015).

Further, although gains on EF may be important outcomes per se, both EF processes and subject-matter proficiencies are required to support continued learning and problem solving. Thus, it is an open question whether it would be more effective to teach EF first and then content such as mathematics, or to develop both simultaneously. A small set of studies, however, suggests that teaching mathematics may develop both mathematics and EF proficiencies (Clements et al., 2017; Farran et al., 2011; Weiland & Yoshikawa, 2013).

In conclusion, developing both EF processes and mathematical proficiencies appear to be essential for young children's development and there is a suggestion in the research corpus that high-quality mathematics education may have the dual benefit of teaching this important content area and developing EF processes. This open question led to the wider DREME project of which this study is a part.

2.3 Preliminary EF Study Work

Given the possibility that mathematics instruction may have the additional benefit of increasing children's EF proficiencies, but little rigorous evaluations of that possibility, we, along with other DREME colleagues, decided to conduct a series of studies to test the causal connection. We began by acquiring and describing all the activities in six research-based preschool mathematics curricula. We then evaluated each of these in an interesting process to identify, and then analyze, a subset that had the potential to develop both substantial mathematics and EF proficiencies. We planned to conduct our research and development activities in four phases.

In phase 1, we described and coded all activities in research-based preschool mathematics curricula. In phase 2, we evaluated all activities coded in phase 1 to reduce these to a small proper subset that had the potential to develop both substantial mathematics and EF proficiencies. We then intensely coded and analyzed these so others could benefit from our descriptions and to identify commonalities among this select subset. In phase 3, we developed and piloted newly developed archetype activities. Phase 4, which is currently underway, involves conducting microgenetic studies to further evaluate and refine the activities.

2.3.1 Phase 1: Description of Activities

Level 0. Six research-based preschool mathematics curricula were identified from the literature as having a strong theoretical and empirical base; from these, 457 activities were identified (Level 0 Coding, Table 2, note that *Everyday Math* was including because of the strong bases, http://everydaymath.uchicago.edu/research, although we know of no study focused only on pre-K). Note that the number of activities is a rudimentary measure, as some activities are simple and short, while others have multiple parts and are conducted over multiple days. We then coded the activities according to their topic content (i.e., adding/subtracting, comparing/ordering, counting, data/graphing, geometry/spatial, measurement, number, numeral, pattern, sorting, subitizing, and other), type of activity (i.e., charts/visuals, game, kinesthetic, math construction, musical, problem solving, read aloud, using tools/manipulatives, etc.), and setting (i.e., individual, pairs, small group, whole group, or some combination). The authors taught two graduate research assistants (GRAs, doctoral candidates), to perform this coding, and then checked the codes for the first 20 activities; thereafter, any questions the GRA coders had were directed to the authors, and were checked via double coding, with consensus reached for final coding.

Table 3 presents the findings by topic, type of activity, and setting. Across all curricula activities, counting was the most frequently represented content area (40%), followed by geometry/spatial (19%) and measurement (13%). The least frequently represented content areas were numeral recognition (4%), subitizing (4%), number (3%), and "other" topics (1%; e.g., division, numeral writing). When coded for type, over one-third of all activities reviewed were categorized as using tools/manipulatives (34%). Games

Table 2 Early Math Curricula Included for Study

Program/ Curriculum	Number of Activities Initially Reviewed: Level 0	Number of Activities: Level 1	Number of Activities: Level 2	Number of Activities: Levels 3–4	Number of Activities: Level 5	References
Big Math for Little Kids (BMLK)	52	5	5	1	1	Ginsburg, Greenes, and Balfanz (2003)
Building Blocks (BB)	94	36	23	9	6	Clements and Sarama (2013)
Developing Math Concepts (DMC)	71	11	5	1	0	Richardson (2008)
Everyday Math (EM)	117	25	9	0	0	Leslie, Audrain, Bell, Bell, and O'Nan Brownell (2012)
Pre-K Mathematics Curriculum (PMC)	32	11	5	3	2	Klein, Starkey, and Ramirez (2002)
Rightstart/ Number Worlds (R/NW)	91	14	11	1	1	Griffin, Clements, and Sarama (2007) and Griffin et al. (1994)
Total activities	457	102	58	15	10	

Table 3 Description of All Activities Across All Curricula (Level 0, 457 Activities)

	BMLK	BB	DMC	EM	PMC	R/NW	Total
Content[a]							
Adding/Subtracting	6	10	0	0	4	1	21
Comparing/Ordering	4	12	3	5	3	19	46
Counting	9	41	27	45	7	53	182
Data/Graphing	1	0	2	10	1	3	17
Geometry/Spatial	15	29	15	18	6	6	89
Measurement	9	5	10	16	4	17	61
Number	0	2	2	8	0	0	12
Numeral	3	2	0	6	1	4	16
Pattern	5	8	7	10	4	3	37
Sorting	0	0	5	10	2	0	17
Subitizing	1	10	0	3	0	2	16
Other	1	0	2	1	1	0	5
Type							
Art	2	0	0	5	0	1	8
Charts/Visuals	3	0	5	12	3	9	32
Discussion of Theme	0	0	0	3	0	0	3
Drawing	1	1	0	0	0	0	2
Game	3	10	4	20	7	22	66
Kinesthetic	7	8	3	8	0	4	30
Language/Vocab	1	1	1	0	1	0	4
Math Construction	0	2	2	5	1	0	10
Model-then-Guide	3	9	1	2	1	4	20
Musical	1	14	3	6	0	4	28
Practice Skills	2	6	12	12	1	23	56
Problem Solving	8	6	1	6	0	2	23
Read Aloud	5	7	0	4	0	0	16

Continued

Table 3 Description of All Activities Across All Curricula (Level 0, 457 Activities)—cont'd

	BMLK	BB	DMC	EM	PMC	R/NW	Total
Role Play	0	1	0	0	0	0	1
Using Tools/Manipulatives	16	28	39	33	18	22	156
Other	0	1	0	1	0	0	2
Setting							
Individual	1	2	22	0	0	2	27
Pairs	3	3	2	4	8	2	22
Small Group	9	18	26	42	24	9	128
Small Group or Pairs	1	0	0	18	0	0	19
Small Group or Whole Group	2	11	0	42	0	0	55
Whole Group	36	60	21	11	0	78	206

[a]Some activities represented more than one content category.

(14%) and activities that focused on practice skills (12%) were also frequently represented, though to a lesser extent. The majority of activities reviewed was identified by curriculum materials as whole group (45%) or small group (28%) activities.

2.3.2 Phase 2: Selecting Fecund Activities

The goal of phase 2 was to rate activities as to their potential to develop both math and EF proficiencies and ultimately to cull, from all activities identified and coded, 10 archetypical instructional activities that were evaluated as having the greatest potential.

Level 1. To ensure the inclusion of activities that included the potential for high-quality mathematics learning and instruction, we first conducted a high-level review of all 457 activities. Three coders, including one of the authors, participated in the Level 1 coding and rated the activities according to a 3-point scale of "yes," "maybe," or "no" for further inclusion and consideration based on the level and content of mathematics learning involved. The same coders trained for Level 0 were trained to perform coding for Level 1. Codes for the first 10 activities were checked; thereafter, any questions the GRAs had were directed to the authors, and were checked via double coding, with consensus reached for final coding.

At this initial level, we used the criteria that the topic had to be central to mathematics learning (counting, subitizing, ordering, comparing, addition/subtraction, geometry, spatial, pattern, or measurement). A sample "yes" activity is *Mr. Mixup* from the Building Blocks curriculum. In this activity, children are introduced to a puppet, Mr. Mixup, that often makes mistakes when practicing various math concepts related to shapes, counting, and measuring. Children help Mr. Mixup count, for example. If he makes a mistake, they need to stop him right away and correct the error. A sample "no" activity is *Number Jump* from the Building Blocks curriculum. In this activity, the teacher holds up a numeral card, children say the numeral, then jump that many times. The activity is repeated with different numerals and other movements.

When the full list of 457 activities was coded for content of math learning at Level 1, a total of 102 "yes" activities (22%) remained for further consideration (Table 4). Owing to the large number of "yes" activities that remained after Level 1 coding, all "maybe" activities were removed from further consideration. Of the remaining 102 activities at this coding level, counting (38%), geometry/spatial (28%), and comparing/ordering (20%) were the most often represented content areas; other areas were much less frequently represented. Activities using tools/manipulatives (36%) and in the form of games (31%) were most common. Most remaining activities were small group activities (39%), with whole group activities (26%) the next most common.

Level 2. To further narrow the field, we conducted a Level 2 triage of the remaining activities. For the mathematics activity per se, we again used the criteria that the topic had to be central to mathematics learning *and* also added the criteria that the activity support child engagement and learning. More specifically, this meant that preschool-aged children should be able to (a) easily comprehend the task (perceive and interpret the materials and verbal interactions of the task and interpret the mathematical requirements, at least intuitively); (b) act on the task in a way that involves engagement with and action upon the ideas/requirements comprehended; and (c) make a response that is then received by others (including acting on materials), often starting a cycle of interaction. The previously established codes of content area, activity type, and setting were considered secondarily to ensure an adequate distribution of activities across these categories the next step. The same coders trained for Levels 0 and 1 were trained to perform coding for Level 2. Similar to previous coding rounds, codes for the first 10 activities were checked; thereafter, any questions the GRAs had were directed to the authors, and were checked via double coding, with consensus reached for final coding.

Table 4 Description of Level 1 Activities Across All Curricula (102 Activities)

	BMLK	BB	DMC	EM	PMC	R/NW	Total
Content[a]							
Adding/Subtracting	6	6	0	0	2	1	15
Comparing/Ordering	3	9	1	2	1	4	20
Counting	1	12	5	11	2	8	39
Data/Graphing	1	0	0	0	1	0	2
Geometry/Spatial	0	16	3	6	3	1	29
Measurement	0	2	1	3	0	2	8
Number	0	1	0	3	0	0	4
Numeral	1	0	0	0	0	0	1
Pattern	0	2	0	2	1	0	5
Sorting	0	0	1	2	0	0	3
Subitizing	0	3	0	3	0	1	7
Other	1	0	0	0	0	0	1
Type							
Art	0	0	0	0	0	0	0
Charts/Visuals	0	0	0	0	1	1	2
Discussion of Theme	0	0	0	0	0	0	0
Drawing	0	1	0	0	0	0	1
Game	0	9	2	7	4	10	32
Kinesthetic	0	0	0	1	0	0	1
Language/Vocab	0	1	0	0	0	0	1
Math Construction	0	1	1	1	0	0	3
Model-then-Guide	0	3	0	2	0	0	5
Musical	0	0	0	0	0	0	0
Practice Skills	0	3	1	1	0	0	5
Problem Solving	3	3	1	6	0	0	13
Read Aloud	1	0	0	0	0	0	1
Role Play	0	1	0	0	0	0	1

Table 4 Description of Level 1 Activities Across All Curricula (102 Activities)—cont'd

	BMLK	BB	DMC	EM	PMC	R/NW	Total
Using Tools/Manipulatives	1	14	6	7	6	3	37
Other	0	0	0	0	0	0	0
Setting							
Individual	0	2	1	0	0	0	3
Pairs	0	2	1	1	3	1	8
Small Group	0	13	7	8	8	4	40
Small Group or Pairs	0	0	0	10	0	0	10
Small Group or Whole Group	1	8	0	5	0	0	14
Whole Group	4	1	2	1	0	9	27

[a]Some activities represented more than one content category.

Level 2 resulted in 58 activities for further consideration (Table 5). At this level of coding, the most frequent content area was counting (47%), followed by geometry/spatial (31%); comparing/ordering (24%) and adding/subtracting (14%) were next, with the other topics much less frequent. The game (36%) and using tools/manipulatives (33%) were the most frequent type, with problem solving (12%) the next most frequent. Similar to previous levels, small group (43%) and whole group (34%) were the most common activity settings.

Level 3. To select activities that had potential to be fecund in simultaneously developing substantial mathematics and EF proficiencies, Level 3 coding was performed. The goal of this level was to narrow the activities to a subset, targeted a priori as a maximum of 15 activities. This followed the same holistic procedures as the previous level with two alterations. First, of course, there was a recalibration of the level of acceptance, as these activities survived the first screening. A "high" rating now required providing substantial opportunity for an activity's EF demands to be able to be adapted, via altering EF demands or providing scaffolding of an activity for children without significantly changing the math content inherent in the activity. Only activities rated as "high" were included for further review and consideration. A sample "high" EF adaptability activity is *Shape Touch* from the Big Math for Little Kids curriculum (also related to the "Feely Box" activity from Building Blocks). In this activity, children are shown a picture of a shape, and they have to find the same shape in a bag of shapes without

Table 5 Description of Level 2 Activities Across All Curricula (58 Activities)

	BMLK	BB	DMC	EM	PMC	R/NW	Total
Content[a]							
Adding/Subtracting	0	5	0	0	2	1	8
Comparing/Ordering	1	8	0	2	0	3	14
Counting	1	9	4	4	1	8	27
Data/Graphing	0	0	0	0	0	0	0
Geometry/Spatial	2	9	1	4	2	0	18
Measurement	0	1	0	1	0	1	3
Number	0	0	0	0	0	0	0
Numeral	0	0	0	0	0	0	0
Pattern	1	1	0	0	0	0	2
Sorting	0	0	0	0	0	0	0
Subitizing	0	0	0	0	0	2	2
Other	0	0	0	0	0	0	0
Type							
Art	0	0	0	0	0	0	0
Charts/Visuals	0	0	0	0	0	0	0
Discussion of Theme	0	0	0	0	0	0	0
Drawing	0	0	0	0	0	0	0
Game	0	6	2	3	1	9	21
Kinesthetic	0	0	0	0	0	0	0
Language/Vocab	0	1	0	0	0	0	1
Math Construction	0	1	0	1	0	0	2
Model-then-Guide	0	1	0	2	0	0	3
Musical	0	0	0	0	0	0	0
Practice Skills	0	3	0	0	0	0	3
Problem Solving	3	3	1	0	0	0	7
Read Aloud	1	0	0	0	0	0	1
Role Play	0	1	0	0	0	0	1

Table 5 Description of Level 2 Activities Across All Curricula (58 Activities)—cont'd

	BMLK	BB	DMC	EM	PMC	R/NW	Total
Using Tools/Manipulatives	1	7	2	3	4	2	19
Other	0	0	0	0	0	0	0
Setting							
Individual	0	1	0	0	0	0	1
Pairs	0	2	0	1	0	1	4
Small Group	0	7	4	6	5	3	25
Small Group or Pairs	0	0	0	1	0	0	1
Small Group or Whole Group	1	6	0	0	0	0	7
Whole Group	4	7	1	1	0	7	20

[a]Some activities represented more than one content category.

looking. A sample "low" EF adaptability activity is the previously described *Mr. Mixup*; this activity was rated as high in math content at Level 2, but was removed for further consideration in Level 3 because it did not include substantial EF adaptability opportunities.

Second, when two or more activities were identified as similar in content and nature, we synthesized them into an *archetype* no longer taken exclusively from one curriculum (rather, combining the advantages gleaned from each) and we conducted all further coding on that archetype. For example, two "high" EF adaptability activities, "Shape Touch" (from the Big Math for Little Kids curriculum) and the related activity, "Feely Box" (from the Building Blocks curriculum), were combined. Both activities have children match shapes, identify and name shapes by touch, and identify and count vertices and sides.

One to two GRAs at each of three project sites were trained on math content/quality and EF adaptability coding. All activities from two math curricula (*Big Math for Little Kids* and *Building Blocks*) were coded by GRAs from two project sites. Similar to previous procedures, for the remaining curricula, coders were trained and the first 10 activities were coded and checked for 100% agreement in coding; thereafter, any questions the GRAs had were directed to the authors, and were checked via double coding, with consensus reached for final coding.

Of the 15 activities remaining after Level 3 coding, geometry/spatial content-focused activities were most frequently represented (40%), followed by adding/subtracting (33%), counting (20%), comparing/ordering (13%), measurement (7%), and subitizing (7%); other content areas were not represented at this level (Table 6). Activities types at this level included those using tools/manipulatives (40%), games (40%), problem solving (13%), and practice skills (7%). A variety of settings were represented at this level: small group (40%), small group or whole group (33%), whole group (13%), and pairs (13%).

Level 4. Level 4 began with two preparatory activities. First, EF benchmarks were created based on work on a previous intervention (see Table 7) (SECURe Intervention—Jones, Bouffard, Bailey, & Jacob, 2011; Philips, 2016). The following EF dimensions and higher-order cognitive skills were coded: working memory, inhibitory control, cognitive flexibility, attention control/focus, planning skills, and self-reflection/error monitoring. Second, to prepare the 15 activities for extensive coding, each was separated into sematic units (e.g., teacher makes a statement or demonstrates a procedure, teacher asks a question, children response) to facilitate the application of the EF benchmarks coding of each significant step. Those accomplished, we completed the EF coding of the 15 archetype activities using this benchmark. For all EF benchmark coding activity, consensus was reached for all activity codes.

Level 4 EF coding was then completed on the remaining 15 activities (Table 8) using the same procedures. All areas of EF and selected higher-order cognitive skills coded were represented to various extents across the 15 activities. Working memory, inhibitory control, and cognitive flexibility were the most frequently coded categories. Perhaps not surprising, all of the activities required children to use their working memory skills, at least to some extent, to successfully complete the activity. When considering the EF coding, it is important to keep in mind that coding at this level only involved coding the scripts for the activities; EF skills actually utilized by children to complete the activity (i.e., behavioral coding of EF) was not completed at this level of coding.

Level 5. At Level 5, all previous codes were used to narrow the field to a total of 10 archetypical activities. The first authors reviewed the remaining 15 activities and 4 members of the team each ranked them after reviewing all associated coding data, including both math and EF codes from various levels. These ratings were then aggregated and consensus reached to determine the final 10 activities to inform new archetype activity development. The specific criteria were as follows:

Table 6 Description of Levels 3 and 4 Activities Across All Curricula (15 Activities)

	BMLK	BB	DMC	EM	PMC	R/NW	Total
Content[a]							
Adding/Subtracting	0	3	0	0	1	1	5
Comparing/Ordering	0	2	0	0	0	0	2
Counting	0	2	1	0	0	0	3
Data/Graphing	0	0	0	0	0	0	0
Geometry/Spatial	1	3	0	0	2	0	6
Measurement	0	1	0	0	0	0	1
Number	0	0	0	0	0	0	0
Numeral	0	0	0	0	0	0	0
Pattern	0	0	0	0	0	0	0
Sorting	0	0	0	0	0	0	0
Subitizing	0	1	0	0	0	0	1
Other	0	0	0	0	0	0	0
Type							
Art	0	0	0	0	0	0	0
Charts/Visuals	0	0	0	0	0	0	0
Discussion of Theme	0	0	0	0	0	0	0
Drawing	0	0	0	0	0	0	0
Game	0	3	1	0	1	1	6
Kinesthetic	0	0	0	0	0	0	0
Language/Vocab	0	0	0	0	0	0	0
Math Construction	0	0	0	0	0	0	0
Model-then-Guide	0	0	0	0	0	0	0
Musical	0	0	0	0	0	0	0
Practice Skills	0	1	0	0	0	0	1
Problem Solving	0	2	0	0	0	0	2
Read Aloud	0	0	0	0	0	0	0
Role Play	0	0	0	0	0	0	0
Using Tools/Manipulatives	1	3	0	0	2	0	6
Other	0	0	0	0	0	0	0

Continued

Table 6 Description of Levels 3 and 4 Activities Across All Curricula (15 Activities)—cont'd

	BMLK	BB	DMC	EM	PMC	R/NW	Total
Setting							
Individual	0	0	0	0	0	0	0
Pairs	0	2	0	0	0	0	2
Small Group	0	2	1	0	3	0	6
Small Group or Pairs	0	0	0	0	0	0	0
Small Group or Whole Group	1	4	0	0	0	0	5
Whole Group	0	1	0	0	0	1	2

[a]Some activities represented more than one content category.

Table 7 Executive Function Benchmark Categories and Operational Definitions

EF Category	Operational Definition
Attention Control/Focus	Maintaining focus while selecting and attending to relevant information and sustaining progress toward a goal (presumably to solve a math problem). This includes *actively* attending to instructions, the activity itself, feedback on the activity, or to peers' math problem-solving behavior in a group work context. "Actively attending" does not require action on the child's part, because it might be indicated by the child watching or listening
Planning Skills	Identifying and organizing the steps or sequence of events needed to complete an activity and achieve a desired goal
Working Memory	Maintaining information in memory during the activity (monitoring working memory) or manipulating/revising information in working memory during the activity (updating working memory)
Inhibitory Control	The ability to suppress or modify a (distracting) behavioral response in the service of engaging behaviors required to attain a longer-term goal (i.e., inhibiting automatic reactions: $4+5=6$ as in counting), while initiating more "effortful" yet appropriate responses (i.e., $4+5=9$)
Cognitive Flexibility	The mental ability to switch between thinking about two different concepts and to think about multiple concepts simultaneously. Additionally, the ability to redirect or shift one's focus of attention away from one salient object, instruction, or strategy to another
Self-Reflection/Error Monitoring	The ability to review work either independently or with the help of peers or teachers for accuracy. The ability to rehearse the steps taken or answers given/chosen and/or evaluate them

Table 8 Executive Function Benchmark Categorizations (Levels 4–5 and Phase 4 Archetype Activities)

		EF Dimensions & Higher-Order Cognitive Skills					
Activity	**Curriculum**	**Attention Control/Focus**	**Planning Skills**	**Working Memory**	**Inhibitory Control**	**Cognitive Flexibility**	**Self-Reflection/Error Monitoring**
Level 4 Activities							
Shape Touch[a]	BMLK	✓	✓	✓	✓	★	✓
Plus Minus Game[a]	R/NW	✓	✓	★	✓	✓	
X-Ray Vision[a]	BB			✓	★	✓	✓
Gone Fishing[a]	BB			★	✓		
Pizza Game[a]	BB		✓	★	✓	✓	✓
Let's Make Tangram Pictures[a]	PMC		✓	✓	✓	★	✓
Guess My Rule[a]	BB			★	✓	✓	✓
Shape Step[a]	BB	✓		★	✓	✓	✓
Is It or Not?[a]	BB	✓		★	✓	✓	✓
Bears in the Cave[a]	PMC	✓		★	✓		✓
Compare Game	BB		✓	★		✓	
Concentration	DMC		✓	★	✓	✓	✓
Imagine a Shape	PMC		✓	✓	✓	★	✓

Continued

Table 8 Executive Function Benchmark Categorizations (Levels 4–5 and Phase 4 Archetype Activities)—cont'd

		EF Dimensions & Higher-Order Cognitive Skills					
Activity	**Curriculum**	**Attention Control/Focus**	**Planning Skills**	**Working Memory**	**Inhibitory Control**	**Cognitive Flexibility**	**Self-Reflection/Error Monitoring**
Length Riddles	BB	✓		★	✓	✓	✓
Adding Board Game	BB		✓	★	✓	✓	✓
Phase 4: Archetype Activities							
Big Fish Story				★	✓		✓
Change Game			✓	★	✓	✓	✓
Cookie Game		✓	✓	★	✓	✓	✓
Count to Add and Subtract		✓		★	✓	✓	✓
Don't Burn Your Feet!		✓		✓	★		✓
Magician's Tricks		✓		★	★	✓	✓
Shape Puzzles				✓	✓	★	★
True or False?		✓		✓	✓	✓	★
What Am I Thinking?		✓		★	✓	✓	✓
What Shape Am I Touching?		✓		✓	★	✓	✓

[a]Represents activities that were selected to move forward for Level 5 archetype activity development and coding.
★ Most frequently represented EF/Higher-Order Cognitive Skills Category based on activity script coding.
✓ Indicates EF/Higher-Order Cognitive Skills Category was represented based on activity script coding.

1. Math content and activity structure
   a. Representation of four primary activity types identified in Levels 0–3 activity selection (game, problem solving, practice skills, using tools/manipulatives).
   b. Representation of five primary content focus areas identified in Levels 0–3 activity selection (addition/subtraction, comparing/ordering, counting, geometry/spatial, measurement).
   c. Representation of activities that can be introduced and conducted in small group or pairs setting.
2. EF demands
   a. Representation of main components of EF identified in the literature, as informed by EF benchmark coding (Level 4).
3. Applicability for future microgenetic study
   a. Representation of activities that are amenable to small group instruction.
   b. Representation of activities that are amenable to eventual independent use by children in centers.

The Level 5 ten finalist activities are described in Tables 8 and 9. The topics of geometry/spatial (50%) and arithmetic (30%) dominated the selected activities, with counting (20%), subitizing, (10%), and comparing/ordering (10%) also represented. Half used tools or manipulatives, 30% were games, 10% were involved problem solving, and 10% were practice skills. Half of the activities were coded as small or whole group, 20% as whole group (but were modified to also work as small group), 20% as small group, and 10% as pairs.

2.3.3 Phase 3: Developing Archetype Activities

Initial activity development. The 10 activities remaining after coding served as the basis for creating the same number of archetype activities (Table 10), which were defined as activities that are representative of a particular activity type/setting/content-focus that include high-quality math content and EF demands. Following the lead of the Big Math for Little Kids and the Building Blocks curricula, activities were also developed so that they included multiple subactivities and phases that built upon one another in terms of content and complexity. This design will allow for opportunities to support children at various levels and across time as they learn math skills and practice EF skills (while also keeping the activities fresh and exciting). Each activity was designed and written such that activities could be scaffolded up or down to make content more or less challenging for children depending on their individual needs; two sets of scaffolds were developed for every phase of each

Table 9 Description of Level 5 Activities Across All Curricula (10 Activities)

	BMLK	BB	DMC	EM	PMC	R/NW	Total
Content[a]							
Adding/Subtracting	0	1	0	0	1	1	3
Comparing/Ordering	0	1	0	0	0	0	1
Counting	0	2	0	0	0	0	2
Data/Graphing	0	0	0	0	0	0	0
Geometry/Spatial	1	3	0	0	1	0	5
Measurement	0	0	0	0	0	0	0
Number	0	0	0	0	0	0	0
Numeral	0	0	0	0	0	0	0
Pattern	0	0	0	0	0	0	0
Sorting	0	0	0	0	0	0	0
Subitizing	0	1	0	0	0	0	1
Other	0	0	0	0	0	0	0
Type							
Art	0	0	0	0	0	0	0
Charts/Visuals	0	0	0	0	0	0	0
Discussion of Theme	0	0	0	0	0	0	0
Drawing	0	0	0	0	0	0	0
Game	0	1	0	0	1	1	3
Kinesthetic	0	0	0	0	0	0	0
Language/Vocab	0	0	0	0	0	0	0
Math Construction	0	0	0	0	0	0	0
Model-then-Guide	0	0	0	0	0	0	0
Musical	0	0	0	0	0	0	0
Practice Skills	0	1	0	0	0	0	1
Problem Solving	0	1	0	0	0	0	1
Read Aloud	0	0	0	0	0	0	0
Role Play	0	0	0	0	0	0	0

Table 9 Description of Level 5 Activities Across All Curricula (10 Activities)—cont'd

	BMLK	BB	DMC	EM	PMC	R/NW	Total
Using Tools/Manipulatives	1	3	0	0	1	0	5
Other	0	0	0	0	0	0	0
Setting							
Individual	0	0	0	0	0	0	0
Pairs	0	1	0	0	0	0	1
Small Group	0	0	0	0	2	0	2
Small Group or Pairs	0	0	0	0	0	0	0
Small Group or Whole Group	1	4	0	0	0	0	5
Whole Group	0	1	0	0	0	1	2

[a]Some activities represented more than one content category.

activity—one for mathematics and one for EF. The activities were then supplemented by relevant handouts for teachers with additional information to assist with content presentation (for example, a handout was created that provided guidance on shape descriptions to use with children).

Prepilot of activities. After developing these new activities, we conducted pilot evaluations of the resulting 10 activities across multiple sites, with each team providing qualitative feedback to the development team. As a convenience sample for the prepilot, sites implemented the activities with preschool-aged children attending schools affiliated with their respective universities. More diverse samples are currently being included in the full pilot phase. The questions the pilots were designed to answer the following questions: Does the implementation reflect adequate and similar fidelity across sites and teachers? Are the materials and structure of the text adequate? Do children appear to be learning the math concepts/skills as postulated? Is EF use observed and is it supported by the activities as postulated? What improvements can be made? We also asked several master teachers for their reactions to the archetypes as written.

The sites involved in prepiloting activities remained in contact during implementation to address any questions or concerns related to the activities. Each site also completed a form that included specific questions related to the activities to inform ongoing and future activity development and refinement. Videos and feedback were provided on a daily basis from the teacher–researchers, and changes were made in tight revision cycles

Table 10 Final Archetype Activities Developed for Further Study (Phase 3)

Archetype Name	Original Curriculum (Activity)	Related Activities	Content Focus	Activity Type	Brief Descriptions
Count to Add and Subtract	Pre-K Math (Bears in the Cave)	How Many Now (Building Blocks), Mouse in the Cookie Jar (Number Worlds)	Counting, Adding/ Subtracting	Game	Relating counting (the successor principle) to arithmetic. Addition and subtraction using hidden objects
Change Game	Number Worlds (Plus–Minus Game)	Number Line Team Game, Race to Ten, Race to Zero, Ice be Nice to my Skating Party (all Number Worlds), Adding: Board Game (Building Blocks)	Adding/ Subtracting	Game	Students roll dot cube, count, and request that many counters, which are put on the game board. +/− cards are drawn to determine movement of counters
Cookie Game	Building Blocks (Pizza Game 1–2)		Counting, Subitizing	Game	Using cookie activity sheets, players roll number cube and put that many counters on their sheet, asking the other player if they are correct or not. Other player must agree. Continues until a player has all spaces covered
Big Fish Story	Building Blocks (Gone Fishing 2)		Adding/ Subtracting	Using Tools/ Manips	Children are given fish crackers and are told to add/subtract various amounts of fish to their lake, and asked how many after adding and subtracting.
Magician's Tricks	Building Blocks (X-Ray Vision)		Ordering/ Comparing, Counting	Practice Skills	Counting cards 1–10 (or 1–20) placed in numerical order. Children point to any card, and other child uses their "magician's tricks" (counting) to tell which card it is without looking

True or False?	Building Blocks (Is It Or Not?)		Geometry/ Spatial	Using Tools/ Manips	Review with children what makes a triangle, for example, and then try to fool them, asking why a shape or drawing presented is or is not a triangle
What Shape Am I Touching?	Big Math for Little Kids (Shape Touch)	Feely Box (Building Blocks)	Geometry/ Spatial	Using Tools/ Manips	Children identify and name shapes by touch, and match shapes to pictures. Also identify and count corners and straight sides
What Am I Thinking?	Building Blocks (Guess My Rule)		Geometry/ Spatial	Problem Solving	Sort shape sets into piles based on something that makes them alike. Children guess the rule, and explain their answer
Shape Puzzles	Pre-K Math (Let's Make Tangram Pictures)		Geometry/ Spatial	Using Tools/ Manips	Children are given shapes to use to fill in a picture puzzle
Don't Burn Your Feet!	Building Blocks (Shape Step)		Geometry/ Spatial	Using Tools/ Manips	Make large shapes on floor/ground. Show several triangles, for example, and tell children to step on triangles only. Ask children how they know it is a triangle

(e.g., daily edits). Regular meetings were also held to further discuss specific activity reflections and observations from the multisite teams, as well as "big picture" topics (e.g., format of activities, process of activity implementation, etc.) to help inform steps moving forward with the full pilot and later microgenetic testing preparation. Initial reactions and observations included the following: the activities were engaging and fun for children; the math content was either just right or too easy for the children (for the convenience sample selected); the scaffolding sections in the activities were helpful in adjusting the activities to meet children's needs and levels; and that additional information or direction needs to be included in certain activities around supporting EF or increasing EF demands.

Teacher feedback and activity refinement. The final act in the development of the archetype activities was to share them with the entire research team and a group of five teachers identified by each research group as excellent teachers of preschool mathematics, with particular experience working with children from low-SES backgrounds, with high needs, and/or dual language learners. Comments were overwhelmingly positive in all areas of feedback (e.g., appropriateness for age-group and special populations served, child engagement, tactics and structures to support math and EF, etc.). Suggestions included ways to make the activities more appropriate for different cultures and for children with special needs, every one of which was incorporated.

2.3.4 Phase 4: Microgenetic Testing of Select Archetype Activities

We are planning to use a subset of the archetype math activities selected during phase 3 to conduct microgenetic studies, a methodological approach that involves repeated observations over time to capture developmental change, uncover the processes that may cause change, and conditions that facilitate change or hinder it from happening. As part of the microgenetic studies, children will receive pretests on math and EF, multiple math instructional sessions, and posttests. Inclusion of multiple data collection sites will afford our use of several math activities and ensure inclusion of diverse participant groups. Coding systems will be developed to measure relevant math and EF behaviors.

We will also conduct complementary experimental studies to further understand how these activities support math and EF development and learning. For example, one subset of studies will focus on select, isolated features of math activities, including features that appear in some of the microgenetic study activities. These studies will test how specific features of instructions or materials (e.g., variation in the size of shapes being compared in the shape identification task) influence children's performance and learning, and if that influence depends on child factors such as EF level. Another set of studies will examine activities that are designed to require low levels of adult direction and, like the microgenetic study activities, manipulate EF demands. Using these activities, we will study how children who vary in their levels of EF benefit from these activities in terms of both math learning and EF development. Once the manipulations involved in such experimental studies are optimized, we will take some of the activities that result into classrooms to observe how children interact with the materials in authentic learning environments.

3. CONCLUSION

Early childhood mathematics education has received increasing attention from researchers, practitioner, and policy makers. Over the next several years, the DREME Network will conduct research and development activities that will have implications for interventions, from the state and district levels, to professional development, families, and teaching. As an immediate contribution, we report on the initial work of one of the projects, a comprehensive review of the instructional activities in six research-based curricula. These data will provide researchers, curriculum developers and specialists, and other educators with new information and new perspectives when creating, evaluating, and choosing curricula and instructional activities in early mathematics.

ACKNOWLEDGMENTS

This chapter and the research reported were supported in part by a grant from the Heising-Simons Foundation, Grant #2015-156. The opinions expressed are those of the authors and do not represent views of the Heising-Simons Foundation. We wish to thank Deborah Stipek for her leadership on all aspects of the DREME Network and to the Network members for their advice on this manuscript, especially Michèle Mazzocco, the lead on the project described in more depth, and other members of her team, Dale Farran, Susan Levine, Deborah Phillips, and Julie Sarama. It should be noted that several curricula described in this article were published by one of the authors and/or members of the DREME Network, who thus could have a vested interest in the results. However, an external auditor oversaw the research design and coding, and five researchers independently confirmed findings and procedures.

REFERENCES

Alloway, T. P., & Alloway, R. G. (2010). Investigating the predictive roles of working memory and IQ in academic attainment. *Journal of Experimental Child Psychology*, *106*(1), 124–131. http://dx.doi.org/10.1016/j.jecp.2009.11.003.

Alloway, T. P., Alloway, R. G., & Wootan, S. (2014). Home sweet home: Does where you live matter to working memory and other cognitive skills? *Journal of Experimental Child Psychology*, *124*, 124–131. http://dx.doi.org/10.1016/j.jecp.2013.11.012.

ATME. (2017). *Standards for preparing teachers of mathematics*. Raleigh, NC: Association of Mathematics Teacher Educators.

Best, J. R., Miller, P. H., & Naglieri, J. A. (2011). Relations between executive function and academic achievement from ages 5 to 17 in a large, representative national sample. *Learning and Individual Differences*, *21*(4), 327–336. http://dx.doi.org/10.1016/j.lindif.2011.01.007.

Bierman, K. L., Domitrovich, C. E., Nix, R. L., Gest, S. D., Welsh, J. A., Greenberg, M. T., ... Gill, S. (2008a). Promoting academic and social-emotional school readiness: The Head Start REDI program. *Child Development*, *79*(6), 1802–1817. http://dx.doi.org/10.1111/j.1467-8624.2008.01227.x.

Bierman, K. L., Nix, R. L., Greenberg, M. T., Blair, C. B., & Domitrovich, C. E. (2008b). Executive functions and school readiness intervention: Impact, moderation, and mediation in the Head Start REDI program. *Development and Psychopathology*, *20*(3), 821–843. http://dx.doi.org/10.1017/S0954579408000394.

Blair, C. B., & Razza, R. P. (2007). Relating effortful control, executive function, and false belief understanding to emerging math and literacy ability in kindergarten. *Child Development*, *78*, 647–663.

Blair, C. B., Protzko, J., & Ursache, A. (2011). Self-regulation and early literacy. In S. B. Neuman & D. K. Dickinson (Eds.), *Vol. 3. Handbook of early literacy research* (pp. 20–35). New York, NY: Guilford.

Bull, R., Johnston, R. S., & Roy, J. A. (1999). Exploring the roles of the visual-spatial sketch pad and central executive in children's arithmetical skills: Views from cognition and developmental neuropsychology. *Developmental Neuropsychology*, *15*(3), 421–442. http://dx.doi.org/10.1080/87565649909540759.

Bull, R., Espy, K. A., Wiebe, S. A., Sheffield, T. D., & Nelson, J. M. (2011). Using confirmatory factor analysis to understand executive control in preschool children: Sources of variation in emergent mathematic achievement. *Developmental Science*, *14*(4), 679–692. http://dx.doi.org/10.1111/j.1467-7687.2010.01012.x.

Cameron, C. E., Brock, L. L., Murrah, W. M., Bell, L. H., Worzalla, S. L., Grissmer, D., & Morrison, F. J. (2012). Fine motor skills and executive function both contribute to kindergarten achievement. *Child Development*, *83*(4), 1229–1244. http://dx.doi.org/10.1111/j.1467-8624.2012.01768.x.

Case, R., Griffin, S., & Kelly, W. M. (1999). Socieconomic gradients in mathematical ability and their responsiveness to intervention during early childhood. In D. P. Keating & C. Hertzman (Eds.), *Developmental health and the wealth of nations* (pp. 125–149). New York, NY: Guilford.

Cerda, C. A., Im, M. H., & Hughes, J. N. (2014). Learning-related skills and academic achievement in academically at-risk first graders. *Journal of Applied Developmental Psychology*, *35*(5), 433–443. http://dx.doi.org/10.1016/j.appdev.2014.08.001.

Clark, C. A. C., Pritchard, V. E., & Woodward, L. J. (2010). Preschool executive functioning abilities predict early mathematics achievement. *Developmental Psychology*, *46*(5), 1176–1191. http://dx.doi.org/10.1037/a0019672.

Clements, D. H. (2008). Linking research and curriculum development. In L. D. English (Ed.), *Handbook of international research in mathematics education* (2nd ed., pp. 589–625). New York: Taylor & Francis.

Clements, D. H., & Sarama, J. (2007). Effects of a preschool mathematics curriculum: Summative research on the *Building Blocks* project. *Journal for Research in Mathematics Education*, *38*(2), 136–163.

Clements, D. H., & Sarama, J. (2008). Experimental evaluation of the effects of a research-based preschool mathematics curriculum. *American Educational Research Journal, 45*, 443–494. http://dx.doi.org/10.3102/0002831207312908.

Clements, D. H., & Sarama, J. (2009). *Learning and teaching early math: The learning trajectories approach*. New York, NY: Routledge.

Clements, D. H., & Sarama, J. (2011). Early childhood mathematics intervention. *Science, 333*(6045), 968–970. http://dx.doi.org/10.1126/science.1204537.

Clements, D. H., & Sarama, J. (2013). Building blocks. In (Vols. 1–2)Columbus, OH: McGraw-Hill Education.

Clements, D. H., & Sarama, J. (2014). *Learning and teaching early math: The learning trajectories approach* (2nd ed.). New York, NY: Routledge.

Clements, D. H., Sarama, J., Spitler, M. E., Lange, A. A., & Wolfe, C. B. (2011). Mathematics learned by young children in an intervention based on learning trajectories: A large-scale cluster randomized trial. *Journal for Research in Mathematics Education, 42*(2), 127–166.

Clements, D. H., Sarama, J., Layzer, C., Unlu, F., & Germeroth, C. (2012, March). The efficacy of an intervention synthesizing scaffolding designed to promote self regulation with an early mathematics curriculum: Effects on executive function. In D. Farran (Chair), *Enhancing executive function and achievement in prekindergarten classrooms: the effectiveness of tools of the mind. Symposium conducted at the meeting of Society for Research on Educational Effectiveness, Washington, DC*

Clements, D. H., Sarama, J., Wolfe, C. B., & Spitler, M. E. (2013). Longitudinal evaluation of a scale-up model for teaching mathematics with trajectories and technologies: Persistence of effects in the third year. *American Educational Research Journal, 50*(4), 812–850. http://dx.doi.org/10.3102/0002831212469270.

Clements, D. H., Sarama, J., & Germeroth, C. (2016). Learning executive function and early mathematics: Directions of causal relations. *Early Childhood Research Quarterly, 36*, 79–90. http://dx.doi.org/10.1016/j.ecresq.2015.12.009.

Clements, D. H., Sarama, J., Layzer, C., Unlu, F., Germeroth, C., & Fesler, L. (2017). *Effects on executive function and mathematics learning of an early mathematics curriculum synthesized with scaffolded play designed to promote self-regulation versus the mathematics curriculum alone*. Manuscript submitted for publication.

Clements, D. H., Fuson, K. C., & Sarama, J. (2017). The research-based balance in early childhood mathematics: A response to Common Core criticisms. *Early Childhood Research Quarterly, 40*, 150–162. http://dx.doi.org/10.1016/j.ecresq.2017.03.005.

Cortina, K. S., Miller, K. F., Hua, S., Peist, E., Huang, Y., & Wei, Y. (2013). Behavioral regulation of preschoolers in China and the US: The role of attention. In N. McElvany & H. G. Holtappels (Eds.), *Empirische Bildungsforschung: Theorien, Methoden, Befunde und Perspektiven* (pp. 125–133). New York, NY: Waxmann.

Cronbach, L. J., Ambron, S. R., Dornbusch, S. M., Hess, R. D., Hornik, R. C., Phillips, D. C., & Weiner, S. S. (Eds.), (1980). *Toward reform of program evaluation: Aims, methods, and institutional arrangements*. San Francisco, CA: Jossey-Bass.

Denton, K., & West, J. (2002). *Children's reading and mathematics achievement in kindergarten and first grade*. Retrieved from U.S. Department of Education, National Center for Education Statistics website: http://nces.ed.gov/pubsearch/pubsinfo.asp?pubid=2002125.

Diamond, A., & Lee, K. (2011). Interventions shown to aid executive function development in children 4 to 12 years old. *Science, 333*(6045), 959–964. http://dx.doi.org/10.1126/science.1204529.

Diamond, A., Barnett, W. S., Thomas, J., & Munro, S. (2007). Preschool program improves cognitive control. *Science, 318*, 1387–1388. http://dx.doi.org/10.1126/science.1151148.

Dowsett, S. M., & Livesey, D. J. (2000). The development of inhibitory control in preschool children: Effects of "executive skills" training. *Developmental Psychobiology, 36*, 161–174.

Duncan, G. J., Claessens, A., & Engel, M. (2004). *The contributions of hard skills and socio-emotional behavior to school readiness Evanston.* IL: Northwestern University.
Duncan, G. J., Dowsett, C. J., Claessens, A., Magnuson, K., Huston, A. C., Klebanov, P., ... Japel, C. (2007). School readiness and later achievement. *Developmental Psychology, 43*(6), 1428–1446. http://dx.doi.org/10.1037/0012-1649.43.6.1428.
Edens, K. M., & Potter, E. F. (2013). An exploratory look at the relationships among math skills, motivational factors and activity choice. *Early Childhood Education Journal, 41*(3), 235–243. http://dx.doi.org/10.1007/s10643-012-0540-y.
Espinet, S. D., Anderson, J. E., & Zelazo, P. D. (2012). Reflection training improves executive function in preschool-age children: Behavioral and neural effects. *Developmental Cognitive Neuroscience.* http://dx.doi.org/10.1016/j.dcn.2012.11.009.
Farran, D. C., Lipsey, M. W., & Wilson, S. J. (2011, November). In *Curriculum and pedagogy: Effective math instruction and curricula. Paper presented at the early childhood math conference, Berkeley, CA.*
Fuson, K. C., Clements, D. H., & Sarama, J. (2015). Making early math education work for all children. *Phi Delta Kappan, 97*(3), 63–68.
Gathercole, S. E., & Pickering, S. J. (2000). Assessment of working memory in six- and seven-year old children. *Journal of Educational Psychology, 92,* 377–390. http://dx.doi.org/10.1037/0022-0663.92.2.377.
Gawrilow, C., Fäsche, A., Guderjahn, L., Gunzenhauser, C., Merkt, J., & von Suchodoletz, A. (2014). The impact of self-regulation on preschool mathematical achievement. *Child Indicators Research,* 7(4), 805–820. http://dx.doi.org/10.1007/s12187-013-9201-y.
Geary, D. C. (2013). Early foundations for mathematics learning and their relations to learning disabilities. *Current Directions in Psychological Science, 22*(1), 23–27. http://dx.doi.org/10.1177/0963721412469398.
Geary, D. C., & Liu, F. (1996). Development of arithmetical competence in Chinese and American children: Influence of age, language, and schooling. *Child Development, 67*(5), 2022–2044.
Gersten, R., Chard, D. J., Jayanthi, M., Baker, M. S., Morpy, S. K., & Flojo, J. R. (2008). *Teaching mathematics to students with learning disabilities: A meta-analysis of the intervention research.* Retrieved from RMC Research Corporation, Center on Instruction website: http://www.centeroninstruction.org/files/Teaching Math to SLD Meta-analysis.pdf.
Ginsburg, H. P., Greenes, C., & Balfanz, R. (2003). *Big math for little kids.* Parsippany, NJ: Dale Seymour.
Griffin, S., & Case, R. (1997). Re-thinking the primary school math curriculum: An approach based on cognitive science. *Issues in Education, 3,* 1–49.
Griffin, S., Case, R., & Siegler, R. S. (1994). Rightstart: Providing the central conceptual prerequisites for first formal learning of arithmetic to students at risk for school failure. In K. McGilly (Ed.), *Classroom lessons: Integrating cognitive theory and classroom practice* (pp. 25–49). Cambridge, MA: MIT Press.
Griffin, S., Clements, D. H., & Sarama, J. (2007). *Number worlds/building block: A prevention/intervention program: Teacher edition level B.* Columbus, OH: SRA/McGraw-Hill.
Huizinga, M., Dolan, C. V., & van der Molen, M. W. (2006). Age-related change in executive function: Developmental trends and a latent variable analysis. *Neuropsychologia, 44,* 2017–2036. http://dx.doi.org/10.1016/j.neuropsychologia.2006.01.010.
Jacob, R., & Parkinson, J. (2015). The potential for school-based interventions that target executive function to improve academic achievement: A review. *Review of Educational Research, 85,* 512–552. http://dx.doi.org/10.3102/0034654314561338.
Jones, S. M., Bouffard, S., Bailey, R., & Jacob, R. (2011). *SECURe coding guidelines and materials.* Cambridge, MA: Harvard Graduate School of Education, Harvard University.

Klein, A., Starkey, P., & Ramirez, A. B. (2002). *Pre-K mathematics curriculum*. Glenview, IL: Scott Foresman.

Klein, A., Starkey, P., Clements, D. H., Sarama, J., & Iyer, R. (2008). Effects of a pre-kindergarten mathematics intervention: A randomized experiment. *Journal of Research on Educational Effectiveness*, *1*, 155–178.

Klingberg, T. (2009). *The overflowing brain: Information overload and the limits of working memory*. New York, NY: Oxford University Press.

Kovas, Y., Haworth, C. M. A., Dale, P. S., & Plomin, R. (2007). The genetic and environmental origins of learning abilities and disabilities in the early school years. *Monographs of the Society for Research in Child Development*, *72(3)*, 1–144.

LeFevre, J.-A., Berrigan, L., Vendetti, C., Kamawar, D., Bisanz, J., Skwarchuk, S.-L., & Smith-Chant, B. L. (2013). The role of executive attention in the acquisition of mathematical skills for children in grades 2 through 4. *Journal of Experimental Child Psychology*, *114*(2), 243–261. http://dx.doi.org/10.1016/j.jecp.2012.10.005.

Lehto, J. E., Juujarvi, P., Kooistra, L., & Pulkkinen, L. (2003). Dimensions of executive functioning: Evidence from children. *The British Journal of Developmental Psychology*, *21*, 59–80. http://dx.doi.org/10.1348/026151003321164627.

Leslie, D. A., Audrain, A. E., Bell, J., Bell, M., & O'Nan Brownell, J. (2012). *Everyday mathematics teacher's guide to activities: Pre-K* (3rd ed.). Chicago, IL: McGraw-Hill & University of Chicago.

Lewis Presser, A., Clements, M., Ginsburg, H., & Ertle, B. (2015). Big math for little kids: The effectiveness of a preschool and kindergarten mathematics curriculum. *Early Education and Development*, *26*(3), 399–426. http://dx.doi.org/10.1080/10409289.2015.994451.

Lillard, A. S., & Else-Quest, N. (2007). Evaluating Montessori education. *Science*, *313*, 1893–1894. http://dx.doi.org/10.1126/science.1132362.

Lonigan, C. J., & Phillips, B. M. (2012, March). *Comparing skills-focused and self-regulation focused preschool curricula: Impacts on academic and self-regulatory skills*. Paper presented at the society for research on educational effectiveness, Washington, DC.

Lyons, K. E., & Zelazo, P. D. (2011). Monitoring, metacognition, and executive function. Elucidating the role of self-reflection in the development of self-regulation. *Advances in Child Development and Behavior*, *40*, 379–380. http://dx.doi.org/10.1016/B978-0-12-386491-8.00010-4.

McClelland, M. M., Cameron, C. E., Duncan, R., Bowles, R. P., Acock, A. C., Miao, A., & Pratt, M. E. (2014). Predictors of early growth in academic achievement: The head-toes-knees-shoulders task. *Frontiers in Psychology*, *5*, 599. http://dx.doi.org/10.3389/fpsyg.2014.00599.

Miyake, A., Friedman, N. P., Emerson, M. J., Witzki, A. H., & Howerter, A. (2000). The unity and diversity of executive functions and their contributions to complex 'frontal lobe' tasks: A latent variable analysis. *Cognitive Psychology*, *41*, 49–100. http://dx.doi.org/10.1006/cogp.1999.0734.

Morgan, P. L., Farkas, G., Hillemeier, M. M., & Maczuga, S. (2014). Who is at risk for persistent mathematics difficulties in the United States? *Journal of Learning Disabilities*, *49*, 305–319. http://dx.doi.org/10.1177/0022219414553849.

Morris, P. A., Mattera, S. K., Castells, N., Bangser, M., Bierman, K. L., & Raver, C. C. (2014). *Impact findings from the Head Start CARES demonstration: National evaluation of three approaches to improving preschoolers' social and emotional competence (OPRE report 2014–44)*. Washington, DC: Office of Planning, Research and Evaluation, Administration for Children and Families, U.S. Department of Health and Human Services.

National Research Council. (2009). *Mathematics learning in early childhood: Paths toward excellence and equity*. Washington, DC: National Academy Press.

Ng, F. F.-Y., Tamis-LeMonda, C. S., Yoshikawa, H., & Sze, I. N.-L. (2014). Inhibitory control in preschool predicts early math skills in first grade: Evidence from an ethnically diverse sample. *International Journal of Behavioral Development, 39*(2), 139–149. http://dx.doi.org/10.1177/0165025414538558.

Olkun, S., Altun, A., Göçer Şahin, S., & Akkurt Denizli, Z. (2015). Deficits in basic number competencies may cause low numeracy in primary school children. *Ted EĞİtİm Ve BİlİM, 40*(177), 141–159. http://dx.doi.org/10.15390/eb.2015.3287.

Otero, T. M., Barker, L. A., & Naglieri, J. A. (2014). Executive function treatment and intervention in schools. *Applied Neuropsychology: Child, 3*(3), 205–214. http://dx.doi.org/10.1080/21622965.2014.897903.

Perels, F., Merget-Kullmann, M., Wende, M., Schmitz, B., & Buchbinder, C. (2009). Improving self-regulated learning of preschool children: Evaluation of training for kindergarten teachers. *The British Journal of Educational Psychology, 79*(2), 311–327. http://dx.doi.org/10.1348/000709908X322875.

Perry, B., MacDonald, A., & Gervasoni, A. (2015). Mathematics and transition to school: Theoretical frameworks and practical implications. In B. Perry, A. MacDonald, & A. Gervasoni (Eds.), *Mathematics and transition to school: International perspectives* (pp. 1–14). New York, NY: Springer.

Philips, D. (2016, July). Executive function mapping: Making more of math. In: Jones, S. (Chair), *Using the executive function (EF) mapping project to inform executive function and regulation-related programs, policies, and assessments. Symposium conducted at the meeting of the National Research Conference on Early Childhood, Washington, DC.*

Preschool Curriculum Evaluation Research Consortium. (2008). *Effects of preschool curriculum programs on school readiness (NCER 2008–2009).* Retrieved from Government Printing Office website http://ncer.ed.gov/.

Raver, C. C. (2013, September). In *Targeting self-regulation through intervention: Lessons from RCTs Paper presented at the society of research on educational effectiveness fall conference, Washington, DC.*

Raver, C. C., Jones, S. M., Li-Grining, C., Zhai, F., Bub, K., & Pressler, E. (2011). CRSP's impact on low-income preschoolers' preacademic skills: Self-regulation as a mediating mechanism. *Child Development, 82*(1), 362–378. http://dx.doi.org/10.1111/j.1467-8624.2010.01561.x.

Razza, R. P., & Raymond, K. (2015). Executive functions and school readiness. In S. Robson & S. F. Quinn (Eds.), *The Routledge international handbook of young children's thinking and understanding* (pp. 133–149). New York, NY: Routledge.

Richardson, K. (2008). *Developing math concepts in pre-kindergarten.* Bellingham, WA: Math Perspectives.

Riggs, N. R., Greenberg, M. T., Kusché, C. A., & Pentz, M. A. (2006). The mediational role of neurocognition in the behavioral outcomes of a social-emotional prevention program in elementary school students: Effects of the PATHS curriculum. *Prevention Science,* 7(1), 91–102. http://dx.doi.org/10.1007/s11121-005-0022-1.

Roebers, C. M., Cimeli, P., Röthlisberger, M., & Neuenschwander, R. (2012). Executive functioning, metacognition, and self-perceived competence in elementary school children: An explorative study on their interrelations and their role for school achievement. *Metacognition and Learning,* 7(3), 151–173. http://dx.doi.org/10.1007/s11409-012-9089-9.

Röthlisberger, M., Neuenschwander, R., Cimelia, P., Michel, E., & Roebers, C. M. (2012). Improving executive functions in 5- and 6-year-olds: Evaluation of a small group intervention in prekindergarten and kindergarten children. *Infant and Child Development, 1,* 411–429. http://dx.doi.org/10.1002/icd.752.

Rouse, C., Brooks-Gunn, J., & McLanahan, S. (2005). Introducing the issue. *The Future of Children, 15,* 5–14.

Rueda, M. R., Rothbart, M. K., McCandliss, B. D., Saccomanno, L., & Posner, M. I. (2008). Training, maturation, and genetic influences on the development of executive attention. *Proceedings of the National Academy of Sciences, 102*, 14931–14936. http://dx.doi.org/10.1073/pnas.0506897102.

Sarama, J., & Clements, D. H. (2009a). Building blocks and cognitive building blocks: Playing to know the world mathematically. *American Journal of Play, 1*, 313–337.

Sarama, J., & Clements, D. H. (2009b). *Early childhood mathematics education research: Learning trajectories for young children*. New York, NY: Routledge.

Sarama, J., & DiBiase, A.-M. (2004). The professional development challenge in preschool mathematics. In D. H. Clements, J. Sarama, & A.-M. DiBiase (Eds.), *Engaging young children in mathematics: Standards for early childhood mathematics education* (pp. 415–446). Mahwah, NJ: Erlbaum.

Sarama, J., Clements, D. H., Wolfe, C. B., & Spitler, M. E. (2012a). Longitudinal evaluation of a scale-up model for teaching mathematics with trajectories and technologies. *Journal of Research on Educational Effectiveness, 5*(2), 105–135. ERIC number: EJ961450.

Sarama, J., Lange, A., Clements, D. H., & Wolfe, C. B. (2012b). The impacts of an early mathematics curriculum on emerging literacy and language. *Early Childhood Research Quarterly, 27*, 489–502. http://dx.doi.org/10.1016/j.ecresq.2011.12.002.

Schoemaker, K., Bunte, T., Espy, K. A., Deković, M., & Matthys, W. (2014). Executive functions in preschool children with ADHD and DBD: An 18-month longitudinal study. *Developmental Neuropsychology, 39*(4), 302–315. http://dx.doi.org/10.1080/87565641.2014.911875.

Seo, K.-H., & Ginsburg, H. P. (2004). What is developmentally appropriate in early childhood mathematics education? In D. H. Clements, J. Sarama, & A.-M. DiBiase (Eds.), *Engaging young children in mathematics: Standards for early childhood mathematics education* (pp. 91–104). Mahwah, NJ: Erlbaum.

Starkey, P., Klein, A., & Wakeley, A. (2004). Enhancing young children's mathematical knowledge through a pre-kindergarten mathematics intervention. *Early Childhood Research Quarterly, 19*, 99–120.

Thomson, S., Rowe, K., Underwood, C., & Peck, R. (2005). *Numeracy in the early years: Project good Start*. Retrieved from Australian Council for Educational Research website: http://www.acer.edu.au/documents/GOODSTART_FinalReport.pdf.

Watts, T. W., Duncan, G. J., Siegler, R. S., & Davis-Kean, P. E. (2014). What's past is prologue: Relations between early mathematics knowledge and high school achievement. *Educational Researcher, 43*, 352–360. http://dx.doi.org/10.3102/0013189X14553660.

Weiland, C., & Yoshikawa, H. (2013). Impacts of a prekindergarten program on children's mathematics, language, literacy, executive function, and emotional skills. *Child Development, 84*(6), 2112–2130. http://dx.doi.org/10.1111/cdev.12099.

Weiland, C., Ulvestad, K., Sachs, J., & Yoshikawa, H. (2013). Associations between classroom quality and children's vocabulary and executive function skills in an urban public prekindergarten program. *Early Childhood Research Quarterly, 28*(2), 199–209. http://dx.doi.org/10.1016/j.ecresq.2012.12.002.

Welsh, J. A., Nix, R. L., Blair, C. B., Bierman, K. L., & Nelson, K. E. (2010). The development of cognitive skills and gains in academic school readiness for children from low-income families. *Journal of Educational Psychology, 102*(1), 43–53. http://dx.doi.org/10.1037/a0016738.

Zelazo, P. D., Craik, F. I. M., & Booth, L. (2004). Executive function across the life span. *Acta Psychologica, 115*(2–3), 167–183. http://dx.doi.org/10.1016/j.actpsy.2003.12.005.

CHAPTER TWO

The Use of Concrete Experiences in Early Childhood Mathematics Instruction

Arthur J. Baroody[*,†,1]
[*]University of Illinois at Urbana-Champaign, Champaign, IL, United States
[†]University of Denver, Denver, CO, United States
[1]Corresponding author: e-mail address: baroody@illinois.edu

Contents

Abstract

Addressed are four key issues regarding concrete instruction: What is concrete? What is a worthwhile concrete experience? How can concrete experiences be used effectively in early childhood mathematics instruction? Is there evidence such experiences work? I argue that concrete experiences are those that build on what is familiar to a child and can involve objects, verbal analogies, or virtual images. The use of manipulatives or computer games, for instance, does not in itself guarantee an educational experience. Such experiences are worthwhile *if* they target and further learning (e.g., help children extend their informal knowledge or use their informal knowledge to understand and learn formal knowledge). A crucial guideline for the effective use of concrete experience is Dewey's principle of interaction—external factors (e.g., instructional activities) need to mesh with internal factors (readiness, interest). Cognitive views of concrete materials, such as the cognitive alignment perspective and dual-representation hypothesis, provide useful guidance about external factors

Advances in Child Development and Behavior, Volume 53
ISSN 0065-2407
http://dx.doi.org/10.1016/bs.acdb.2017.03.001

but do not adequately take into account internal factors and their interaction with external factors. Research on the effectiveness of concrete experience is inconclusive because it frequently overlooks internal factors.

Piaget's (1936) stages of development helped popularize the use of concrete experiences and manipulatives in early childhood education. If young children were capable of only concrete thinking, then it made sense to focus on "real math" (everyday experiences or objects), not symbolic mathematics that required abstract thinking. Manipulatives—objects purposely structured to more transparently embody a mathematical idea (e.g., relation or procedure) than real objects—were seen as useful tool for introducing mathematics in a developmentally appropriate manner (Mix, 2010). Although young children are now generally considered more capable of abstract thinking and concepts, manipulatives remain a mainstay of early childhood and elementary mathematics education (McNeil, Uttal, Jarvin, & Sternberg, 2009). This chapter addresses four questions: What is concrete? What is a worthwhile concrete experience? How can concrete experiences be used effectively in early childhood mathematics instruction? Is there empirical evidence that such experiences work?

1. WHAT IS CONCRETE?

Theorists have proposed that mathematical thinking progresses from concrete to abstract along basically two different dimensions. One dimension can broadly be characterized as moving from perceptual-motor actions to mental actions. Bruner (1966), for instance, proposed that development involved adding new modes of thinking to previous capabilities. Young children can represent knowledge in terms of motor actions (enactive phase). More advanced children add the capability of representing knowledge as mental pictures involving key characteristics only (iconic phase). In time, children represent ideas using words or written symbols that do not resemble the idea (symbolic phase). Steffe and Cobb (1988) proposed that mathematical thinking moves from a dependence on representations of physical objects or movements triggered only by perceptual input to a reliance on mental objects or actions (concepts evoked independently of perceptions). The educational implication of such perceptual-motor to mental action views is that instruction should start with concrete experiences that involve physically manipulating objects (e.g., create "muscle memory").

An alternative way of conceptualizing the concrete-to-abstract progression is from specific to general thinking. For instance, Resnick (1992) conceptualized concrete-to-abstract development as moving from local (e.g., context- or object-specific) concepts to general concepts (e.g., broad generalizations applied or regardless of context). Put differently, whereas concrete knowledge, which is based on personnel experience, typically can be applied to familiar situations, abstract knowledge, which is based on generalizations, is applicable to unfamiliar problems, as well as familiar situations (Bisanz, Watchorn, Piatt, & Sherman, 2009; Prather & Alibali, 2009). From this perspective, concrete experiences are those that build on young children's existing informal and familiar knowledge gleaned largely from meaningful everyday experiences. Abstract is the formal symbolic knowledge that is general and applicable even beyond personally meaningful contexts.

Each of these perspectives of concrete-to-abstract development seems to apply to particular aspects or levels of development. That is, in some cases, physically acting on objects or physical movement may be needed to promote more advanced thinking (Mix, 2010). In other cases, the particular mode for prompting reflection, insight, and generalization—whether it be physical object, virtual image, verbal analogy, or formal (written) symbols—may be unimportant as long as the context can be related to familiar knowledge. Consider the development of early addition.

- Level 1. Initially, children may need to literally see, for example, one item added to two to make three items to construct a basic understanding of composition or addition. Once such a basic understanding of addition is constructed, children may look for additional examples such as decomposing ••• into •• and • or noticing that two objects and two more make four (Baroody & Rosu, 2006).
- Level 2. About 2.5 years of age, children can solve nonverbal addition (e.g., two items placed on a mat and then hidden and one more item placed next to the mat and then moved under the mat) using a mental representation involving, perhaps, a mental or iconic image (Huttenlocher, Jordan, & Levine, 1994).
- Level 3. Several years later, children solve equivalent word problems—symbolic problems in a verbal format set in a familiar context (Jordan, Hanich, & Uberti, 2003).

- Level 4. Later yet, children solve equivalent formal symbolic problems in a verbal format ("How much is two and one"; Jordan et al., 2003) or a written format (e.g., 2 + 1) without a context.[a]

Children at Levels 1 and 2 could be considered as having *sensory-concrete knowledge* and in need of sensory input to make sense of addition or solve for the sum (Clements, 1999; Sarama & Clements, 2009a). Level 3 children, who can solve word problems but not symbolic problems, would have *integrated-concrete knowledge* in they can relate (assimilate) the symbolic language of a word problem to familiar knowledge. Level 4 children are operating at the integrated-concrete knowledge level *if* they must relate the formal expression 2 + 1 to familiar knowledge to make sense of the problem and devise or select a strategy for solving it. Level 4 children who do not have to relate the symbolism to a meaningful context to do so are operating at an entirely symbolic level and as having *symbolic knowledge.*

The distinction between sensory-concrete and integrated-concrete knowledge is important because "concrete" is often equated exclusively with physical objects or movements. Consistent with the perceptual-motor to mental actions perspective, physically real materials or experiences may be necessary for novice learners or developmentally less advanced children—especially presymbolic or prelanguage infants and toddlers. However, consistent with specific to general perspective, the same may not true of developmentally more advanced children. For such children, an interpretation of concrete experiences would *not* necessarily exclude using verbal analogies, pictures/diagrams, or virtual-based activities as a basis for instruction, *if* it built on a child's existing knowledge (Baroody, 1989; Clements & McMillen, 1996). There is substantial evidence that children's familiar informal knowledge (including verbal counting-based representations) plays a key role in understanding and learning formal systematic knowledge

[a] Doug Clements (Personal Communication, 1/5/2017) indicated, "We get children to solve verbal (but still abstract) 'two and one more' problems the same time as word problems." According to the specific-to-general hypothesis, the order in which the word problems (WP) and verbal symbolic problems (VSP; e.g., "two and one more") are presented should matter. Specifically, presenting WP should induce better performance on VSP than in VSP-first condition, because the relatively sterile VSP context could be related to familiar and meaningful situations evoked by the WP (i.e., the previously introduced meaningful contexts would give meaning to relatively abstract statements such as "How much is two and one more?"). Moreover, the difference would be particularly pronounced if VSP were even more sterile as those used by Jordan and her colleagues (e.g., "How much is two and one?"). That is, the presence or absence of the term "more" may be important because its presence may better connect a VSP with children's existing informal change add-to understanding of addition (e.g., an initial collection of two is made larger by one *more* item). Clearly, though, these two hypotheses need to be empirically tested.

(Ginsburg, 1977; Ginsburg, Klein, & Starkey, 1998). Verbal analogies—relating unfamiliar symbols to familiar experiences—can be invaluable in understanding formal symbols (e.g., relating the minus sign to "taking away," relating the idea of angle to "the amount of turn," or introducing fractions such as 2/3 in terms of the familiar experience of fair sharing: two pizzas shared fairly among three children; Baroody & Coslick, 1998). Diagrams can be useful in distinguishing types of addition and subtraction and choosing an appropriate strategy (Jitendra et al., 2013). Young children clearly benefit from, for example, the careful use of virtual or computer-based instruction (Clements & Nastasi, 1993; Clements & Sarama, 2003, 2004, 2012). Indeed, Sarama and Clements (2009a) detail seven ways virtual manipulatives afford greater integrated-concrete learning than do physical manipulatives.

However the concrete-to-abstract continuum is conceptualized, the evidence does not support the view that children's thinking is concrete across mathematical domains and adults' thinking is abstract (McNeil et al., 2009). A child's thinking may be relatively abstract in one area such as cardinal numbers (numbers that represent the total of a collection) but relatively concrete in another such as subtraction or the composition of shapes.[b] Indeed, a 3-year old may have a relatively abstract understanding of (the cardinal number) "two" but a relatively limited and inexact understanding of "ten." This perspective underscores the need for educators to understand a child's developmental level within a particular domain when considering whether an instructional activity is sufficiently concrete. This point is underscored in the next two sections of this chapter.

2. WHAT IS A WORTHWHILE CONCRETE ACTIVITY?

In his analysis of why Progressive Education was unsuccessful, Dewey (1963) concluded that instruction should not consist of a conglomeration of activities without clear educational purposes. For example, concrete experiences such as everyday calendar activities are often used without a clear, justifiable goal, or good effect (Clements, Baroody, & Sarama, 2013; National Research Council, 2009). Similarly, using the same M&M activity to teach probability for four consecutive grade levels without modification

[b] The flipside is that even adults can frequently benefit from having unfamiliar aspects of formal mathematics related to their familiar knowledge and can benefit from reflecting on manipulative models (McNeil et al., 2009).

or extension is not likely to result in new learning. Manipulatives used in isolation (e.g., left in a learning center without guidance or feedback) are not likely to achieve worthwhile instructional goals (Uttal, 2003).

Dewey (1963) argued that teachers must avoid *miseducative experiences* (activities for the sake of activity and that may actually impede development) and strive to provide educative experiences (experiences that lead to worthwhile learning or a basis for later learning; see also Sophian, 2004). Concrete experiences and manipulative use need to be guided by a plan for achieving useful instructional goals. According to Dewey, educative experiences result "from an interaction of external factors, such as the nature of the subject matter and teaching practices, and internal factors, such as a child's [developmental readiness] and interests" (Baroody, 1987, p. 37). Teachers, then, must ensure that external and internal factors mesh.

Piaget's principles of meaningful learning illustrate how the external factor of introducing formal symbols needs to fit the internal factor of a child's developmental level. His principle of assimilation specifies that new information is understood and interpreted in terms of existing knowledge. For example, introducing the formal symbolism for subtraction by relating $5-3=?$ to "What is five minus three?" may not make sense to children who do not know the meaning of the formal term "minus." As a result, even though they might be able to solve an analogous word problem, students may be unresponsive, simply guess, or assimilate the formal expression to what do they understand (i.e., assume it represents addition and answer "8"). In contrast, relating $5-3=?$ to their informal knowledge of subtraction (e.g., "Five candies take away three candies leaves what?") can enable them to understand and learn that the minus sign means subtract. Piaget's moderate novelty principle prescribes that information that is neither too familiar nor unfamiliar will pique interest and engagement. For instance, for a child who can already count to thirty-one and read the written numbers to 31, the task of counting up to today's date on a calendar can be readily assimilated but is uninspiring. In contrast, the question "How many tens does 31 show?" asked of the same child without grouping and place-value knowledge cannot be assimilated to prior knowledge and is, thus, incomprehensible and uninteresting. However, a novel compare-type problem that builds on the child's familiar knowledge (e.g., "If today is December 5 and our field trip is December 11, how many more days until our trip?") might make sense and spark interest in solving it. Dewey's principle of interaction is perhaps even more closely akin to Vygotsky's (1962) zone of proximal development: The domain-specific—or even activity-specific—competence a child can exhibit with careful and minimal assistance.

Hypothetical learning trajectories (HTLs) can serve as an invaluable tool for implementing Dewey's (1963) principle of interaction—for ensuring external factors (e.g., instructional goals and activities) effectively interact with the internal factor of a child's developmental level. HLTs—which include theoretically and empirically based developmental levels or steps—(a) define goals for meaningful instruction; (b) provide assessments for identifying a pupil's current developmental level and the next (developmentally appropriate) instructional step; and (c) detail instructional activities to help the pupil achieve this next level (Daro, Mosher, Corcoran, & Barrett, 2011; Sarama & Clements, 2009b, 2009c). HTLs are invaluable in readiness or remedial instruction in that they lay out in sequential order the developmental prerequisites and accompanying instructional activities needed to achieve a target level of instruction. In effect, they can provide guidance in what scaffolding children at a particular developmental level need.

Consider, for instance, the case of 3-year-old Alison (Baroody, 1987). Her father asked, "What number comes after nine when we count?" Stumped, Alison did not reply. Her mother intervened by asking, "What comes after "one, two, three, four, five, six, seven, eight, nine." Alison quickly responded, "Ten." This scenario makes sense in terms of a verbal-counting HLT summarized in Table 1. Alison was at Level 2 in Table 1 (the unbreakable chain level) and had to use a "running start" (count from "one") to indicate the next number in the counting sequence (e.g., What comes after nine?). As children become familiar with the counting sequence, they achieve the breakable chain level (Level 4A in Table 1) and no longer need a "running start" to determine the number that follows another. Instead they can access the counting sequence at the given number and state its successor (e.g., immediately indicate that "ten" comes after "nine"). As her father's question (What comes after nine?) was two levels above Alison's developmental level, she may not have understood the question and certainly did not have the means for answering. Her mother's question (What comes after one, two...nine?) was at her developmental level and thus permitted her to apply what she knew to solving the novel task of answering a number-after question. According to the HLT in Table 1, instruction that might help a child move from Level 2 (the unbreakable chain level) to achieving Level 4 (the breakable chain level) is the transitional Level 3 (determining the number-after with increasingly abbreviated running starts).

Table 1 An Example of a HLT: Verbal-Counting Development

HLT Level	**Conceptual Basis (Fuson, 1988)/Comments**
Level 1: Sing-songer	*String level*: Knows the first few number words as an undifferentiated string of sounds (multisyllabic sound): "onetwothree."
Level 2: Counter from "one"	*Unbreakable chain level*: Differentiates among number words of the sequence but does the have the flexibility to reproduce the sequence other than starting with one (i.e., cannot enter the sequence at any point and count). Thus, to determine the number after five, a child would have to use a running start—that is, count from 1 to 5 (e.g., count: "1–2–3–4–5, 6").
Level 3: Number-after citer with abbreviated running starts	*Transition between unbreakable chain level and breakable chain level*: In order to help children at Level 2, a teacher or parent could pose number-after questions with an increasingly abbreviated running start: "What comes after 2, 3, 4, 5?"; then "What comes after 3, 4, 5?"; and finally "What comes after 4, 5?"
Level 4A: Number-after citer without a running start	*Breakable chain*: Familiarity with the counting sequence permits the child to enter the counting sequence at any point and specify the next number. Thus, a child no longer needs a running start or even an abbreviated running start to determine what comes after any given number (e.g., a child answer the question "What comes after 5?" by saying "5, 6," or "6").
Level 4B: Counter from *n*	Once a child can determine the number after another, they can continue a count from any *n* (e.g., in response to the request "Start with 5 and count," a child counts "5, 6, 7, 8, 9..." or "6, 7, 8, 9...").
Level 5: Counter-on using patterns	*Transition between breakable chain level and numerable chain level*: At this level, a child can count on a specified number of times intuitively—usually only a few times—by means of a pattern (e.g., in response to start with 5 and count two more times, a child counts, "Five, SIX, seven").
Level 6: Counter-on keeping track	*Numerable chain level*: Realizes that counting words themselves can be counted. Requires a keeping-track process to determine when to stop the counting-on process (e.g., for start with five and count four more times: "5, 6 is 1 time, 7 is 2 times, 8 is 3 times, *9* is *4 times*").

The careful use of HLTs can better ensure that concrete experiences at the preschool (age 3–5 years) and primary (grades K to 3) levels are educative. A key cause of mathematical learning difficulties in school, especially among children at risk due to poverty, are gaps in their readiness (informal) knowledge. For preschoolers without rich learning opportunities, informal knowledge can be seriously incomplete or narrow (Baroody, 1987). Identifying where a child is on a HLT can help preschool or primary-level teachers provide the educative experiences that remedy informal deficiencies and extend existing informal knowledge strengths. Another key cause of mathematical learning difficulties in school is a gap between school mathematics instruction and children's existing knowledge (Ginsburg, 1977). Identifying where a child is on an HLT can help preschool to primary-level teachers select the concrete experiences that connect formal mathematics to the child's informal knowledge and enable the child to assimilate symbolic school instruction successfully.

3. HOW CAN CONCRETE EXPERIENCES BE USED EFFECTIVELY IN EARLY CHILDHOOD MATHEMATICS INSTRUCTION?

Since at least the time of Resnick's (1982; Resnick & Omanson, 1987) failed attempt to use base-ten blocks to foster an understanding of base-ten, place-value concepts and skills, it has been clear that manipulatives do not guarantee meaningful learning (Ball, 1992; Baroody, 1989; Clements & McMillen, 1996). That is, children may not spontaneously link concrete models or what they learn from working with concrete objects to other representations, particularly written representations (Uttal, O'Doherty, Newland, Hand, & DeLoache, 2009). How to link a concrete model to symbolic mathematics to effectively illuminate an intended referent remains a critically important challenge for early childhood mathematics instruction. Discussed next are cognitive perspectives, a perspective based on Dewey's interaction principle and learning trajectories, and instructional strategies for ensuring an educative experience.

3.1 Perspectives From Cognitive Psychology

3.1.1 Cognitive Alignment Framework

A key factor in linking concrete models effectively to targeted referent knowledge may be how well the model embodies the relations, steps, or other attributes of the referent knowledge. As Laski and Siegler (2014,

p. 861) noted, "The cognitive alignment framework proposes that the more precise the alignment among the desired mental representation, the physical materials being used to promote learning, and the activities that direct learners' thoughts and actions during the acquisition process, the greater the learning is likely to be." See Clements and Battista (2000) for a discussion on cognitively aligning software.

Consider, for example, children's relative magnitude judgments (see Baroody & Purpura, 2017, for a detailed discussion). Research indicates that initially children do not recognize that number words represent relative magnitude. Theoretically, this would result in assigning all the numbers 1–10 the same level of magnitude or point on a number line (Frame A in Fig. 1). In time, they recognize that smaller, familiar numbers have distinct relative magnitudes but their sense of relative magnitude with larger, unfamiliar numbers is fuzzy. For instance, in finding out that her father was 42, 5.5-year old, Arianne asked in all innocence, "Is that close to a hundred, Daddy?" For younger children, the first few numbers may have distinct relative magnitudes, but even somewhat larger numbers such "six" or "seven" may simply be viewed as "many" or "a lot" and almost indistinguishable from "ten." The result is a response pattern that is logarithmic even for numbers 1–10 (see Frame B in Fig. 1). Children then see each succeeding number in the counting sequence from 1 to 10 as representing a larger amount (a linear representation of magnitude comparisons; Frame C in Fig. 1) and eventually as exactly one larger than its predecessor (i.e., a linear representation with a slope of 1; Frame D in Fig. 1). Research also has indicated that playing a linear numerical board game (*The Great Race*), which directly models a mental left-to-right number list for counting from 1 to 10 (see Fig. 2; Resnick, 1983) significantly improved the linearity of children's magnitude judgments and other aspects of early numeracy (Ramani & Siegler, 2008; Siegler & Ramani, 2008). Indeed, playing *The Great Race* with a linear board significantly produced better results than doing so with a circular board (Ramani & Siegler, 2011; Siegler & Ramani, 2009). Unfortunately, methodological issues cloud what conclusions can be drawn from such research (again see Baroody & Purpura, 2017, for a detailed discussion).

Similarly, playing a semilinear numerical board game using a 0-to-100 hundreds chart (*Race to the Moon*) significantly improved the linearity of children's magnitude judgments to 100 with a counting-on, but not a counting from "one" process (Laski & Siegler, 2014). This was attributed to the fact that, unlike the count-from-1 training, the count-on training directs attention to the numbers in the squares by requiring a child to explicitly name

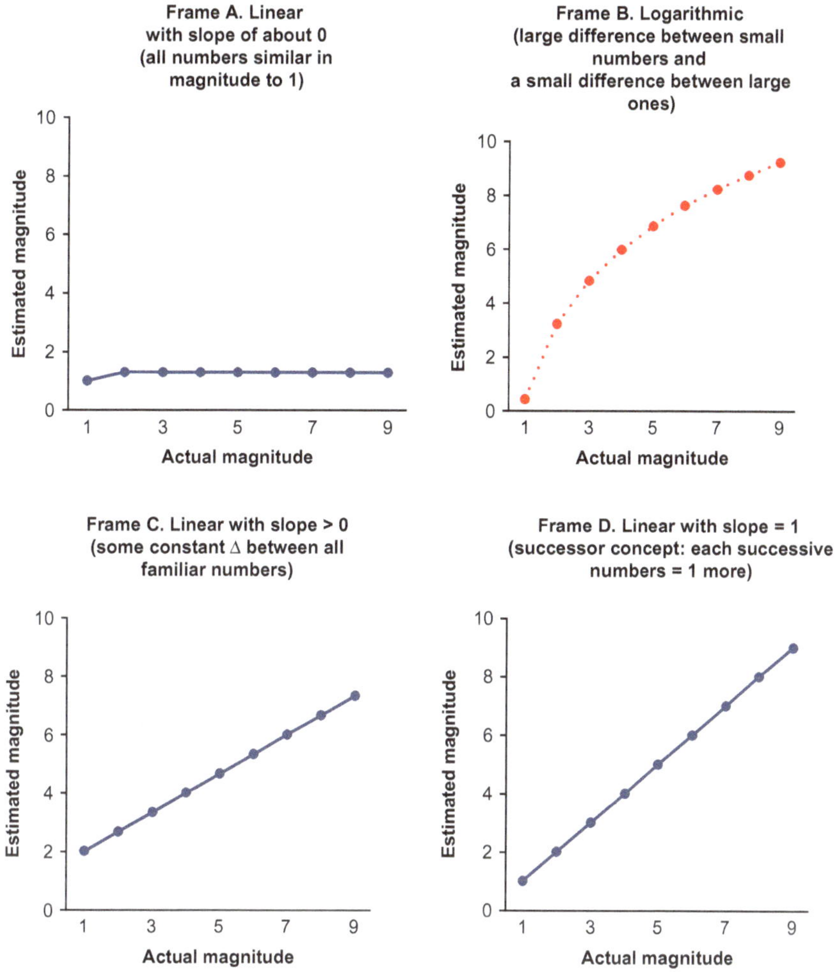

Fig. 1 Different response patterns for magnitude estimation task (where does *n* fit on 1–10 number line?).

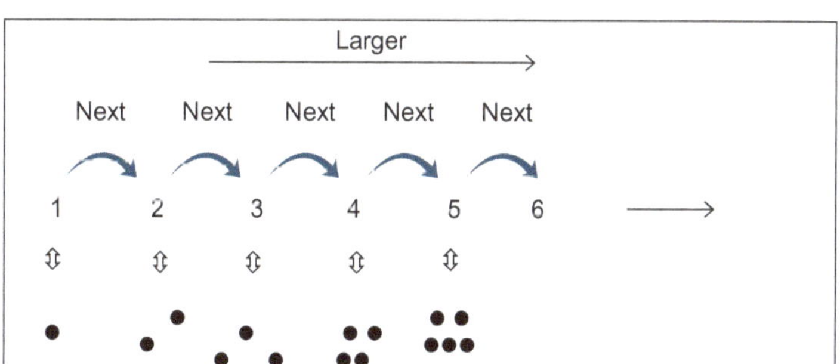

Note. As depicted, the figure represents a child who associates specific quantities with the number words "one" to "six" and who understands the increasing magnitude principle—a number further along in the sequence represents a larger quantity. The representation for a child who has discover the successor principle would include the label "next is one more" (instead of "next").

Fig. 2 A representation of magnitude relations of the counting sequence: a "mental number list" (based on fig. 3.1 in Resnick, 1983).

them. The resulting encoding process "provides the data needed to correlate numbers in different squares with the visuospatial, auditory, kinesthetic, and temporal cues that accompany reaching that square. In contrast, the counting-from-1 procedure does not require, or even encourage, attention to the numbers in the squares" (p. 861). Although counting-on 1 or 2 especially (e.g., moving two spaces from 37 by simply counting "38, 39") may not require an explicit keeping-track process, Baroody and Purpura (2017) hypothesized adding a process (e.g., counting-on "38 is 1 more, 39 is 2 more") might significantly enhance the efficacy of *Race to the Moon*.

Researchers and educators need to consider carefully what instructional goals they are trying to achieve with particular concrete materials and, as needed, adapt the process by which children interact with the materials. An adaptation of Wynroth's (1986) *Slides and Ladders*, the *Grid Race* (Baroody & Coslick, 1998)—like *Race to the Moon*—utilizes a 0–100 hundreds chart. The goal of *Slides and Ladders* or the *Grid Race*, though, is to foster discovery of mental-addition (and subtraction) rules, such as the decade-after rule (adding 10 to a decade results in the next decade; e.g., $50+10=60$) or the add-10 to a two-digit number rule (the sum is the next decade with the same ones digit; $57+10=67$), and a partial-decomposition mental-addition strategy (e.g., $57+36=57+30+6=87+6=93$). For such goals, the 0-to-5 spinner used in *Race to the Moon* must be replaced by one of several different decks of cards designed to emphasize a particular regularity. For fostering the decade-after rule, the deck would be composed of 0 or 10 cards. For promoting the add-10 to a 2-digit number rule, the deck would be composed of some cards with a number from 1 to 9 and an abundance of cards with 10. For promoting the partial-decomposition strategy, the deck would consist of cards with two-digit numerals (e.g., 14 and 23). For all three versions, the card a player draws would determine the number of spaces a child moves on the hundreds board. For instance, if a child is on space 57 and draws a 10, the child would count by ones 10 spaces to move to space 67. Wynroth's (1986) rationale was that children would discover a shortcut for the laborious process of counting 10 spaces from 57 to 67, namely you can simply move vertically up one row in the table whenever you add 10. Theoretically, this single vertical-move shortcut provides a basis for discovering the decade-change rule for adding 10 to multidigit numbers (e.g., $57+10$ is the next decade, 60, plus the same number of units, seven).

If Laski and Siegler's (2014) hypothesis that count-on procedure requires an encoding process that facilitates attention to number and numerical relations is correct, then, a version of the *Grid Race* that requires a child to specify the problem and outcome (e.g., "57 plus 10 is 67") should be more efficacious in helping children discover mental-addition (and subtraction) shortcuts than a version in which children do not relate their move to a verbal equation. Indeed, encouraging children to use a keeping track process (e.g., "57, 58 is 1 more, 59 is 2 more, 60 is 3 more…67 is 10 more") might be particularly motivating for finding the vertical-jump shortcut or the same ones—next decade rule.

3.1.2 The Dual-Representation Hypothesis

A rationale for using hands-on manipulatives is that the activity of physical exploration can be progressively internalized to form conceptual understanding. Intuitively making manipulatives more perceptually rich should make them more attractive and interesting and help children focus better on a task. Recent research raises questions about these assumptions and even indicates that such physical manipulation and perceptual enrichment of manipulatives may, in many cases, distract from learning (Kaminski, Sloutsky, & Heckler, 2008, 2009, 2013; McNeil & Jarvin, 2007; McNeil et al., 2009).

According to the dual-representation hypothesis, successful use of a symbolic artifact entails mentally representing both its physical and abstract nature. However, such a dual representation is challenging to young children because they are less likely to recognize that a new artifact can have symbolic import. Moreover, they have difficulty simultaneously formulating two active (concrete and abstract) representations of an artifact. "Manipulations that increase the salience of a [model] as a concrete object should decrease children's appreciation and use of it as a [representation], whereas manipulations that decrease children's focus on the object properties of the model should increase their ability to focus on what the [stand-in] represents" (Uttal et al., 2009, p. 157). Consider four examples.

1. DeLoache (2000) hid a small toy dog in a miniature replication of a room and then asked a child to use this model to find a large stuffed dog hidden in the regular size room. In one study, 3-year olds allowed to physically interact with the model during a free play prelude were less successful than those who did not play with the model in using the representation of the hiding place and regular size room to find the hidden larger dog. In another study, 2.5-year olds who viewed the model dog being hid from behind a window were more successful than their peers who had physically interacted with the small toy dog. Marzolf and

DeLoache (1994) found that using a picture as a "map" was superior to using a three-dimensional model. Uttal et al. (2009) concluded that physical interaction interferes with viewing a manipulative as a model.

2. Sarama and Clements (2009a) concluded that computer-based manipulatives may facilitate generalization and transfer more effectively than physical manipulatives. Uttal et al. (2009) concluded, "Using computer-based manipulatives reduces the demands of dual representation, enabling children to focus less on the on-screen objects themselves and more on the connections between the manipulatives and mathematical representations" (p. 138).
3. The results of Uttal et al. (2013) appear to support the view that children have difficulty connecting concrete and abstract representations. In Experiment 1a, these researchers found that manipulative-based instruction on multidigit subtraction with second graders did not transfer to using a written algorithm and instruction on a written algorithm did not transfer to block models.
4. Experiment 1b by Uttal et al. (2013) corroborated the view that highly attractive manipulatives or concrete objects may make it particularly difficult to link a concrete model with an abstract referent. Specifically, second graders who used regular Digi-Blocks designed to teach the base-ten system and how to solve two-digit subtraction seldom used the manipulative inappropriately, whereas most of their peers who used brightly colored and patterned blocks were side tracked.
5. Kaminski et al. (2009) had 11-year olds learn a mathematical concept either concretely with perceptually rich symbols or abstractly with symbolic models. Although the concrete model made learning easier, it resulted in less transfer, whereas the symbolic model made learning harder but resulted in greater transfer. The implication is that concrete representations may limit the generalizations pupils can derive.

Clearly, educators need to use manipulatives thoughtfully, not laden representations with frivolous or unnecessary details, and consider carefully the advantages and disadvantages of manipulatives in various forms. Barmby, Harries, Higgins, and Suggate (2009) reasoned that groups-of or number-line models of multiplication obscure the operation's commutative property. Indeed, it is plausible (based on the authors' informal observations) that children may also better exhibit understanding of multiplicative commutativity in the sterile context of symbolic expressions (e.g., 9×7 and 7×9) than with even rectangular-array models (e.g., 9 rows of 7 items or 7 rows of 9 items) or area models (e.g., rectangles 9- × 7-linear units and

7- × 9-linear units), because the concrete model of 9 × 7 and 7 × 9 in each case look different.

However, the evidence supporting the dual-representation hypothesis is not clear-cut.

1. A maxim in mathematics education is: If manipulatives are to be used effectively as a teaching tool, provide the manipulatives to children for a period of free exploration. With their curiosity satisfied, they will then be ready to use the manipulatives as intended by the educator—as a tool for prompting reflection and learning. Children in the DeLoache (2000) study were given up to 10 min of free play with the model materials "or any time after 5 min [if] the child seemed to be losing interest" (p. 334). It is unclear whether—consistent with the maxim just described and inconsistent with the dual-representation hypothesis—the children who lost interest performed better than those who did not. It also seems important to distinguish between child-initiated activities and adult-imposed tasks, because children may exhibit a competence on the former not revealed on the latter (Baroody, Lai, & Mix, 2006). Requiring a young child to engage in an adult-imposed task is challenging enough without doing so while the child is still engaged in an interesting child-initiated activity. A way of making adult-imposed testing (or instruction) more palatable and motivating to young children is to make it as similar as possible to child-initiated activities (e.g., present it as a game). As a transition from free play to the experimental task in the DeLoache (2000) study, "the experimenter indicated they would 'do something different now'" followed by an "extensive orientation to the [experimental] task" (p. 334). A perhaps more engaging and potentially more productive transition might be introducing the experimental task as the "Hiding Game." The tester could first show the large room, explain that a large stuffed dog is hidden somewhere in the room, and then offer a clue or hint to the hiding place with toy dog and miniature room. Such a transition would help evaluate whether failure on the standard task when preceded by free play has more to with motivational issues than cognitive limitations. Even if dual representation proves to be a challenge to 2.5-year olds over a variety of conditions (including motivating situations), symbolic representation ability and the ability to discern adult expectations improves with age. Thus, it is not clear that the dual-representation hypothesis is (as) applicable to 7-year olds encouraged to show how base-ten blocks could be used to solve a two-digit subtraction problem involving renaming in the Uttal et al. (2013) study.

2. Direct comparisons of the use of virtual and physical manipulatives are relatively rare, and the advantage of virtual manipulatives may be due to a number of factors other than that they are less distracting than physical manipulatives, which can be touched. See Sarama and Clements (2009a) or Clements and Sarama (2012) for a list of advantages of virtual manipulatives over real manipulatives.
3. A plausible alternative explanation for the results of Uttal et al.'s (2013) Experiment 1a is that the lack of transfer between modes was due to the ineffective nature of the instruction. Both manipulative- and written-based instruction focused on procedures, not the conceptual rationale for the multidigit subtraction algorithm and spanned only two sessions. Brief, meaningless instruction—whether with concrete materials or written symbols—is not likely to promote flexible application of knowledge (Baroody, 1989).
4. Moreover, the key issue may not be as simple as "gaudy vs plain" (as suggested by the result of Uttal et al.'s (2013) Experiment 1b. Petersen and McNeil (2013), for example, found that perceptually rich objects facilitate performance when counting tasks involved unfamiliar objects but hindered performance when the tasks involved familiar objects. Although it would seem that the dual-representation hypothesis would predict that novelty and perceptual richness would combine to render a jazzy, unfamiliar object relatively useless as a learning tool, the authors speculated that the increased attention such objects attracted was channeled into understanding the task, whereas, with familiar objects, it activated existing knowledge of the object and served to draw attention away from the task.

   Perhaps more importantly, there may be value in making a model perceptually rich, interesting, or attractive *if* it helps draw attention to a critical relation or otherwise promotes meaningful learning. For example, the bright colors and patterns of Digi-Blocks in the Uttal et al. (2013) study did not serve any pedagogically useful purpose (e.g., the blocks did not help underscore the grouping or place value ideas underlying the two-digit subtraction procedure). In contrast, using one bright color to represent groups of 10 and a different bright to represent units might help to underscore a grouping concept and the conceptual rationale for regrouping.
5. There are plausible alternative explanations to Kaminski et al.'s (2008, 2009, 2013) evidence. One is the common problem in instructional/curriculum evaluation of using a dependent measure (the test of transfer) that better matched the abstract learning condition. The concrete

training involved showing participants two measuring cups and asking them to determine what would be "left over" after combining the content of one cup of the cups (e.g., 1/3 + 1/3 results in 2/3 leftover; 2/3 + 2/3 results in 1/3 leftover[c]). The abstract training involved generic shapes (e.g., ● + ● results in ◆; ◆ + ◆ results in ●). The transfer task involved "intentionally concrete and contextually rich" displays (e.g.,  +  results in  ;  +  results in ), which theoretically should favor the concrete trainees (Kaminski et al., 2009, p. 153). However, whereas the concrete training involved logical reasoning about fraction addition, the abstract training might simply elicit memorization of rules (two circles makes a diamond and vice versa or simply two of a kind make one of a different kind). Whereas applying knowledge of fraction addition is readily apparent in the concrete condition, inducing arbitrary rules might be more challenging. This would account for why the concrete condition resulted in better learning during training. However, on the transfer task, fraction addition is irrelevant but the rule/rules memorized by rote in the abstract condition can be readily applied to two of the four test trials. Indeed, De Bock, Deprez, Van Dooren, Roelens, and Verschaffel (2011) found that the concrete learning context transferred to a similar concrete context and their qualitative analysis raised questions about Kaminski et al.'s (2013) interpretation of what students learned from the abstract training.

It is essential to keep in mind Piaget's caution that it is not the physical activity itself that is important but a child's reflection on the physical activity (Baroody, 1989; Sarama & Clements, 2009a). For research on learning beyond the sensory-concrete level, a focus on whether children touch objects or whether objects are perceptually rich is of secondary, or perhaps little, importance. The key issue is whether a manipulative is pedagogically meaningful—whether it engages children's thinking in ways that prompt reflection on ideas or issues targeted by an educator. That is, a "concrete" model is one that connects with what children already know and prompts them to reconsider and extend this existing knowledge.

---

[c] The concrete training seems confusing. For example, if the analogy is interpreted as 1/3 in the first cup is poured into the second cup containing 1/3, nothing is left in the first cup and 2/3 is left in the second cup. So, in this case, the answer (what is leftover) seems to reference what is in the *second cup* after combining the two cups. With 2/3 + 2/3, 1/3 would be left in the first cup and the second cup would have 3/3. So in this case, the answer (what is leftover) seems to refer to the *first cup*, not the second cup. An inconsistent analogy stacks the deck against finding a regularity.

Although children's reflection can and should be guided by a teacher or activity, it probably cannot be forced or imposed. As with instruction in general, imposing how to use manipulatives may be no more effective than imposing symbols and written algorithms on children (Baroody, 1989; Gravemeijer, 2002). As McNeil et al. (2009, p. 139) elegantly noted: Some evidence supports that "students may not actively process the consequences of actions on concrete objects unless they, themselves, are the directors of that activity (Martin, 2009) ..." Future research must investigate the best ways for teachers to strike a balance between providing direction and allowing students to regulate and direct their own activity.

3.2 Perspective Based on Dewey's Interaction Principle and Learning Trajectories

Cognitive perspectives (the cognitive alignment framework and dual-representation hypothesis) focus on external factors. To better understand the effective use of concrete materials, it is essential to take into account the interaction of external and internal factors and children's development or readiness as embodied by HLTs. However closely a concrete representation aligns with a mathematical idea, it does not guarantee that it will be an appropriate or effective model for a child *if* the formal idea is developmentally too advanced. For this reason, children may need to begin with a model that is aligned with their existing developmental level of mathematical understanding and thinking and only approximates the formal mathematical idea. In effect, educational researchers and practitioners might better consider a series of concrete models that build on children's informal understanding and thinking and increasingly becomes aligned with the formal mathematical idea. In this subsection, the following five areas of early childhood mathematics education are analyzed in terms of Dewey's interaction principle and HLTs: patterning, a linear representation of the counting numbers and its hypothesized developmental prerequisites, using number lines to formally introduce magnitude comparisons and arithmetic, and base-ten place-value concepts and skills.

3.2.1 Patterning

In *Workjobs* published in 1972, Baratta-Lorton recommended using letters to label, identify, and discuss repeating patterns. For example, she suggested labeling ○□○□○□, up-down-up-down-up-down, blue-red-blue-red-blue-red, and long-short-long-short-long-short all ABABAB or, for short, AB patterns and ○□□○□□○□□, up-down-down-up-down-down-up-down-down, blue-red-red-blue-red-red-blue-red-red, and long-short-short-long-short-short-long-short-short all ABBABBABB or, for short, ABB patterns. Note that the short version serves to identify a repeating pattern's "core"—the basic elements that are repeated or part that repeats.

Doug Clements, who was a kindergarten teacher at the time, recalls being highly skeptical of Baratta-Lorton's (1972) recommendation to label patterns with letters because it struck him as too abstract. Nevertheless, he tried the letter-labeling method with his classes and became convinced of its value. The method is now a popular feature of early childhood mathematics programs, such as the *Building Blocks* program (Clements & Sarama, 2013).

Baroody (1993, p. 2–84) offered the following explanation for why the letter-labeling method may be effective:

> *"Labeling patterns can help children find repeating patterns and discover commonalities among such patterns ... using letters can help students analyze and identify patterns. Moreover, coding patterns with letters gives children a convenient way of explicitly describing patterns ... such letter codes can help children see that patterns constructed of different materials can* share the same structure *(emphasis added). Understanding that a pattern can be embodied in various, even different-looking, ways represents an important advance in children's mathematical thinking. There comes a dawning recognition that mathematics is a search for underlying structure that transcends appearances."*

Note that the last insight could arguably be classified as a really "big idea."

Recently, Fyfe, McNeil, and Rittle-Johnson (2015) adduced evidence that labeling concrete material abstractly with letters (e.g., "ABABAB") is significantly more powerful than doing so concretely with a physical characteristic such as color names (e.g., "blue-red-blue-red-blue-red") in fostering patterning among preschoolers. They concluded: "Using abstract language to describe patterns facilitates children's pattern abstraction" (p. 5) and was particularly beneficial when children correctly adopted the letter-labeling method.

Fyfe et al.'s (2015) results reinforce the point that manipulatives themselves do not guarantee learning (Ball, 1992; Baroody, 1989; Clements & McMillen, 1996) but that what is critical is how concrete materials are used. Specifically, they concluded that their results are consistent with the cognitive alignment framework (Laski & Siegler, 2014) and the dual-representation hypothesis (Kaminski et al., 2013; Uttal et al., 2009). Fyfe et al.'s (2015) research, though, raises four interesting questions that underscore the importance of Dewey's interaction principle.

1. Is there a critical age or developmental level for successfully introducing the letter-labeling method?

   Fyfe et al. (2015) found a significant age effect—older, but not younger, children tended to benefit. (Participants ranged in age from 3.6 to 4.9 years, average = age 4.4, SD = 0.4.) Perhaps there is a critical age that is necessary to benefit from the letter-labeling training. Research needs to address this potentially critical internal factor. Perhaps more likely, though, children may need to reach a certain developmental level before the letter-labeling method makes sense. Research is needed to ascertain the parameters of this critical internal factor—including any developmental prerequisites for understanding and using the letter-labeling method.
2. Where in a patterning HLT should the letter-labeling method be introduced?

   The learning trajectory detailed by Sarama and Clements (2009b) indicates that the ability to translate a pattern in one medium into another (e.g., translate a pattern comprised of geometric shapes into numbers or letters) develops relatively late (Level 4 in Table 2) and seems to imply teaching the letter-labeling method should not be taught before 6 years of age. Unclear, though, is whether Levels 1–3 are developmental prerequisites for learning the letter-labeling method or whether these levels might benefit from introducing this method first. Note that the Level 1 (Pattern Recognizer) activity of describing the pattern on a strip (e.g., square, circle, square, circle, square, circle, etc.) may not necessarily involve recognizing a pattern. A child could just name each shape in turn. Note that the Level 2 (Pattern Duplicator) activity of fixing or duplicating a pattern such as square, circle, square, circle, square, and circle may simply involve matching a shape on the strip with the shape of a block. So in a real sense, a child could duplicate a pattern by using a nonpattern-based process. The Level 2 activity of extending a pattern would seem to require at least an intuitive understanding of a repeating core. As the next question suggests, it may be particularly useful to introduce the letter-labeling method before introducing more complex patterns (Level 3 in Table 2).[d]

[d] Nicole McNeil (Personal Communication, 3/23/2017) noted that it would be interesting to compare musical sounds, such as da-dum, with the letter-labeling method to see if the former might help young children even more in identifying pattern cores.

Table 2 Selected Levels From the Patterning Learning Trajectory From Sarama and Clements (2009a)

Level Name	Description of Developmental Level
0. Preexplicit Patterner	Implicit pattern detection and use. Labels a visual, rhythmic, or other regularity a "pattern" (e.g., "My shirt has pattern")
1. Pattern Recognizer	Explicitly recognizes and labels a simple sequential repeating pattern (e.g., "I am wearing a blue, red, blue, red blue, red pattern")
2. AB Pattern Fixer, Duplicator, Extender	Can fill in a missing element of a ABABAB pattern, then copy it, and finally, extend it
3. Pattern Fixer, Duplicator, Extender of More Complex Repeating Patterns	Can fill in a missing element of, for example, ABBABBABB pattern; then copy it, and finally, extend it
4. Pattern Unit Recognizer	Identifies the core (smallest unit) of a repeating pattern and can translate patterns into new media (e.g., identify ○□○□○□ as an AB pattern)

3. Is the letter-labeling method effective with all types of repeating patterns?

As Fig. 3 indicates, the impact of the abstract letter-labeling method varied with the type of pattern. Specifically, it had the least impact with AB and ABC patterns, substantial impact with ABB and AAB patterns, and the greatest impact with the AABB pattern. With the concrete method of labeling patterns with physical characteristics, the AB pattern was at least somewhat more salient than the other patterns and, thus, provided less room for improvement. An AB pattern's saliency may be due to its simplicity—a core of two different elements that alternate. In contrast, ABB, AAB, and AABB patterns have a core of two different elements that repeat in a more complex manner, and the ABC pattern have a core of three different elements. The letter-labeling method may have a limited impact on ABC pattern recognition perhaps for the reasons discussed next.

4. Are the results sustainable and generally applicable?

The Fyfe et al. (2015) study involved a single training session of 20 min. So it is unclear whether the results would persist over time and how much instruction is needed to maintain gains. Importantly,

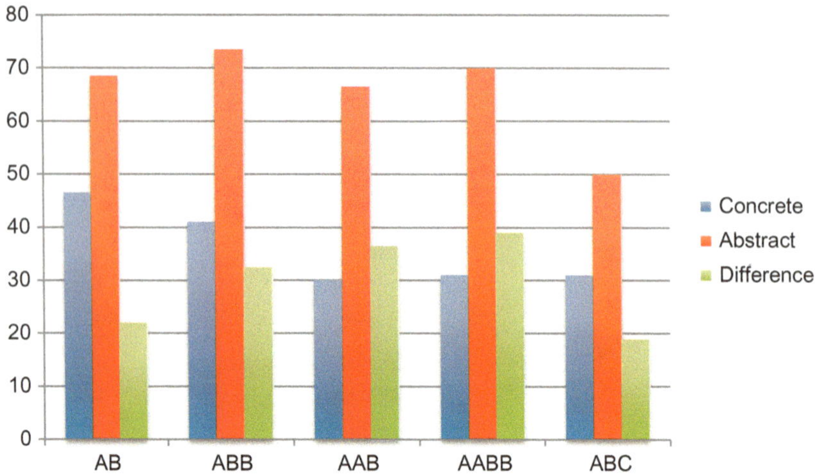

Fig. 3 Summary of Fyfe et al.'s (2015) training experiment on patterns.

unclear is whether the brief training would achieve the hallmark of meaningful instruction, namely spontaneous and successful application of the letter-labeling method to a new pattern, especially less salient regularities such ABC patterns, and how much instruction is needed to promote such transfer. As the small sample ($n=64$) involved middle class children, will the abstract letter-labeling method be as successful with other less-advantaged populations of children (e.g., preschoolers from low-income families, or ELL children)?

3.2.2 A Linear Representation and Its Hypothesized Developmental Prerequisites

Table 3 illustrates a HLT for children's relative magnitude judgments of the number words 1–10. A series of increasing abstract models that take into account developmental level may promote the levels of this HLT more effectively than relatively abstract linear numerical board games such as *The Great Race*.

3.2.2.1 Increasing Magnitude Principle

A plausible developmental prerequisite for making linear magnitude judgments is the increasing magnitude principle: numbers coming later in the counting sequence represent larger quantities than predecessors (e.g., nine is more than four because it comes after four when we count; Level 2 in Table 3). A cardinality chart—separate numeral cards with interlocking blocks above each numeral to represent relative quantity—might be even more effective than the number-list model alone (e.g., *The Great Race*) in

Table 3 HLT of Mental Magnitude Comparisons

Level 0: Number words without magnitude	Children learn the cardinal meaning of the first few number words—provides a basis for *verbal subitizing* (e.g., "two" represents pairs of objects) but does not realize that "two" is more than "one").
Level 1. Subitizing-based relative magnitude	Verbal subitizing provides a basis for recognizing that "two is more than one" and "three is more than two" (ordinal meaning of the first few number words), because the child can directly see the differences in the relative magnitude to two small collections.
Level 2. Increasing magnitude principle—relative magnitude of distal numbers	The insight that the order of the number words for small numbers (one, two, three) reflect increasing magnitude provides a basis for the generalization that the counting sequence represents increasingly larger quantities—the increasing magnitude principle. The principle and *familiarity with the counting sequence* and *the increasing magnitude concept* provide a basis for comparing two numbers at obviously different positions in the counting sequence (i.e., to make gross comparisons of 2 or 7, 10 or 3, 9 or 5, and 4 or 8).
Level 3. Mental comparisons of neighboring or successive numbers (number after equals more)	Once children become sufficiently familiar with the counting sequence, they can enter the sequence at any point and *specify the next number* instead of always counting from one. This permits children to apply the *increasing magnitude principle* to the task of *mentally comparing even close numbers*, such as the larger of two neighboring numbers (e.g., immediately indicate that eight is more than seven because it follows seven in the counting sequence).
Level 4. Successor principle (number after equals 1 more) and a linear representation of the counting sequence	Verbal subitizing enables children to see that "two" is exactly one more than "one" item and that "three" is exactly one more than "two" items, and this can help them induce the *successor principle* (each successive number in the counting sequence is exactly one more than the previous number). The *successor principle*, in turn, enables children to view the counting sequence as n, $n+1$, $[n+1]+1$, ... (the *positive integer sequence*)—a linear representation of number.

Most concrete (most directly models an informal counting-based view of the number sequence) ←		→ Most abstract (least directly models an informal counting-based view of the number sequence)
Discrete-quantity model (e.g., Cardinality chart)	Semidiscrete-quantity model (e.g., Number list)	Continuous-quantity model (e.g., Number line)
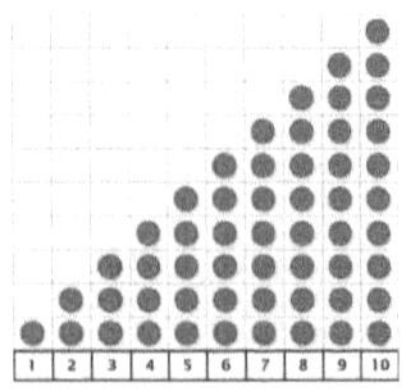	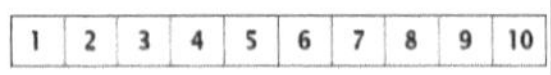	

Note: The three models of number embody different meanings of number:

- A discrete-quantity model such as a cardinality chart represent what she called a *cardinal meaning of number* and corresponds to children's informal counting-based view of the number sequence.
- A continuous-quantity model such as a number line (linear length) represents what Fuson (1988) called a *measurement meaning of number*, which presumbaly requires a formal understanding of the concept of a unit of measurement.
- A semidiscrete-model such as a number list approximates both measurement and cardinal meanings.

Fig. 4 Three models of number.

fostering the principle, because—in terms of young children's existing informal (discrete-quantity-based) view of the counting sequence—it is a concrete and more direct model of relative magnitude (see Fig. 4; Baroody & Purpura, 2017; Frye et al., 2013).

3.2.2.2 Successor Principle

A plausible conceptual basis for a linear representation of the counting sequence and hence linear magnitude judgments is the successor principle (Level 4 in Table 3). This principle entails the insight that each successive number in the counting sequence is exactly one more than its predecessor (e.g., nine which immediately follows eight when we count is one more than eight; Izard, Pica, Spelke, & Dehaene, 2008). Currently, almost no research has been reported on how to promote the successor principle and whether such instruction also fosters a linear representation. Reid, Baroody, and Purpura (2013) compared computer-based training that focused directly on successor principle and that involving *The Great Race*. The successor training consisted of children playing three games involving a hungry monkey swinging through tree branches on which a number line was superimposed to retrieve a banana. The games targeted the number-after knowledge (*what is the number after 3?* Level 4A in Table 1), successor understanding (*how many more branches must the monkey swing to get from 3 to 4? Level 4 in Table* 3), and identifying correct application of the successor principle (*the monkey thinks he needs to swing one more branch to get from 3 to 4, is he correct?*). The direct successor training was significantly more effective than *The Great Race* in promoting successor learning but not linear magnitude comparisons.

3.2.3 Introducing Addition and Subtraction Using a Number-Line Model

Whether a number-line model should be used to formally introduce children to operations on numbers needs to take into account the internal factor of children's understanding of numbers. In recent years, it has become popular to recommend using a number line as a basis for formally introducing young children to addition and subtraction (Bell et al., 2004; National Mathematics Advisory Panel, 2008). For instance, the supplemental *Bridges in Mathematics* (Math Learning Center, 2009) program introduces an addition item such as 5+3=8 as hopping to the right 5 and then 3 more. The inverse of this—the related subtraction item 8 − 3=5—is shown next by hopping from 8 in the opposite direction (to the left) three times.

Although it is crucial that children see addition and subtraction as related operations (e.g., that addition and subtraction of the same number undo each other), the number-line model used by *Bridges in Mathematics* may be too abstract for young children for two reasons (Baroody, 2016a). Consistent with physical to mental embodiments view, one is that it initially connects symbolic equations such as 5+3=8 and 8 − 3=5 to a relatively sterile iconic representation (number line) instead of an enactive representative (e.g., modeling inversion the addition and subtraction with objects) or a "transitional" representation (e.g., modeling inversion the addition and subtraction of the same number with pictures of objects).

Consistent specific to general (familiarity) view, a second reason is that a number line may be inconsistent with young children's informal view of numbers. As Fig. 4 suggests, a number line represents a linear extent (a continuous quantity) and thus embodies what Fuson (1988) identified as a *measurement meaning of number*. Evidence indicates that young children view the number sequence as representing a series of increasingly larger discrete quantities—as embodying what Fuson (1988) identified as a *cardinal meaning of number*. For example, they typically (a) do not spontaneously break up continuous quantities into countable units (Fuson, 1988; Huntley-Fenner, 2001) and (b) misconceive number lines as representing discrete quantities, as evidenced by them counting the numbers or hash marks on the number line, not the linear extents between hash marks (e.g., counting 0 or its hash mark as "one"; Lehrer, 2003). A key source of the difficulties is that children construct a concept of linear unit relatively late. Indeed, Saxe et al. (2010) found that fifth graders struggled with number-line-based problems that required the application of such a concept. For instance, shown a number line with 8 and 10 marked and labeled, many participants incorrectly considered the distance between 8 and 10 as one unit and located 11 at the position for 12. For such reasons, mathematics educators have long recommended against using number lines as basis for introducing formal mathematics instruction (Ernest, 1985; Fuson, 2009), and the National Research Council (2001) recommend against using these models as instructional tool before grade 2.

Two qualifications are in order. *If* instruction helps students construct a solid concept of linear units, the use of a number line for teaching addition and subtraction would be consistent with Dewey's (1963) interaction principle. Using the number line to introduce two-digit mental addition and subtraction as done in Realistic Math may also be consistent with this principle and a potentially useful technique. As Frame A of Fig. 5 illustrates, the number line in this program is initially related to children's informal discrete-quantity view of numbers by showing 10 circles between each hash mark (Blöte, Klein, & Beishuizen, 2000). As Frame B of Fig. 5 illustrates, the empty number line is essentially used as a notation system for aiding students' calculations that can be based on their discrete-quantity view of numbers (Blöte, Van der Burg, & Klein, 2001).

3.2.4 Base-Ten and Place-Value Concepts and Skills

Like the previous section, this subsection illustrates why it is crucial to consider internal (psychological) factors when considering the use of concrete experiences and progressively adapting these experiences as children achieve higher levels on a HLT.

3.2.4.1 Understanding Multidigit Numbers in Terms of Base-Ten/Place-Value Concepts

Children informally view multidigit numbers such as 23 in terms of 23 (countable) units. They must construct base-ten concepts (e.g., 10 ones can be grouped together to form a larger unit called "ten") and place-value concepts (e.g., the value of the digit 2 in 23, by virtue of its position in the tens place, represents 2 tens (2×10) and 3, by virtue of its position in the ones place, represents 3 ones (3×1).

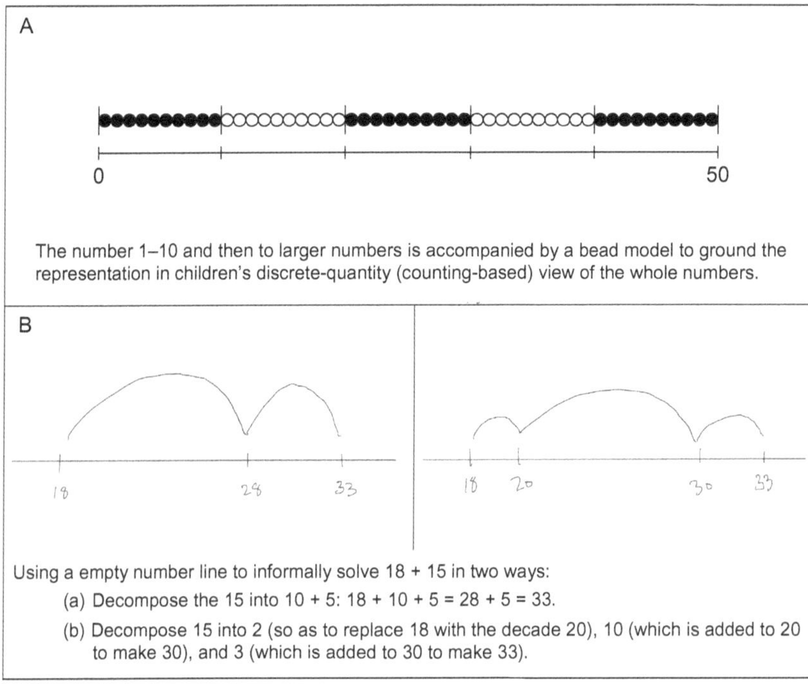

Fig. 5 Use of the number line in Realistic Math to teach two-digit addition and subtraction.

Baroody (1990) adduced a sequence of increasingly abstract manipulative- or picture-based models for fostering the meaningful learning of grouping-by-ten (base-ten) and place-value concepts. Frame A of Fig. 6 illustrates the most concrete multidigit number model (e.g., interlocking blocks) in the sense that it builds directly on children's informal counting-based view of multidigit numbers in two ways: (a) an interlocking-block model is *proportional* (e.g., the representation of a ten is 10 times larger than that for one) and (b) it requires children to *physically put together* 10 ones to create a ten. Although also *proportional*, Frame B models (e.g., base-ten blocks) are somewhat more abstract in that they entail trading 10 embodiments of ones for a *pregrouped embodiment of ten*. That is, unlike using interlocking blocks, which literally entails the experience of constructing a ten from 10 ones, using base-ten blocks entails a child recognizing that a 10-stick or "long" could theoretically be constructed by gluing 10 units together. Conversely, a "ten stick" composed of interlocking blocks can physically be decomposed into 10 unit blocks, whereas a long can only be traded for 10 cubes.

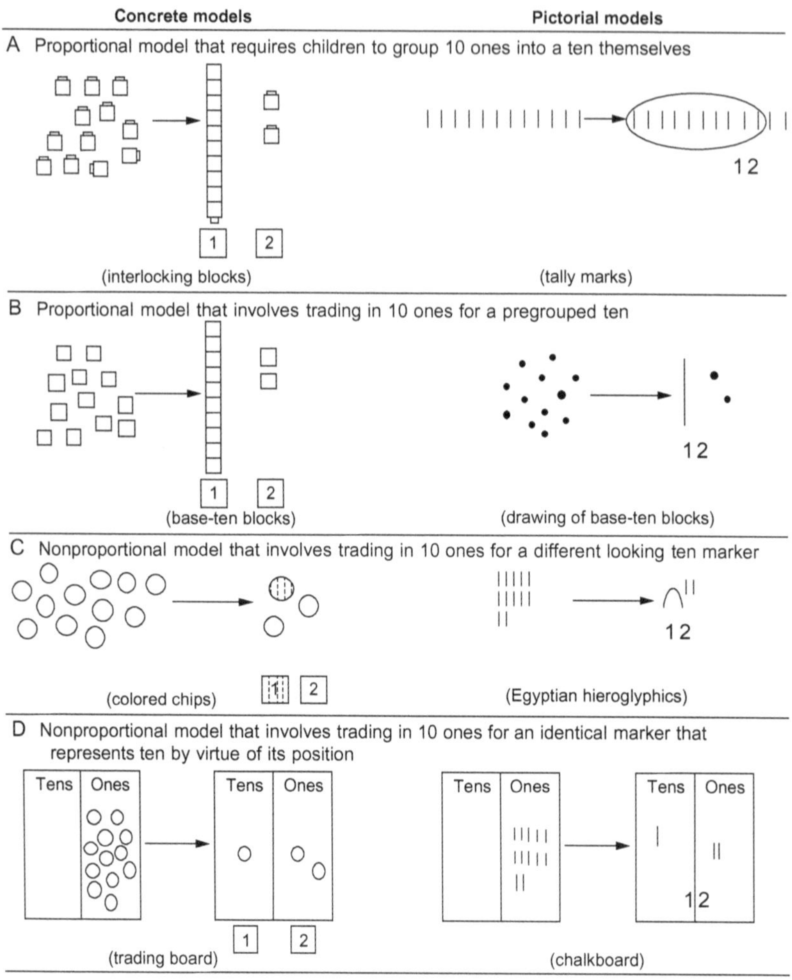

Note. Montessori Beads are between Frame A and B in abstractness. Although 10 discrete beads are strung together to form a ten string (and 10 strings of ten are combined to form block of 100), it may not be practical for children to do this. Yet, the ten string is more clearly made up of 10 discrete units than the long (ten-stick) of a base-ten model.

Fig. 6 Concrete-to-abstract models of a base-ten and place-value meaning of two-digit numbers.

Both Frame A models (e.g., interlocking blocks) and Frame B models (e.g., base-ten blocks) are more concrete than *nonproportional* models shown in Frames C and D of Fig. 6. Frame C models, which involve *trading 10 embodiments of one for a single different-looking embodiment of 10*) are less abstract than Frame D models, which involve *trading 10-for-1 embodiments that are all identical save for the position on say a place-value trading board*. Note that while Frame A, B, C, and D models all model grouping (albeit at increasingly abstract levels), only Frame D models represent an effort to directly embody a place concept (i.e., the position of a digit determines it value; e.g., whether it is a ones or tens).

The key unresolved pedagogical issue is whether it is more efficacious to use a series of increasingly abstract models that start by building directly on children's informal counting-based view of numbers or—as suggested by cognitive alignment perspective—start with a model that directly represents both the grouping and place-value characteristics of multidigit numbers. Other unresolved questions include: Consistent with an interaction perspective and contrary to the dual-representation hypothesis (DeLoache, 2000; Kaminski et al., 2013; Uttal et al., 2009), is using Frame A models in which, for instance, children physically construct a ten from units and then Frame B models (e.g., trading 10 unit blocks a long) more efficacious than starting with Frame B models? Similarly, are using proportional models and then nonproportional models more efficacious than starting with nonproportional models? Are using less abstract models and then the relatively abstract Model D more powerful than starting with Model D? Do the results for all of the previous questions with advantaged children also apply to various at-risk populations, such as children from low-income families or those with mathematical learning difficulties?

3.2.4.2 Magnitude Comparisons and a Linear Representation to 100

Even after children have constructed an understanding of the successor principle (e.g., recognize in principle that a number is one more than its predecessor in the counting sequence), they continue to exhibit a logarithmic response pattern on magnitude comparison tasks involving larger and relatively unfamiliar numbers. Various factors have been adduced to account for such a nonlinear response pattern (Anobile, Cicchini, & Burr, 2012; Barth & Paladino, 2011; Slusser, Santiago, & Barth, 2013). Landy, Charlesworth, and Ottmar (2016, p. 21) concluded:

> *It appears, however, that human representations of natural numbers, at least beyond a paltry few hundred thousand iterations, rely on resources quite distinct from successorship or even a metric "number line," and impose substantial additional structure given by the numeral systems. A fundamental mistake made by classical empiricism was to assume that the inner representations were iconic—that they were like the outer represented. When reasoning about large numbers, we appear to rely on representations that are fundamentally unlike the numbers themselves.*

Perhaps unlike smaller familiar numbers, larger numbers may not be represented in the form of a mental number line as suggested by Resnick (1983).

A million—a billion, what's the difference?

If the left-hand endpoint below = 0 and the right-hand endpoint = a billion and

$a < 1/100$, $b = 1/100$, $c = 10/100$, $d = 50/100$, $e = 90/100$, $f = 99/100$, and $g > 99/100$,

What letter best represents the position that a million would occupy on the number line below?

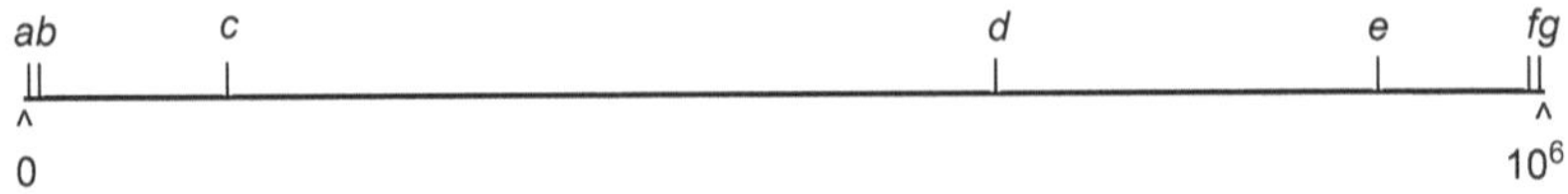

Fig. 7 The million vs a billion problem.

Instead, their representation may be based on an understanding of the structure of the number system. For example, correctly judging the relative magnitude of a million and a billion (see Fig. 7) depends on understanding that a billion is 1000 times larger than a million.

For children who are just becoming familiar with the numbers to 100, the *Grid Race* might be more efficacious than *Race to the Moon* in exploring the structure of the number from 0 to 100 and, thus, have a greater impact on various aspects of numeracy, including a linear representation of numbers. Specifically, whereas counting-on 1 or 2 from a two-digit number is effectively geared toward discovering the (already known) successor principle, it—unlike counting 10 spaces from a decade or other two-digit number—is less likely to reinforce the repetitive structure of the numbers from 0 to 100 (e.g., each decade is 10 more than the preceding decade or repeatedly subtracting 10 changes only the decade not the number of ones). Whereas reinforcing the successor principle is not likely to improve the linearity of magnitude comparisons, a better grasp of how two-digit are interrelated might.

3.3 Instructional Strategies for Ensuring Educative Experiences

3.3.1 Eight Guidelines

Mix (2010) cited five (largely external) dimensions or variables that might impact the effectiveness of manipulatives: variations in structure, the amount of contact, the relation of a model to written symbols, the amount of exposure, and the number of models. These variables are related to Points 1, 2, 5, 6, and 7, respectively. In addition, Points 3, 4, and 8 are important applications of Dewey's (1963) interaction principle. Note that for helping children overcome the long-recognized difficulty of connecting physical models of mathematical ideas or processes to formal written representations (Ginsburg, 1977; Resnick, 1982; Uttal, Liu, & DeLoache, 2006), Points 3, 4, and 5 are particularly important.

1. *Structured use.* A Piagetian approach based on free play, once the mantra of radical constructivists and many early childhood educators, is now generally not considered a viable instructional approach. Although free play is important in its own right, the presence of manipulatives alone in a free play context does not guarantee an educative experience (Baroody, Clements, & Sarama, in press; Fuson, 2009). For example, Clements, Copple, and Hyson (2002) concluded that it is unlikely incidental experiences will foster educative experiences and that young children are ready for organized, sequenced experiences embedded in specific activities (e.g., reading children's literature, music or art activities), play (e.g., math games or physical activities), or projects. Frye et al. (2013) further concluded that, in addition to embedded mathematics activities, early childhood instruction should include daily, targeted mathematics instruction (i.e., dedicated time for structured, mathematics-focused activities). In brief, ensuring educative experience entails using manipulatives with a purpose (e.g., clear goals), direction (e.g., scaffolding), and feedback (including questions that prompt reflection such as "Why do you think that makes sense?" or "Does everyone agree?"). As Brown, McNeil, and Glenberg (2009, p. 161), noted, "Without appropriate structure, learners may fail to discover the target concept."
2. *Active use.* However, not all structured instruction is equally effective. Chi (2009) hypothesized that constructive activities (which elicit responses involving ideas that go beyond the information provided) are more effective than active activities (physically doing something), which in turn are more effective than passive activities (e.g., listening or watching). Consider how this general framework applies to the case of using concrete materials. Using manipulatives to actively involve children in discovery learning—exploring a phenomenon, noticing patterns or relations, and generalizing such regularities—is more effective than either their unguided use or adult-directed use in advancing children's understanding (see Alfieri, Brooks, Aldrich, & Tenenbaum, 2011). Prompting the imitation of a teacher-taught manipulative-based procedure or simply watching a teacher model a procedure with manipulatives may be meaningful to children and promote their successful adoption and application of the procedure when it builds on existing conceptual

understanding but when new conceptual insight is needed (Baroody, Tiilikainen, & Tai, 2006).

The use of concrete experiences involving structured or guided discovery has a number of advantages over direct instruction (see, e.g., Alfieri et al., 2011; Ambrose, Baek, & Carpenter, 2003; DeCaro & Rittle-Johnson, 2012; Fyfe, DeCaro, & Rittle-Johnson, 2014; Rittle-Johnson, Fyfe, Loehr, & Miller, 2015; Schwartz, Chase, Chin, & Oppezzo, 2011; Shafto, 2014). One is that first discovering concepts or inventing self-styled solution strategies is more likely to result in understanding a concrete model and, thus, greater learning or more flexible application of procedures than first learning procedures. For instance, Thompson (1992) found that students who had already learned teacher-taught renaming procedures did not improve in computational accuracy after an intervention with either base-ten blocks or virtual base-ten blocks (Blocksworld). Learning concepts (e.g., via "meaningful" manipulatives) after learning a standard procedure is inconsistent with the moderate novelty principle for many students because of the belief they already know how or what to do.

For a thoughtful discussion of how Froebel and Montessori artfully used carefully designed protocols to guide the manipulatives-based discovery of mathematical ideas, see Balfanz (1999) and Mix (2010). One effective way of involving children in guided discovery learning is to provide them materials that embody examples and nonexamples of a concept so that they can discover for themselves the critical (defining) attributes of the idea (Durkin & Rittle-Johnson, 2012; Rittle-Johnson & Star, 2007, 2009, 2011; Schwartz et al., 2011). For example, a game for helping preschoolers construct a concept of "three" (subitize collections of three) is the game *Is It Three?* Children are shown different examples of three *and* nonexamples of three (e.g., collections of 2 and 4) and asked, "Is this three?" The variety of examples aid in helping children in abstracting the defining property of three (a trio) by discounting other properties such as shape, color, and size, and the nonexamples help define the boundaries of the concept (e.g., •••, but not •• [fewer] or •••• [more], is "three"; Baroody, Lai, et al., 2006; Frye et al., 2013). See Baroody and Coslick (1998) for a variety of example lessons that employ examples and nonexamples. In sum, similar to Martin and Schwartz (2005), Brown et al. (2009, p. 161) observed, "educators need to find an appropriate balance between structure and spontaneity," as learners need some freedom to construct meaningful knowledge for themselves.

3. *Reflective use.* A particularly important aspect of the active use of concrete materials and the construction of meaningful knowledge is thoughtful reflection on the experience—mental, as well as physical, engagement (Baroody, 1989; Sarama & Clements, 2009a). Indeed, consistent with Chi's (2009) distinction between active and constructive activities, Piaget (1964) differentiated between a *physical experience* and a *logical-mathematical experience.* The former "consists of acting upon objects and drawing some knowledge about the objects (e.g., noticing that a steel ball is heavier than a rubber ball of the same size; p. 11). A logical-mathematical experience is knowledge that is drawn from thinking about the actions on the objects. As an example, Piaget cited the experience of a roughly 4-year-old child who counted a row of pebbles first in one direction, then in the opposite direction. Next the child arranged the pebbles in a circle and counted them clockwise and then counter clockwise. After experimenting with yet other arrangements, the child discovered that, regardless of order or arrangement, counting the collection (accurately) always had the same result. In effect, by reflecting on his experience, the child discovered the order-irrelevance and number-constancy principles applicable to any collection. In sum, experiences with concrete materials are more likely to be educative if children spontaneously reflect on their physical actions or are prompted to do so by a teacher or peers asking them to explain or justify their actions. Points 4 and 5 are especially important for prompting reflective use.
4. *Connecting to meaningful knowledge (e.g., capitalizing on familiar analogies).* To prompt thoughtful reflection, discovery learning or any form of instruction needs to be meaningful, which requires carefully building on what children know (Bruner, 1961; Fyfe, Rittle-Johnson, & DeCaro, 2012). Encouraging children to use manipulatives to solve mathematics problems with understanding needs to take into account Piaget's principle of assimilation and the moderate novelty principle, and a key tool for doing so with learners of all ages is using meaningful analogies. Building on existing and meaningful knowledge (e.g., a familiar analogy) is particularly important for a guided discovery approach, as it enables children to invent their own informal strategy for solving a problem. Building on children's existing knowledge and strategies is the rationale for the *Everyday Math* program (Bell et al., 2004) and can serve as basis for understanding and reinventing formal procedures (see Baroody & Coslick, 1998). Doing so minimizes the need for

students to memorize a teacher-taught problem-solving procedure for a particular class of problems—an advantage that may be particularly important for children with mathematical learning difficulties.

Consider the case of division. Primary-grade children often readily devise a divvy-up strategy for fair-sharing problems (Amount ÷ number of shares = size of shares; see Table 4; Hiebert & Tonnessen, 1978; Kouba & Franklin, 1993). In contrast, Squire and Bryant (2002) noted they often have difficulty solving more generic partitive problems (Amount ÷ number of groups = size of each group) and especially measurement problems (Amount ÷ size of each group = number of groups; again see Table 4). Capitalizing on children's everyday experience of fair sharing can help them understand, compare, and solve more challenging (i.e., generic partitive and measurement) division word problems. The following guidelines outline how instruction can successfully foster the use of manipulatives to informally solve division word problems:

Table 4 Types of Division Word Problems, Examples, and Children's Informal Modeling Strategy

Type of Problem[a]	Example Word Problem[b]	Informal Strategy for Directly Modeling the Meaning of the Problem
Fair sharing	How much candy did each of three friends get if they shared 12 candies fairly?	*Divvy-up*: • Count out 12 chips, • Then distribute a chip to each of three piles until all the chips are gone, and • Finally, count the chips in one of the piles to determine the answer.
Partitive	Mandy has 12 candies. She puts 4 candies in each bag. How many bags can she fill?	
Modified fair sharing	How many friends can share 12 candies fairly if each share is 3 candies?	*Measure-out*: • Count out 12 chips, • Then create piles of three until all the chips are gone, and • Finally, count the number of piles to determine the answer.
Measurement	Mandy has 3 bags of candies and 12 candies altogether. If there is the same number of candies in each bag, how many candies are in each bag?	

[a]Fair-sharing problems are familiar examples of the more generic partitive class of division problems. Modified fair-sharing problems are familiar examples of the more generic measurement class of division problems.

[b]The examples of the partitive and measurement problems are modeled after those illustrated at http://www.math.niu.edu/courses/math402/packet/packet-4.pdf (examples that had been adapted from Cognitively Guided Instruction, University of Wisconsin-Madison, 1992).

- Introduce division with highly familiar fair-sharing problems. By devising a divvying-up strategy with manipulatives themselves, children will understand, own, and apply effectively apply the strategy.
- Relate less familiar generic partitive problems to fair sharing. For instance, encourage a child who does not understand such problems (e.g., does not spontaneously apply a divvy-up strategy) to translate them into fair-sharing terms: The amount is analogous to knowing the total number of items, the number of groups is comparable to the number of people (the number of shares), and the unknown (the size of each group) corresponds to the size of each share.
- Introduce the moderately novel modified fair-sharing problem (Amount ÷ size of each share = number of shares) to foster the invention of an informal measure-out strategy, see Table 4). Solving such problems before introducing generic measurement problems will provide a basis for understanding and solving the latter, more unfamiliar and challenging problems.
- Relate less familiar generic measurement problems to modified fair sharing: Measurement division is analogous to knowing the total amount and the size of each share and not knowing the number of people with whom the total amount can be shared (number of shares). This will help children recognize that their informal measure-out strategy is applicable to such problems.

As illustrated in Fig. 8, the familiar and meaningful analogy of (modified) fair sharing can also be invaluable in helping students

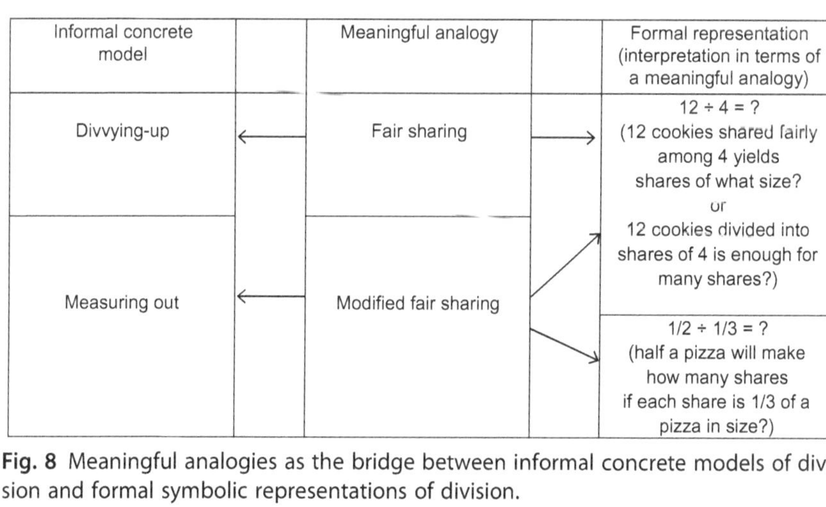

Fig. 8 Meaningful analogies as the bridge between informal concrete models of division and formal symbolic representations of division.

making sense of formal representations of division and bridging the gap between written division expressions (e.g., 12÷3 or 1/2÷1/3) or equations (e.g., 12÷3=? or 1/2÷1/3=?) and their informal models or solution strategies. That is, the analogy provides the missing connection between formal representations of division and children's knowledge of how to use manipulatives to informally solve division problems. With older elementary or even college-level students, the modified fair-sharing analogy for measurement division can provide the crucial basis for making sense of fraction division. For instance, for many students, it is a mystery why 1/2÷1/3 results in the larger number 3/2 or 1–1/6 (not a smaller number as with the division of whole numbers). The result makes sense if you consider that 1/2 of pizza allows can be used to create one full share and a part (1/6) of another share.

5. *Explicit connections.* Many manipulative-based activities do not include corresponding written representations or procedures. Moreover, children often do not spontaneously relate concrete models to abstract formalisms or vice versa (Resnick, 1982; Resnick & Omanson, 1987; Uttal et al., 2013). Frequently, then, concrete materials need to be explicitly linked to the abstract idea such material are intended to represent to ensure thoughtful reflection (Brown et al., 2009; Fyfe et al., 2015). As a first step, self-explanations may be particularly valuable in facilitating learning both concepts and strategies because they prompt a pupil to explicitly summarize new information and thus consciously reflect on it (Berthold & Renkl, 2009; Matthews & Rittle-Johnson, 2009; Rittle-Johnson, 2006; Rittle-Johnson et al., 2015; Star & Rittle-Johnson, 2009). Consistent with the Common Core State Standards Math Practices Standard 3 (http://www.corestandards.org/Math/Practice/) encouraging children to share, justify, and compare their informal strategies can also be valuable in fostering explicit reflection and understanding, including gaps or inconsistencies in their thinking (Baroody & Coslick, 1998; Battista, 2016; Piaget, 1928). So as to foster connections with symbolic procedures, teachers can encourage children to translate informal manipulative-based strategies into a written procedure or vice versa (e.g., by asking them to evaluate a formal procedure in terms of informal concepts and procedures). Finally, teachers can explicitly summarize pupil's discoveries and explanations and, as needed, fill in gaps to ensure coherency. See Brown et al. (2009) for a discussion of how artfully used gestures can assist in guiding children's attention to

the connection between concrete representations or action and abstract ideas.

6. *Extended use.* Uttal (2003, p. 111) noted: "Several lines of research have shown that for manipulatives to be effective, they must be used repeatedly for the same concept." Children may need time using manipulatives to discover regularities or otherwise make sense of their intended purpose. Unfortunately, formal instruction often limits manipulative exposure (Mix, 2010).
7. *Multiple representations.* Using multiple representations can serve to promote abstraction or generalization of a discovered regularity or deepen conceptual understanding of a symbolic representation (Barmby et al., 2009; Kaminski et al., 2009; McNeil et al., 2009; Moseley, 2005; Steinbring, 1997). Moreover, translating formal symbolism into multiple concrete embodiments (as well as the reverse) can further ensure multiple connections and a deeper understanding of an idea (Baroody & Coslick, 1998). However, research has shown that simply accessing multiple representations does not necessarily lead to improved understanding (Ainsworth, 2006). Underscoring the importance of helping students explicitly connects various representations of an idea, Seufert (2003, p. 228) noted: "Learners must interconnect the external representations and actively construct a coherent mental representation in order to benefit from the complementing and constraining functions of multiple representations."

   A concern with multiple models is that it reduces the exposure to any one model and may confuse students (Mix, 2010; Uttal, 2003). For this reason, Asian schools focus on a single concrete model (Stevenson & Stigler, 1992). Concerns with the single-model approach, though, are an overly local concept or inflexible strategy use. This may contribute to an education that undermines pupils' disposition, creativity, autonomy, and initiative (Zhao, 2014).
8. *Purposeful use.* Manipulatives may be particularly engaging when children have a purpose of their own for using them or an adult creates a reason. Math games may be a particularly useful way to create a real need to use and reflect on the use of concrete experiences (for examples, see Baroody, 2016b; Baroody et al., in press; Baroody & Coslick, 1998; Clements & Sarama, 2013). Dramatic play (e.g., a store scenario) also provides a structured yet purposeful way of creating a situation where children have the opportunity to use concrete materials to explore, reflect on, and extend mathematical concepts or apply mathematical skills.

3.3.2 A Case in Point

Consider the research and recommendations by Kaminski and Sloutsky (2013) regarding graph instruction for primary-grade pupils. Participants were randomly assigned to an extraneous information condition, which involved a combination of pictogram and bar graph (see Frame A in Fig. 9), or a no extraneous information condition, which involved a solid bar graph (see Frame B in Fig. 9). All participants were taught how to read a bar graph by relating the *x*-axis (time in weeks) to the *y*-axis (the number of shoes). Children in the extraneous condition were posttested with a combination of a misleading pictogram but accurate bar graph (see Frame C of Fig. 9, and those in no extraneous condition were tested with a solid bar graph (see Frame D of Fig. 9).[e] Whereas, the vast majority of kindergartners and first graders in the extraneous condition counted the items in the misleading pictogram and thus responded incorrectly, the vast majority of those in the no extraneous condition used the taught procedure and responded correctly. On a posttest task involving novel (pattern-filled) bar graphs (see Frames E and F of Fig. 9), kindergarten and first graders in the extraneous information condition used far less counting, and the majority responded correctly. However, those in the no extraneous information condition performed significantly better.

Kaminski and Sloutsky (2013) drew two conclusions:

1. If pictograms are used in primary instruction, then teachers should explicitly direct children's attention to the relation between the *x*-axis and *y*-axis so that they do not have to rely on a counting strategy.
2. Although pictograms are often recommended as a scaffold for helping young child make sense of bar graphs (e.g., Choate & Okey, 1981), such scaffolding is not necessary for successfully learning how to read bar graphs.

If a child repeatedly failed to see the connection between counting the items in a pictogram and the corresponding value on the *y*-axis, then Kaminski and Sloutsky's (2013) first recommendation is consistent Point 5—help make connections explicit. However, the second recommendation should be treated with caution or even skepticism for several reasons. Consistent with Point 4 (connect new instruction to prior knowledge), real graphs, picture graphs, and pictograms (especially with grid) are recommended as scaffolds

[e] The condition-specific task for the Extraneous Information Condition of the Kaminski and Sloutsky (2013) study (see Frame C of Fig. 9) is not ecologically valid, as teachers and textbook publishers (typically) do not use pictograms that depict an incorrect number of items.

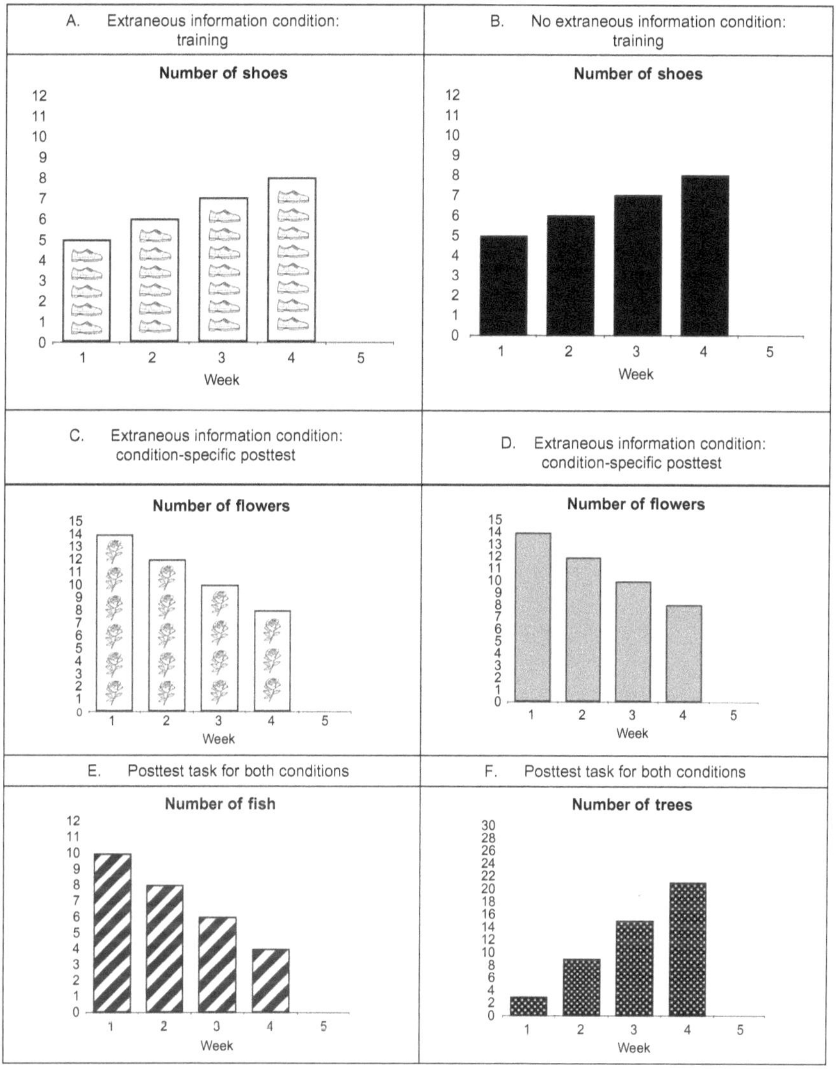

Fig. 9 Graphs similar to those used by Kaminski and Sloutsky (2013) for training and testing (based on their Figs. 1 and 2).

for meaningful introducing bar graphs, in part, because young children have an informal, counting-based (discrete-quantity) view of number and bar graphs entail a measurement (continuous-quantity) model of number. Some primary-grade children, then, may need such scaffolding to *understand* a bar graph and how to read it. Importantly, Kaminski and Sloutsky (2013) apparently did not administer a delayed posttest to assess retention. There are several reasons to believe their intervention involving no extraneous information would not have a lasting impact. Inconsistent with Points 2 and 3, the instruction entailed a passive activity (e.g., listening or watching the trainer read a graph) and an active activity (imitating the trainer's procedure) but not constructive activities (producing responses that entail ideas that go beyond provided information) requiring reflection. Although the instruction refers to common objects and unit of time, it is not clear whether the two variables were related in a meaningful way (e.g., by a familiar story line; Point 4). As a trainer demonstrated a graph-reading procedure four times and a participant practiced it only four times (in same session), the intervention did not involve extended use (Point 6). Finally, the training did not seem purposeful from a child's perspective or particularly engaging (Point 8). All in all, the instruction seems readily forgettable.

4. IS THERE EVIDENCE THAT CONCRETE EXPERIENCES WORK?

Some empirical evidence indicates that, for example, concrete experiences are useful in extending existing informal knowledge by providing young children an opportunity to discover and apply a mathematical regularity or devise and practice an informal strategy—(Boggan, Harper, & Whitmire, 2010; Clements & Sarama, 2012) and that games (Bright, Harvey, & Wheeler, 1985), including computer games (Baroody, Purpura, Eiland, & Reid, 2015; Obersteiner, Reiss, & Ufer, 2013; Shin, Sutherland, Norris, & Soloway, 2012) can be valuable in promoting mathematical learning. However, in light of the preceding discussions on instructional strategies for ensuring effective use of concrete experiences and Dewey's (1963) interaction principle, it should not be surprising that research on the effectiveness of concrete experiences is mixed (see, e.g., reviews by Mix, 2010; Uttal, 2003). In explaining the mixed results of Thompson's (1992) use of the Blocks Microworld program and research on the effectiveness of manipulatives in general, Mix (2010) concluded that whether a model might or might not work depends on such factors as how manipulatives are used, the outcome measure, and the characteristics of the learner.

For instance, Walker, Mickes, Bajic, Nailon, and Rickard (2013) evaluated the relative efficacy of using a conceptual approach (fact triangles) and a drill approach (answer-production [AP] training) to promote fluency with subtraction combinations with grade 1–6 students. Fact triangles

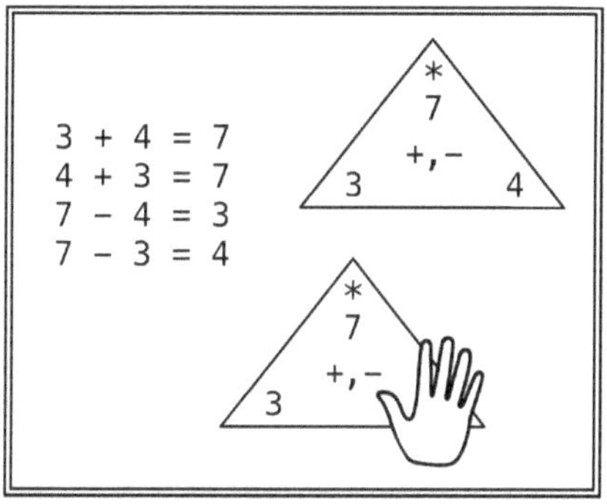

Fig. 10 Example of a fact triangles using 3–4–7.

(e.g., see Fig. 10) are widely used to help children see that subtraction is related to addition and that known sums can be used to reason out unknown differences (e.g., if $3+4=7$, then $7-3=4$). Walker et al. (2013) found that AP training was significantly more efficacious in promoting subtraction fluency with practiced combinations than the fact-triangle intervention but that neither approach promoted transfer of fluency to unpracticed subtraction combinations. These researchers concluded that fact triangles "are not an effective vehicle for fluency training or for establishing flexibly applicable arithmetic skill" and curricula should "deemphasize fact-triangle exercises in favor of more AP training" (p. 30).

However, the fact-triangle intervention in the Walker et al. (2013) study may have failed because of the ineffective manner in which the training was implemented. The model used was (a) perhaps only semiactive (Section 3.3, Point 2), (b) not designed to prompt reflection (Point 3), (c) not particularly meaningful (Point 4), (d) without explicit connections between procedures and concepts (Point 5), (e) relatively short in duration (Point 6), (f) dependent on a single (not multiple) representations (Point 7), and (g) not purposeful and engaging (Point 8). In regard to Point 4, their fact-triangle training involved only two of seven steps in a HLT for fostering the meaningful memorization of subtraction combinations (Baroody, 2016a). In contrast, virtual concrete experiences designed to be consistent with Points 2–4 and 6–8 were significantly efficacious in promoting fluency with unpracticed subtraction combination—transfer, which is a primary goal of education (Baroody, Purpura, Eiland, & Reid, 2014; Baroody, Purpura, Eiland, Reid, & Paliwal, 2016). Moreover, although Walker et al. did attempt to gauge transfer, they did not measure which method was more effective in fostering the conceptual understanding that addition and subtraction are related operations. Finally, aside from a fluency pretest, these researchers did assess internal factors such as developmental readiness to benefit from fact-triangle instruction.

Mix (2010) concluded that manipulatives "play different roles in different situations" (p. 41) and the key question is not "do such educational tools work" but "do these materials used in this particular way activate this particular mechanism in this particular learner?" Moreover, citing Ginsburg and Golbeck (2004), she noted, "almost no research has addressed how or why these materials might help" (p. 41). To address such a question effectively, researchers need to spell out the role manipulatives are presumed to play—a factor that dictates the outcome measures. Importantly, both the theoretical model and intervention effort need to take into account Dewey's (1963) principle of interaction—how external factors are intended to mesh with internal factors. Evaluations of manipulatives involving interventions that violate this principle (e.g., attempt to impose a manipulative procedure via direct instruction quickly) confound instructional approaches with the potential value of the manipulative in a learning environment that honors the principle (e.g., encourages a child to use their existing knowledge to reflect on how to use the manipulative). Furthermore, a fair evaluation of manipulatives should include assessing the developmental readiness of participants along a HLT. To date, such internal factors have been largely overlooked when researchers construct hypotheses regarding why and how manipulatives work and evaluate the impact of manipulatives.

5. CONCLUSIONS

Both educators and researchers need to take into account Dewey's (1963) principle of interaction when considering the use of concrete experiences. The main implication of the interaction perspective is that the choice of concrete experience depends on internal or psychological factors (e.g., a child's developmental level and interests) as well as external factors (e.g., the goal of instruction, the nature of the instructional activity). Indeed, ensuring an educative experience requires that the latter mesh with the former, including a child's developmental level.

What is "concrete" may depend on whether the goal is sensory-concrete, integrated-concrete, or abstract knowledge. "Students may require physically concrete materials to build meaning [sensory-concrete knowledge] initially" (Sarama & Clements, 2009a, p. 146). However, a priori, there is no reason to believe that virtual representations could not provide the sensory input needed to construct mathematical ideas initially. Whether virtual experiences can be designed to promote sensory-concrete knowledge as effectively as (or even more so than) physical materials, is an open and important question. Until such evidence is available, results showing that virtual manipulatives are more effective than physical manipulatives (e.g., Thompson, 1992) are not necessarily inconsistent with perceptual-motor to mental actions perspective (cf. Mix, 2010), *if* the goal is integrated or abstract knowledge. For these goals, concrete is what is meaningful—what can be connected to other familiar ideas or situations—not sensory input from physical objects or movements (Sarama & Clements, 2009a).

A worthwhile concrete experience or activity is educative (e.g., promotes the next level of development in a HLT or lays the groundwork for more advanced levels). "Good [concrete experiences or] manipulatives are those that aid students in building, strengthening, and connecting various representations of mathematical ideas" (Sarama & Clements, 2009a, p. 146). Their good use could benefit from a systematically planned approach to assessment and instruction based on HLTs. Such an approach can help ensure that external mesh with internal factors by identifying a child's level of development and the goal, suggested activities for achieving this goal, and the assessment for gauging whether the goal has been achieved. Although largely overlooked by researchers, an educative experience may also include activities that foster a positive disposition for further learning or problem-solving competence.

The effective use of concrete experiences and manipulatives—whether they work or not—depends on what is arguably the most important and too often overlooked factor in learning, namely internal factors (e.g., a child's developmental level or interests; Fyfe et al., 2012; McNeil & Jarvin, 2007). Key questions both educators and researchers need to ask include: Is the concrete experience or manipulative consistent with the principle of assimilation (e.g., does it fit a child's existing knowledge and is thus meaningful). In particular, does instruction tap a conceptual schema (e.g., a familiar and meaningful analogy) by which children can connect a formal representation or process to their existing informal concrete model or strategy? Is it moderately novel and purposeful and, thus, engaging to children? Is the experience designed to prompt reflection, discovery, or other mental actions, or does it merely entail passively watching and modeling an adult-imposed strategy? Indeed, whether a manipulative is meaningful, engaging, and thought provoking may be appreciably more important factors than its physical appearance (e.g., whether gaudy or sterile). Considering such questions will reduce the chances of concluding that a concrete activity or material is ineffective or attributing its failed use to a child's or children's cognitive limitations when such activities and materials are not meaningful, engaging, or thought provoking.

Taking into account internal factors and how they mesh with external ones may mean using an approach Bruner (1960) called a "spiral curriculum." It may be developmentally inappropriate to start with a model that most closely aligns with an ultimate goal (a formal representation). Instead, increasing abstract and aligned manipulatives may need to be gradually introduced in a stepwise fashion. More specifically, instruction may need to start with a concrete model that is not only *partially* aligned with the ultimate goal of instruction or desired mental representation but is aligned with a child's developmental level. Then a series of increasingly less concrete models can be used to revisit the idea as the child achieves more advanced levels of development. Finally, when the child is developmentally ready, a relatively abstract model that precisely aligns with the ultimate goal of instruction or desired mental representation can be introduced.

For example, an abacus embodies both base-ten and place-value concepts and is a relatively precise concrete model of multidigit numbers, addition, and subtraction. As such, the model meets the criterion for a useful model according to cognitive alignment perspective, especially if related to written representations. However, an abacus corresponds to the most abstract model (Frame D) in Fig. 6. As such, it may not a good tool for introducing a base-ten/place-value meaning of multidigit numbers, whereas bundling 10 items to create a group of ten (a proportional model that entails the physical action of grouping but not place value) may be more concrete (align more closely to children's informal counting-based view of number). Later, though, the abacus (a nonproportional grouping and place-value model)—which aligns more closely with place-value, base-ten meaning of multidigit written numbers—may well be useful.

Clearly, though, the spiral curriculum hypothesis of concrete experience/manipulative use—like so many aspects of the educational use of these tools—needs rigorous empirical verification. Much more research is needed to develop or refine HLTs that guide the effective use of such educational tools. The promise is that concrete experiences can be a valuable means of discovering and generalizing mathematical regularities and extending informal or formal knowledge (for learners of all ages). Such experiences may be a particularly invaluable means of making sense of formal mathematics instruction—of relating the unfamiliar, context-free ideas of school mathematics to familiar, personally meaningful contexts.

ACKNOWLEDGMENTS

Preparation of this chapter was support by Institute of Education Science Grant R305A150243 (Evaluating the Efficacy of Learning Trajectories in Early Mathematics). The opinions expressed are solely those of the author and do not necessarily reflect the position, policy, or endorsement of the Institute of Education Science or the U.S. Department of Education. In addition to the guest editors, I wish to thank Nicole McNeil for her thoughtful comments on the chapter.

REFERENCES

Ainsworth, S. (2006). DeFT: A conceptual framework for considering learning with multiple representations. *Learning and Instruction, 16*, 183–198. http://dx.doi.org/10.1016/j.learninstruc.2006.03.001.

Alfieri, L., Brooks, P. J., Aldrich, N. J., & Tenenbaum, H. R. (2011). Does discovery-based instruction enhance learning? *Journal of Educational Psychology, 103*, 1–18. http://dx.doi.org/10.1037/a0021017.

Ambrose, R., Baek, J.-M., & Carpenter, T. P. (2003). Children's invention of multidigit multiplication and division algorithms. In A. J. Baroody & A. Dowker (Eds.), *The development of arithmetic concepts and skills: Constructing adaptive expertise* (pp. 305–336). Mahwah, NJ: Erlbaum.

Anobile, G., Cicchini, G. M., & Burr, D. C. (2012). Linear mapping of numbers onto space requires attention. *Cognition, 122*, 454–459. http://dx.doi.org/10.1016/j.cognition.2011.11.006.

Balfanz, R. (1999). Why do we teach young children so little mathematics? Some historical considerations. In J. V. Copley (Ed.), *Mathematics in the early years* (pp. 3–10). Reston, VA: National Council of Teachers of Mathematics.

Ball, D. L. (1992). Magical hopes: Manipulatives and the reform of math education. *American Educator, 16*, 14–18.

Baratta-Lorton, M. (1972). *Workjobs: Activity-centered learning for early childhood*. Menlo Park, CA: Addison-Wesley.

Barmby, P., Harries, T., Higgins, S., & Suggate, J. (2009). The array representation and primary children's understanding and reasoning in multiplication. *Educational Studies in Mathematics, 70*, 217–241. http://dx.doi.org/10.1007/s10649-008-9145 1.

Baroody, A. J. (1987). *Children's mathematical thinking: A developmental framework for preschool, primary, and special education teachers*. New York, NY: Teacher's College Press.

Baroody, A. J. (1989). One point of view: Manipulatives don't come with guarantees. *Arithmetic Teacher, 37*(2), 4–5.

Baroody, A. J. (1990). How and when should place-value concepts and skills be taught? *Journal for Research in Mathematics Education, 21*, 281–286.

Baroody, A. J. (1993). *Problem solving, reasoning, and communicating, grades K-8: Helping children think mathematically*. New York, NY: Merrill-MacMillan.

Baroody, A. J. (2016a). Curricular approaches to introducing subtraction and fostering fluency with basic differences in grade 1. In R. Bracho (Ed.), *Vol. 10(3). The development of number sense: From theory to practice. Monograph of the Journal of Pensamiento Numérico y Algebraico* (pp. 161–191). *numerical and algebraic thought: Granada, Spain: University of Granada.*

Baroody, A. J. (2016b). Reasoning and sense making in grades PK-2: Using number and arithmetic instruction as a basis for fostering mathematical reasoning. In M. T. Battista (Ed.), *Reasoning and sense making in the mathematics classroom: Pre-K—Grade 2* (pp. 27–69). Reston, VA: National Council of Teachers of Mathematics.

Baroody A.J., Clements D. and Sarama J. (in press). Teaching and learning mathematics in early childhood programs, In: Brown C., McMullen M. and File N.K., (Eds.), *Handbook of early childhood care and education (section III of curriculum in early childhood care and education)*, Wiley Blackwell Publishing; Hoboken, NJ.

Baroody, A. J., & Coslick, R. T. (1998). *Fostering children's mathematical power: An investigative approach to K–8 mathematics instruction*. Mahwah, NJ: Erlbaum.

Baroody, A. J., Lai, M.-L., & Mix, K. S. (2006a). The development of young children's number and operation sense and its implications for early childhood education. In B. Spodek & O. Saracho (Eds.), *Handbook of research on the education of young children* (pp. 187–221). Mahwah, NJ: Erlbaum.

Baroody, A. J., & Purpura, D. J. (2017). Early number and operations: Whole numbers. In J. Cai (Ed.), *Compendium for research in mathematics education.* National Council of Teachers of Mathematics: Reston, VA.

Baroody, A. J., Purpura, D. J., Eiland, M. D., & Reid, E. E. (2014). Fostering first-graders' fluency with basic subtraction and larger addition combinations via computer-assisted instruction. *Cognition and Instruction, 32*, 159–197. http://dx.doi.org/10.1080/07370008.2014.887084.

Baroody, A. J., Purpura, D. J., Eiland, M. D., & Reid, E. E. (2015). The impact of highly and minimally guided discovery instruction on promoting the learning of reasoning strategies for basic add-1 and doubles combinations. *Early Childhood Research Quarterly, 30*, 93–105. http://dx.doi.org/10.1016/j.ecresq.2014.09.003.

Baroody, A. J., Purpura, D. J., Eiland, M. D., Reid, E. E., & Paliwal, V. (2016). Does fostering reasoning strategies for relatively difficult basic combinations promote transfer? *Journal of Educational Psychology, 108*, 576–591. http://dx.doi.org/10.1037/edu0000067.

Baroody, A. J., & Rosu, L. (2006). Adaptive expertise with basic addition and subtraction combinations: The number sense view. [Chair]. In A. J. Baroody & J. Torbeyns (Eds.), *Developing adaptive expertise in elementary school arithmetic Symposium conducted at the annual meeting of the American Educational Research Association, San Francisco, CA.*

Baroody, A. J., Tiilikainen, S. H., & Tai, Y. (2006b). The application and development of an addition sketch goal. *Cognition and Instruction, 24*, 123–170.

Barth, H. C., & Paladino, A. M. (2011). The development of numerical estimation: Evidence against a representational shift. *Developmental Science, 14*, 125–135. http://dx.doi.org/10.1111/j.1467-7687.2010.00962.x.

Battista, M. T. (2016). *Reasoning and sense making in the mathematics classroom: Pre-K—Grade 2.* Reston, VA: National Council of Teachers of Mathematics.

Bell, M., Bell, J., Bretzlauf, J., Dillard, A., Hartfield, R., Issacs, A., … Saeker, P. (2004). *Everyday mathematics: University of Chicago School Mathematics Project (grade 1).* Chicago, IL: McGraw-Hill.

Berthold, K., & Renkl, A. (2009). Instructional aids to support a conceptual understanding of multiple representations. *Journal of Educational Psychology, 101*, 70–87. http://dx.doi.org/10.1037/a0013247.

Bisanz, J., Watchorn, R. P. D., Piatt, C., & Sherman, J. (2009). On "understanding" children's developing use of inversion. *Mathematical Thinking and Learning, 11*, 10–24.

Blöte, A. W., Klein, A. S., & Beishuizen, M. (2000). Mental computation and conceptual understanding. *Learning and Instruction, 10*, 221–247.

Blöte, A. W., Van der Burg, E., & Klein, A. S. (2001). Students' flexibility in solving two-digit addition and subtraction problems: Instruction effects. *Journal of Educational Psychology, 93*, 627–638.

Boggan, M., Harper, S., & Whitmire, A. (2010). Using manipulatives to teach elementary mathematics. *Journal of Instructional Pedagogies, 3*, 1–6.

Bright, G. W., Harvey, J. G., & Wheeler, M. M. (1985). *Learning and mathematics games (JRME monograph number 1)*. Reston, VA: National Council of Teachers of Mathematics.

Brown, M. C., McNeil, N. M., & Glenberg, A. M. (2009). Using concreteness in education: Real problems, potential solutions. *Child Development Perspectives* ***3***, 160–164.

Bruner, J. S. (1960). *The process of education*. Cambridge, Mass: Harvard University Press.

Bruner, J. S. (1961). The act of discovery. *Harvard Educational Review, 31*, 21–32.

Bruner, J. S. (1966). *Toward a theory of instruction*. Cambridge, MA: Bilknap.

Chi, M. T. H. (2009). Active-constructive-interactive: A conceptual framework for differentiating learning activities. *Topics in Cognitive Science, 1*, 73–105.

Choate, L. D., & Okey, J. K. (1981). Graphically speaking: Primary-level graphing experiences. In A. P. Shulte & J. R. Smart (Eds.), *Teaching statistics and probability–1981 Yearbook* (pp. 33–41). Reston, VA: National Council of Teachers of Mathematics.

Clements, D. H. (1999). "Concrete" manipulatives, concrete ideas. *Contemporary Issues in Early Childhood, 1*, 45–60.

Clements, D. H., Baroody, A. J., & Sarama, J. (2013). *Background research on early mathematics*. Washington, DC: National Governors Association. Retrieved from: http://www.nga.org/files/live/sites/NGA/files/pdf/2013/1311SEME-Background.pdf.

Clements, D. H., & Battista, M. T. (2000). Designing effective software. In A. E. Kelly & R. A. Lesh (Eds.), *Handbook of research design in mathematics and science education* (pp. 761–776). Mahwah, NJ: Erlbaum.

Clements, D. H., Copple, C., & Hyson, M. (Eds.), (2002). *Early childhood mathematics: Promoting good beginnings. A joint position statement of the National Association for the Education of Young Children (NAEYC) and the National Council for Teachers of Mathematics* (Rev. ed.) Washington, DC: National Association for the Education of Young Children/National Council for Teachers of Mathematics.

Clements, D. H., & McMillen, S. (1996). Rethinking "concrete" manipulatives. *Teaching Children Mathematics, 2*, 270–279.

Clements, D. H., & Nastasi, B. K. (1993). Electronic media and early childhood education. In B. Spodek (Ed.), *Handbook of research on the education of young children* (pp. 251–275). New York: Macmillan.

Clements, D. H., & Sarama, J. (2003). Strip mining for gold: Research and policy in educational technology—A response to "Fool's gold" *Educational Technology Review, 11*, 7–69.

Clements, D. H., & Sarama, J. (2004). Technology in early childhood mathematics: Building Blocks™ as an innovative technology-based curriculum. In D. H. Clements, J. Sarama, & A.-M. DiBiase (Eds.), *Engaging young children in mathematics: Standards for early childhood mathematics education* (pp. 361–375). Mahwah, NJ: Erlbaum.

Clements, D. H., & Sarama, J. (2012). Learning and teaching early and elementary mathematics. [Series Eds.]. In J. Carlson & J. Levin (Eds.), *Psychological perspectives on contemporary educational issues: Vol. 3. Instructional strategies for improving students' learning* (pp. 163–175). Information Age: Charlotte, NC.

Clements, D. H., & Sarama, J. (2013). Building blocks. In *Teacher's edition, teacher's resource guide, assessment guide* (Vols. 1–2). Columbus, OH: McGraw-Hill.

Daro, P., Mosher, F. A., Corcoran, T., & Barrett, J. (2011). *Learning trajectories in mathematics: A foundation for standard, curriculum, assessment, and instruction*. Philadelphia: Consortium for Policy Research in Education.

De Bock, D., Deprez, J., Van Dooren, W., Roelens, M., & Verschaffel, L. (2011). Abstract or concrete examples in learning mathematics? A replication and elaboration of Kaminski, Sloutsky, and Heckler's study. *Journal for Research in Mathematics Education, 42*, 109–126. Retrieved from: http://hdl.handle.net/10067/876830151162165141.

DeCaro, M. S., & Rittle-Johnson, B. (2012). Exploring mathematics problems prepares children to learn from instruction. *Journal of Experimental Child Psychology, 113*, 552–568.

DeLoache, J. S. (2000). Dual representation and young children's use of scale models. *Child Development, 71*, 329–338.

Dewey, J. (1963). *Experience and education*. New York: Collier.

Durkin, K., & Rittle-Johnson, B. (2012). The effectiveness of using incorrect examples to support learning about decimal magnitude. *Learning and Instruction, 22*, 206–214.

Ernest, P. (1985). The number line as a teaching aid. *Educational Studies in Mathematics, 16*, 411–424.

Frye, D., Baroody, A. J., Burchinal, M. R., Carver, S., Jordan, N. C., & McDowell, J. (2013). *Teaching math to young children: A practice guide*. Washington, DC: National Center for Education Evaluation and Regional Assistance (NCEE), Institute of Education Sciences, U.S. Department of Education.

Fuson, K. C. (1988). *Children's counting and concepts of number*. New York: Springer-Verlag.

Fuson, K. C. (2009). Avoiding misinterpretations of Piaget and Vygotsky: Mathematical teaching without learning, learning without teaching, or helpful learning-path teaching? *Cognitive Development, 24*, 343–361.

Fyfe, E. R., DeCaro, M. S., & Rittle-Johnson, B. (2014). An alternative time for telling: When conceptual instruction prior to problem solving improves mathematical knowledge. *British Journal of Educational Psychology, 84*, 502–519. http://dx.doi.org/10.1111/bjep.12035.

Fyfe, E. R., McNeil, N. M., & Rittle-Johnson, B. (2015). Easy as ABCABC: Abstract language facilitates performance on a concrete patterning task. *Child Development, 86*, 927–935. http://dx.doi.org/10.1111/cdev.12331.

Fyfe, E. R., Rittle-Johnson, B., & DeCaro, M. S. (2012). The effects of feedback during exploratory mathematics problem solving: Prior knowledge matters. *Journal of Educational Psychology, 104*, 1094–1108.

Ginsburg, H. P. (1977). *Children's arithmetic*. New York, NY: D. Van Nostrand.

Ginsburg, H. P., & Golbeck, S. L. (2004). Thoughts on the future of research on mathematics and science learning and education. *Early Childhood Research Quarterly, 19*, 190–200.

Ginsburg, H. P., Klein, A., & Starkey, P. (1998). The development of children's mathematical knowledge: Connecting research with practice. In I. E. Sigel & K. A. Renninger (Eds.), *Child psychology in practice: Vol. 4. Handbook of child psychology* (pp. 401–476). New York, NY: Wiley & Sons.

Gravemeijer, K. (2002). In *Learning trajectories and local instruction theories as a means of support for teachers in reform mathematics education Paper presented at the annual meeting of the American Educational Research Association, Las Vegas, NV.*

Hiebert, J., & Tonnessen, L. H. (1978). Development of the fraction concept in two physical contexts: An exploratory investigation. *Journal for Research in Mathematics Education, 9*, 374–378.

Huntley-Fenner, G. (2001). Children's understanding of number is similar to adults' and rats': Numerical estimation by 5- to 7-year-olds. *Cognition, 78*, B27–B40. http://dx.doi.org/10.1016/S0010-0277(00)00122-0.

Huttenlocher, J., Jordan, N. C., & Levine, S. C. (1994). A mental model for early arithmetic. *Journal of Experimental Psychology: General, 123*, 284–296. http://dx.doi.org/10.1037/0096-3445.123.3.284.

Izard, V., Pica, P., Spelke, E. S., & Dehaene, S. (2008). Exact equality and successor function: Two key concepts on the path towards understanding exact numbers. *Philosophical Psychology, 21*, 491–505.

Jitendra, A. K., Dupuis, D. N., Rodriguez, M. C., Zaslofsky, A. F., Slater, S., Cozine-Corroy, K., & Church, C. (2013). A randomized controlled trial of the impact of schema-based instruction on mathematical outcomes for third-grade students with mathematics difficulties. *Elementary School Journal, 114*, 252–276. http://dx.doi.org/10.1086/673199.

Jordan, N. C., Hanich, L. B., & Uberti, H. Z. (2003). Mathematical thinking and learning difficulties. In A. J. Baroody & A. Dowker (Eds.), *The development of arithmetic concepts and skills: Constructing adaptive expertise* (pp. 359–383). Mahwah, NJ: Erlbaum.

Kaminski, J. A., & Sloutsky, V. M. (2013). Extraneous perceptual information interferes with children's acquisition of mathematical knowledge. *Journal of Educational Psychology, 105*, 351–363.

Kaminski, J. A., Sloutsky, V. M., & Heckler, A. F. (2008). The advantage of abstract examples in learning math. *Science, 320*, 454–455. http://dx.doi.org/10.1126/science.1154659.

Kaminski, J. A., Sloutsky, V. M., & Heckler, A. (2009). Transfer of mathematical knowledge: The portability of generic instantiations. *Child Development Perspectives, 3*, 151–155.

Kaminski, J. A., Sloutsky, V. M., & Heckler, A. (2013). The cost of concreteness: The effect of nonessential information on analogical transfer. *Journal of Experimental Psychology: Applied, 19*, 14–29.

Kouba, V. L., & Franklin, K. (1993). Multiplication and division: Sense making and meaning. In R. J. Jensen (Ed.), *Research ideas for the classroom: Early childhood mathematics* (pp. 103–126). New York: Macmillan.

Landy, D., Charlesworth, A., & Ottmar, E. (2016). Categories of large numbers in line estimation. *Cognitive Science, 40*, 1–28. http://dx.doi.org/10.1111/cogs.12342.

Laski, E. V., & Siegler, R. S. (2014). Learning from number board games: You learn what you encode. *Developmental Psychology, 50*, 853–864. http://dx.doi.org/10.1037/a0034321.

Lehrer, R. (2003). Developing understanding of measurement. In J. Kilpatrick, W. G. Martin, & D. Schifter (Eds.), *A research companion to principles and standards for school mathematics* (pp. 179–192). Reston, VA: National Council of Teachers of Mathematics.

Martin, T. (2009). A theory of physically distributed learning: How external environments and internal states interact in mathematics learning. *Child Development Perspectives, 3*, 140–144.

Martin, T., & Schwartz, D. L. (2005). Physically distributed learning: Adapting and reinterpreting physical environments in the development of fraction concepts. *Cognitive Science, 29*, 587–625.

Marzolf, D. P., & DeLoache, J. S. (1994). Transfer in young children's understanding of spatial representations. *Child Development, 64*, 1–15.

Math Learning Center. (2009). *Bridges in mathematics*. Salem, OR: Math Learning Center.

Matthews, P., & Rittle-Johnson, B. (2009). In pursuit of knowledge: Comparing self-explanations, concepts, and procedures as pedagogical tools. *Journal of Experimental Child Psychology, 104*, 1–21.

McNeil, N. M., & Jarvin, L. (2007). When theories don't add up: Disentangling the manipulatives debate. *Theory into Practice, 46*, 309–316.

McNeil, N. M., Uttal, D. H., Jarvin, L., & Sternberg, R. J. (2009). Should you show me the money? Concrete objects both hurt and help performance on mathematics problems. *Learning and Instruction, 19*, 171–184.

Mix, K. S. (2010). Spatial tools for mathematical thought. In K. S. Mix, L. B. Smith, & M. Gasser (Eds.), *The spatial foundations of language and cognition*. New York, NY: Oxford University Press.

Moseley, B. (2005). Students' early mathematical representation knowledge: The effects of emphasizing single or multiple perspectives of the rational number domain in problem solving. *Educational Studies in Mathematics, 60*, 37–69.

National Mathematics Advisory Panel (NMAP). (2008). *Foundations for success: The final report of the National Mathematics Advisory Panel.* Washington, DC: U.S. Department of Education.
National Research Council (NRC). (2001). In J. Kilpatrick, J. Swafford, & B. Findell (Eds.), *Adding it up: Helping children learn mathematics.* Washington, DC: National Academy Press.
National Research Council (NRC). (2009). In C. T. Cross, T. A. Woods, & H. Schweingruber (Eds.), *Mathematics in early childhood: Learning paths toward excellence and equity.* Washington, DC: National Academy Press.
Obersteiner, A., Reiss, K., & Ufer, S. (2013). How training on exact or approximate mental representations of number can enhance first-grade students' basic number processing and arithmetic skills. *Learning and Instruction, 23,* 125–135. http://dx.doi.org/10.1016/j.learninstruc.2012.08.004.
Petersen, L. A., & McNeil, N. (2013). Effects of perceptually rich manipulatives on preschoolers' counting performance: Established knowledge counts. *Child Development, 84,* 1020–1033. http://dx.doi.org/10.1111/cdev.12028.
Piaget, J. (1928). *Judgment and reasoning in the child.* New York, NY: Harcourt, Brace, and Co.
Piaget, J. (1936). *Origins of intelligence in the child.* London, UK: Routledge & Kegan Paul.
Piaget, J. (1964). Development and learning. In R. E. Ripple & V. N. Rockcastle (Eds.), *Piaget rediscovered* (pp. 7–20). Ithaca, NY: Cornell University.
Prather, R. W., & Alibali, M. W. (2009). The development of arithmetic principle knowledge: How do we know what learners know? *Developmental Review, 29,* 221–248. http://dx.doi.org/10.1016/j.dr.2009.09.001.
Ramani, G. B., & Siegler, R. S. (2008). Promoting broad and stable improvements in low-income children's numerical knowledge through playing number board games. *Child Development, 79,* 375–394. http://dx.doi.org/10.1111/j.1467-8624.2007.01131.x.
Ramani, G. B., & Siegler, R. S. (2011). Reducing the gap in numerical knowledge between low- and middle-income preschoolers. *Journal of Applied Developmental Psychology, 32,* 146–159. http://dx.doi.org/10.1016/j.appdev.2011.02.005.
Reid, E. E., Baroody, A. J., & Purpura, D. J. (2013). Impact of interventions on preschoolers' successor principle understanding and linear representation of number[Chair] In J. Stigler (Ed.), *Developing mathematical thinking Symposium presented at the biennial meeting of the Society for Research in Child Development, Seattle, WA.*
Resnick, L. B. (1982). Syntax and semantics in learning to subtract. In T. R. Carpenter, J. M. Moser, & T. A. Romberg (Eds.), *Addition and subtraction: A cognitive perspective* (pp. 136–155). Hillsdale, NJ: Erlbaum.
Resnick, L. B. (1983). A developmental theory of number understanding. In H. P. Ginsburg (Ed.), *The development of mathematical thinking* (pp. 110–151). New York, NY: Academic Press.
Resnick, L. B. (1992). From protoquantities to operators: Building mathematical competence on a foundation of everyday knowledge. In G. Leinhardt, R. Putnam, & R. A. Hattrup (Eds.), *Analysis of arithmetic for mathematics teaching* (pp. 373–425). Hillsdale, NJ: Erlbaum.
Resnick, L. B., & Omanson, S. F. (1987). Learning to understand arithmetic. In R. Glaser (Ed.), *Vol. 3. Advances in instructional psychology.* Hillsdale, NJ: Erlbuam
Rittle-Johnson, B. (2006). Promoting transfer: Effects of self-explanation and direct instruction. *Child Development,* 77, 1–15. http://dx.doi.org/10.1111/j.1467-8624.2006.00852.x.
Rittle-Johnson, B., Fyfe, E. R., Loehr, A. M., & Miller, M. R. (2015). Beyond numeracy in preschool: Adding patterns to the equation. *Early Childhood Research Quarterly, 31,* 101–112.
Rittle-Johnson, B., & Star, J. R. (2007). Does comparing solution methods facilitate conceptual and procedural knowledge? An experimental study on learning to solve equations.

Journal of Educational Psychology, *99*, 561–574. http://dx.doi.org/10.1037/0022-0663.99.3.561.

Rittle-Johnson, B., & Star, J. R. (2009). Compared with what? The effects of different comparisons on conceptual knowledge and procedural flexibility for equation solving. *Journal of Educational Psychology*, *101*, 529–544. http://dx.doi.org/10.1037/a0014224.

Rittle-Johnson, B., & Star, J. R. (2011). The power of comparison in learning and instruction: Learning outcomes supported by different types of comparisons. In J. P. Mestre & B. H. Ross (Eds.), *Vol. 55. The psychology of learning and motivation: Cognition in education* (pp. 199–225). Oxford: Academic Press.

Sarama, J., & Clements, D. H. (2009a). "Concrete" computer manipulatives in mathematics education. *Child Development Perspectives*, *3*, 145–150.

Sarama, J., & Clements, D. H. (2009b). *Early childhood mathematics education research: Learning trajectories for young children*. New York, NY: Routledge.

Sarama, J., & Clements, D. H. (2009c). *Learning and teaching early math: The learning trajectories approach*. New York, NY: Routledge.

Saxe, G. B., Earnest, D., Sitabkhan, Y., Haldar, L. C., Lewis, K. E., & Zheng, Y. (2010). Supporting generative thinking about integers on number lines in elementary mathematics. *Cognition & Instruction*, *28*, 433–474. http://dx.doi.org/10.1080/07370008.2010.511569.

Schwartz, D. L., Chase, C. C., Chin, D. B., & Oppezzo, M. (2011). Practicing versus inventing with contrasting cases: The effects of telling first on learning and transfer. *Journal of Educational Psychology*, *103*, 759–775. http://dx.doi.org/10.1037/a0025140.

Seufert, T. (2003). Supporting coherence formation in learning from multiple representations. *Learning and Instruction*, *13*, 227–237.

Shafto, P. (2014). Direct instruction: More than meets the eye[Chair]. In B. Rittle-Johnson (Ed.), *Different perspectives on the role of explanation and exploration in learning A symposium presented at the annual meeting of the American Educational Research Association, Philadelphia, PA.*

Shin, N., Sutherland, L. M., Norris, C. A., & Soloway, E. (2012). Effects of game technology on elementary student learning in mathematics. *British Journal of Educational Technology*, *43*, 540–560. http://dx.doi.org/10.1111/j.1467-8535.2011.01197.x.

Siegler, R. S., & Ramani, G. B. (2008). Playing linear numerical board games promotes low-income children's numerical development. *Developmental Science*, *11*, 655–661. http://dx.doi.org/10.1111/j.1467-7687.2008.00714.x.

Siegler, R. S., & Ramani, G. B. (2009). Playing linear number board games—But not circular ones—Improves low-income preschoolers' numerical understanding. *Journal of Educational Psychology*, *101*, 545–560. http://dx.doi.org/10.1037/a0014239.

Slusser, E. B., Santiago, R. T., & Barth, H. C. (2013). Developmental change in numerical estimation. *Journal of Experimental Psychology: General*, *142*, 193–208. http://dx.doi.org/10.1037/a0028560.

Sophian, C. (2004). Mathematics for the future: Developing a head start curriculum to support mathematics learning. *Early Childhood Research Quarterly*, *19*, 59–81. http://dx.doi.org/10.1016/j.ecresq.2004.01.015.

Squire, S., & Bryant, P. (2002). From sharing to dividing: Young children's understanding of division. *Developmental Science*, *5*, 452–466.

Star, J. R., & Rittle-Johnson, B. (2009). It pays to compare: An experimental study on computational estimation. *Journal of Experimental Child Psychology*, *101*, 408–426. http://dx.doi.org/10.1016/j.jecp.2008.11.004.

Steffe, L., & Cobb, P. (1988). *Construction of arithmetical meanings and strategies*. New York, NY: Springer-Verlag.

Steinbring, H. (1997). Epistemological investigation of classroom interaction in elementary mathematics teaching. *Educational Studies in Mathematics*, *32*, 49–92.

Stevenson, H. W., & Stigler, J. W. (1992). *The learning gap: Why our schools are failing and what we can learn from Japanese and Chinese education*. New York: Summit Books.
Thompson, P. (1992). Notations, conventions, and constraints: Contributions to effective uses of concrete materials in elementary mathematics. *Journal for Research in Mathematics Education, 23*, 123–147.
Uttal, D. H. (2003). On the relation between play and symbolic thought: The case of mathematics manipulatives. In B. Spodek (Ed.), *Contemporary perspectives on play in early childhood education* (pp. 97–114). Charotte, NC: Information Age Publishing.
Uttal, D. H., Amaya, M., Maita, M. R., Hand, L. L., Cohen, C. A., O'Doherty, K., & DeLoach, J. S. (2013). It works both ways: Transfer difficulties between manipulatives and written subtraction solutions. *Child Development Research, 2013*, 1–13. 216367. http://dx.doi.org/10.1155/2013/216367.
Uttal, D. H., Liu, L. L., & DeLoache, J. S. (2006). Concreteness and symbolic development. In L. Balter & C. S. Tamis-LeMonda (Eds.), *Child psychology: A handbook of contemporary issues* (2nd ed., pp. 167–184). Philadelphia: Psychology Press.
Uttal, D., O'Doherty, K., Newland, R., Hand, L. L., & DeLoache, J. (2009). Dual representation and the linking of concrete and symbolic representations. *Child Development Perspectives, 3*, 156–159.
Vygotsky, L. S. (1962). Thought and language (E. Hanfmann & G. Vakar, Eds., Trans.). Cambridge, MA: MIT Press.
Walker, D., Mickes, L., Bajic, D., Nailon, C. R., & Rickard, T. C. (2013). A test of two methods of arithmetic fluency training and implications for educational practice. *Journal of Applied Research in Memory and Cognition, 2*, 25–32. http://dx.doi.org/10.1016/j.jarmac.2013.02.001.
Wynroth, L. (1986). *Wynroth math program—The natural numbers sequence*. Ithaca, NY: Wynroth Math Program.
Zhao, Y. (2014). *Who's afraid of the big bad dragon: Why China has the best (and worst) education system in the world*. San Francisco: Jossey-Bass.

CHAPTER THREE

Interventions in Early Mathematics: Avoiding Pollution and Dilution

Julie Sarama[1], Douglas H. Clements
University of Denver, Denver, CO, United States
[1]Corresponding author e-mail address: Julie.Sarama@du.edu

Contents

Abstract

Although specific interventions in early mathematics have been successful, few have been brought to scale successfully, especially across the challenging diversity of populations and contexts in the early childhood system in the United States. In this chapter, we analyze a theoretically based scale-up model for early mathematics that was designed to avoid the pollution and dilution that often plagues efforts to achieve broad success. We elaborate the theoretical framework by noting the junctures that are susceptible to dilution or pollution. Then we expatiate the model's guidelines to describe specifically how they were designed and implemented to mitigate pollution and dilution. Finally, we provide evidence regarding the success of these efforts.

Although specific interventions in early mathematics have been successful (Clements & Sarama, 2011), few have been brought to scale successfully, especially across the challenging diversity of populations and contexts in

Advances in Child Development and Behavior, Volume 53
ISSN 0065-2407
http://dx.doi.org/10.1016/bs.acdb.2017.03.003

the early childhood system in the United States. In this chapter, we describe a theoretically based scale-up model for early mathematics that has avoided the pollution and dilution that often plagues efforts to achieve broad success (Clements & Sarama, 2011).

1. BACKGROUND

Taking promising interventions to scale means implementing them successfully in larger settings. Moreover, though, such scale-up goes beyond larger numbers to confront the complexity of such an enterprise. For example, there is an increase in not just the number of students and teachers, but also in both the number of categories of stakeholders and the number of different and often conflicting perspectives they hold. Therefore, we define scale-up as instantiation of an intervention in varied settings with diverse populations, addressing the needs of multiple sociopolitical stakeholders, to achieve satisfactory fidelity of implementation and, thus, intervention goals. Although this already is a substantial challenge, consider the complexity of addressing the perspectives, needs, and desires of the different categories of stakeholders, including parents, various community groups, the professional teaching community, educational leadership in early childhood and in mathematics (often these are distinct groups), and higher-level administration (Sarama & Clements, 2013). Achieving an adequate fidelity of implementation during implementation presents other challenges, such as sufficient materials, technology, professional development, and in-class support.

Also, interventions are not useful if they are not maintained after the initial thrust for the implementation. Thus, attention must be given to including transfer to local ownership and sustainability. We define *sustainability* as the length of time an innovation continues to be implemented with fidelity (cf. Baker, 2007).

Implementation, fidelity, and sustainability are threatened by both pollution and dilution. *Pollution* is the introduction of elements into the intervention environment or teaching practices that vitiate or are inimical to the intervention. *Dilution* is the gradual replacement of components of the intervention with other aspects that do not fulfill the same role as the replaced components.

The next section describes the theoretical framework we developed for our research and development work on implementation and wider scale-up, and the model we derived from it. After we define each of the 10 guidelines that constitute the TRIAD model, we describe how they were designed to mitigate the usual forces of pollution and dilution.

2. THE TRIAD MODEL

2.1 Theoretical Framework

Our theoretical framework (Sarama, Clements, Starkey, Klein, & Wakeley, 2008) is an elaboration of the *Network of Influences* framework (Sarama, Clements, & Henry, 1998). We consider successful implementation of an intervention at scale to involve multiple coordinated efforts to maintain the integrity of the vision and practices of an innovation through increasingly numerous and complex socially mediated filters.

2.1.1 Interactions

The TRIAD model goes beyond solely adopting new curricula. Instead, it scales up the support of "interactions among teachers and children around educational material" (Ball & Cohen, 1999, p. 3). This strategy creates extensive opportunities for teachers to focus on math, goals, and children's thinking and learning, which improves teachers' knowledge of subject matter, teaching, and learning, and increases child achievement (Ball & Cohen, 1999; Cohen, 1996, p. 98; Schoen, Cebulla, Finn, & Fi, 2003; Sowder, 2007). The depiction of the *Network of Influences* framework in Fig. 1 illustrates the hypothesized influences of context and implementation variables on outcomes such as teacher knowledge, child achievement, and sustainability.

2.1.2 Administrators and Other School Leaders (Fig. 1, Factors K and I)

Principal leadership is strongly correlated with levels of implementation and effectiveness (Berends, Kirby, Naftel, & McKelvey, 2001; Bodilly, 1998; Bryk, Sebring, Allensworth, Suppescu, & Easton, 2010; Fullan, 1992; Heck, Weiss, Boyd, & Howard, 2002; Kaser, Bourexis, Loucks-Horsley, & Raizen, 1999; Klingner, Ahwee, Pilonieta, & Menendez, 2003; Teddlie & Stringfield, 1993). Effective administrators provide the time for teachers to experiment, discuss, and, in general, construct their own meanings of the innovation. They communicate continuing commitment, not just in verbal form, but also in other ways, such as resource allocation (Bodilly, 1998). School mathematics and early childhood leaders can serve as essential bridges between administrators and teachers (Cobb, McClain, de Silva, & Dean, 2003). District level leaders

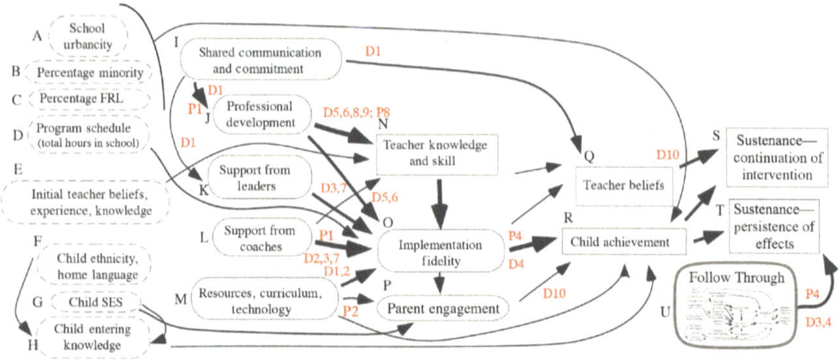

Fig. 1 Revised *Network of Influences* theoretical framework elaborated with junctures susceptible to dilution or pollution. The shaded text such as "D1" or "P2" refer to hypothesized threats (D = dilution, P = pollution) to implementation addressed by guidelines (#1, #2, etc.) of the TRIAD model. The "Follow Though" model at the bottom right is simple a microcosm of the framework. Contextual variables in *dotted* ovals include the school (A–D), teacher (E), and child (F—H) factors from the revised Network of Influences framework. For example, child socioeconomic status, or SES (G), impacts children's initial mathematics knowledge (H), which influences children's achievement (R)—an outcome variable indicated by the *solid rectangle*. Implementation variables in *solid ovals* are features that the project can encourage and support, but cannot control absolutely. For example, *heavy arrows* from professional development (J), to teacher knowledge (N), to implementation fidelity (O), to child achievement (R), indicate the strong effects in that path. Support from coaches (L) also has a strong effect on implementation fidelity, while other factors (J, K, and M) are influential, but to a moderate degree (not all small effects are depicted).

and their decisions also impact implementation and ultimately child achievement (Bodilly, 1998; Klingner et al., 2003; Snipes, Doolittle, & Herlihy, 2002; Spillane, 2000). Communication between principals and all other groups is particularly essential. Principals forgetting the study and their involvement in it and similar communication lapses can be a huge challenge, which increases with the size of the district (Foorman, Santi, & Berger, 2007).

2.1.3 Communication

Communication, collaboration, and agreement among all groups are essential. Our own previous research revealed multiple missed opportunities for facilitation of innovations due to the divergent beliefs of social groups, even about ostensibly observable "facts," such as "there is adequate technology available" (Sarama et al., 1998).

2.1.4 Teachers and Professional Development (Fig. 1, Factors E, N, and Q)

Research suggests that the most critical feature of a high-quality educational environment is a knowledgeable and responsive adult and that high-quality professional development is essential to innovation (Fig. 1J and N; Darling-Hammond, 1997; Ferguson, 1991; National Research Council, 2007; Sarama & DiBiase, 2004; Schoen et al., 2003; Sowder, 2007). Scaling up such professional development has special challenges and opportunities given the early childhood setting, as noted previously. Our model specifies only a weak effect of *initial* teacher expertise (Fig. 1, dotted oval E) because of the low level of mathematics content and pedagogical content knowledge of most pre-K teachers, regardless of background (Copley, 2004; Sarama, 2002; Sarama & DiBiase, 2004), which is consistent with previous research (e.g., Bryk et al., 2010). Changes in beliefs follow changes in practice (boxes N, Q, Showers, Joyce, & Bennett, 1987); moreover, we believe that changes in beliefs help *sustain* teacher practices (Fig. 1, boxes Q and S).

Research-based solutions to these challenges can be found (Corcoran, 2007; Klingner et al., 2003; NAEYC, 2002; National Research Council, 2007; Peisner-Feinberg et al., 1999; Sarama, 2002; US Department of Education, 1999). Use of theory, demonstrations, practice, and feedback, especially from coaches, quadruples the positive effects of information-only training (e.g., strong effects from J and L to N and O in Fig. 1, Foorman et al., 2007; Pellegrino, 2007; Showers et al., 1987). Effective professional development eschews "one-shot" interventions, begins with a specific curriculum (Fig. 1M), embodies the kind of flexible, interactive teaching styles that work well with children, and weaves together math content, pedagogy, and knowledge of child development and family relationships (Baroody & Coslick, 1998; Schoen et al., 2003; Sowder, 2007). The professional development in TRIAD provides a promising path for developing teachers' understanding of learning, teaching, curriculum, and assessment by focusing on research-based models of children's thinking and learning. Research indicates the efficacy of such cognitive models (Bredekamp, 2004; Carpenter & Franke, 2004; Hiebert, 1999; Klingner et al., 2003), especially compared with other approaches such as process–product models (Lawless & Pellegrino, 2007). Research-based *learning trajectories* are TRIAD's core (Clements, Sarama, & DiBiase, 2003). We defined learning trajectories as "descriptions of children's thinking and learning…and a related, conjectured route through a set of instructional tasks" (Clements & Sarama, 2004, p. 83). Thus, learning trajectories have three components: a goal (that is, an aspect of a mathematical domain children should learn), a developmental progression, or learning path through which children move through levels of thinking, and instruction that helps them move along that path (Fig. 2 provides an example). Learning trajectories are at the heart of both TRIAD's math curriculum and its professional development. Learning trajectories help teachers

The *goal* connects learning trajectories to the important *big ideas* in mathematics. The relevant big idea is that numbers can be used to tell how many, describe order, and measure; they involve numerous relations and can be represented in various ways. The specific goals include understanding that subitizing can be used to tell how many (perceptual subitizing), that a quantity can consist of parts and can be broken apart (decomposed) into the parts, and that those parts can be combined (composed) to form the whole (conceptual subitizing). Specific objectives are for children to achieve those understandings and fluency in perceptual and conceptual subitizing.

Age (years)	Developmental progression	Sample instructional tasks
0–1	**Preexplicit number.** Within the first year, the child is sensitive to small numbers, but does not have explicit, intentional knowledge of number.	*Noticing collections.* Provide a rich sensory environment, use words such as *more,* and use actions of adding objects.
1–2	**Small-collection namer.** Names groups of 1 to 2, sometimes 3. Shown a pair of shoes, says, "Two shoes."	*Board games–small numbers.* Play board games with a special die (number cube) or spinner that shows only 1, 2, and 3 dots (then add 0 to it).
3	**Maker of small collections.** Nonverbally makes a small collection (no more than 4, usually 1–3) with the same number as another collection via mental model (i.e., not necessarily by matching). Might also be verbal. When shown a collection of 3, makes another collection of 3.	*Get the number.* Ask children to get the right number of crackers or some other item for a small number of children.
4	**Perceptual subitizer to 4.** Instantly recognizes collections up to 4 briefly shown, and verbally names the number of items. When shown objects briefly, says, "Four."	*Snapshots.* Play "Snapshots" with collections of one to four objects, arranged in line or other simple arrangement, asking children to respond verbally with the number name. Start with the smaller quantities and easier arrangements, moving to those of moderate difficulty only when children are fully competent and confident. 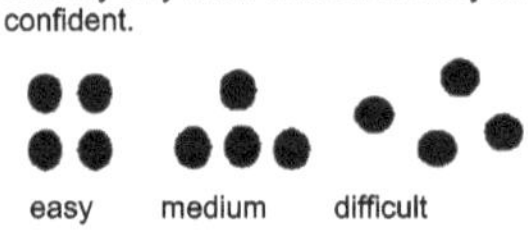
5	**Perceptual subitizer to 5.** Instantly recognizes briefly shown collections up to 5 items and verbally names the number of items. Recognizes and uses spatial and numeric structures beyond the situations in which they were already experienced. Shown 5 objects briefly, says, "Five."	*Snapshots*: Play "Snapshots" with dot cards, starting with easy arrangements, moving to more difficult arrangements when children are able to do so. 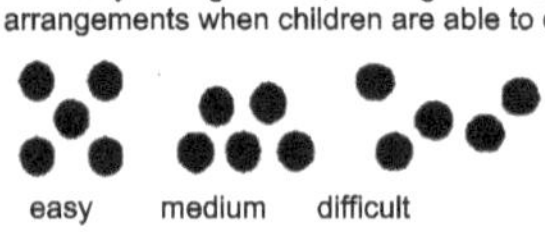
5	**Conceptual subitizer to 5.** Verbally labels all arrangements to about 5, when shown only briefly. Asked "Why?" says, "Because I saw 3 and 2, and so I said 5."	*Snapshots.* Use different arrangements of the various modifications of "Snapshots" to develop conceptual subitizing and ideas about addition and subtraction. The goal is to encourage children to see the addends and

Fig. 2 A learning trajectory for recognition of number and subitizing.

		the sum. 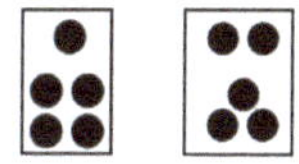
5	**Conceptual subitizer to 10**. Verbally labels most briefly shown arrangements to 6, then up to 10, using groups. "In my mind, I made two groups of 3 and one more, so 7."	*Snapshots*. Play "Snapshots" with larger quantities to develop ideas about addition and subtraction. 
6	**Conceptual subitizer to 20**. Verbally labels structured arrangements up to 20, shown only briefly, using groups. "I saw three 5s, so 5, 10, 15."	*Ten frame addition snapshots*: Briefly show 2 ten frames to help children visualize addition combinations. 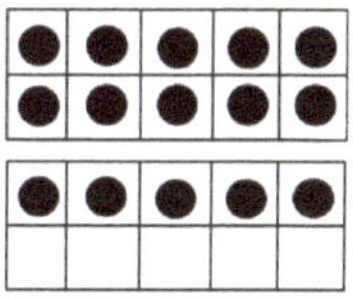
7	**Conceptual subitizer with place value and skip counting**. Verbally labels structured arrangements, shown only briefly, using groups, skip counting, and place value. "I saw groups of 10s and 2s, so 10, 20, 30, 40, 42, 44,. . . 44!"	*Ten frame addition snapshots*. Briefly show several ten frames. 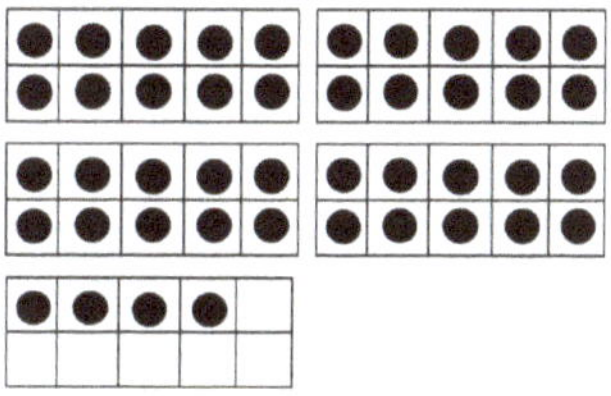
8	**Conceptual subitizer with place value and multiplication.** Verbally labels structured arrangements shown only briefly, using groups, multiplication, and place value. "I saw groups of 10s and 3s, so I thought, 3 tens is 30 and 4 threes is 12, so 42 in all."	*Snapshots with dots*. Play "Snapshots" with structured groups that support the use of increasingly sophisticated mental strategies and operations. 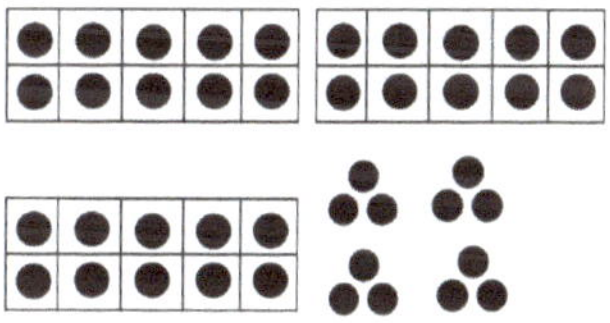

(Adapted from Clements & Sarama, 2014, pp. 17–20; copyright 2014, *Learning and teaching early math: The learning trajectories approach* (2nd ed.) by Clements & Sarama. Adapted with permission of Taylor and Francis Group, LLC, a division of Informa plc.)

Fig. 2—Cont'd

focus on the "conceptual storyline" of the curriculum, a critical element that is often missed (Heck et al., 2002; Weiss, 2002). They facilitate teachers' learning about mathematics, how children think about and learn this math, and how such learning is supported by the curriculum and its teaching strategies. They address domain-specific components of learning and teaching that have the strongest impact on cognitive outcomes (Lawless & Pellegrino, 2007). By illuminating potential developmental progressions, they bring coherence and consistency to goals, curricula, and assessments (for a comprehensive description and review of research on these learning trajectories, see Sarama and Clements (2009a), and for the instructional component).

2.1.5 Children and Their Families (Fig. 1, Factors F, G, and P)

Interventions are more effective if they involve parents, especially by providing activities to do with their children (Bryk et al., 2010; Galindo & Sonnenschein, 2015; Halpern, 2004; Ramey & Ramey, 1998). As with pre-K teachers, parents have a limited view of the breadth of math appropriate for young children (Sarama, 2002). Low-income parents, compared to middle-income parents, believe that math education is the responsibility of the preschool and that children cannot learn aspects of math that research indicates they can learn (Starkey et al., 1999).

2.1.6 Resources, Curriculum, and Technology

The print curriculum, research-based learning trajectories that link developmental progressions to connected instruction, manipulatives, supporting manuals, the *Building Blocks* software activities and management system, and the professional development website, all components of the mathematics intervention, provided a central focus around which the reform revolved. Scale-up is more likely to be successful if resources are targeted to implementing a single curriculum well (Snipes et al., 2002).

2.2 The TRIAD Model's 10 Guidelines

Building on our theoretical framework and review of research, we created a scale-up model for scale-up called TRIAD (technology-enhanced, research-based, instruction, assessment, and professional development, Sarama et al., 2008). The TRIAD model scales up the support of "interactions among teachers and children around educational material" (Ball & Cohen, 1999, p. 3). This strategy creates extensive opportunities for teachers to focus on mathematics, goals, and students' thinking and learning, which improves teachers' knowledge of subject matter, teaching, and learning, and increases student achievement (Ball & Cohen, 1999; Cohen, 1996, p. 98; Schoen et al., 2003; Sowder, 2007). This section summarizes 10 guidelines we followed in creating the TRIAD intervention (Sarama et al., 2008) and the following section describes how each was expected to minimize pollution and dilution.

1. *Involve, and promote communication among, key groups around a shared vision of the innovation* (Hall & Hord, 2001). Emphasize connections between the project's goals, national and state standards, and greater societal need. Promote clarity of these goals and of all participants' responsibilities. School and project staff must share goals and a vision of the intervention (Bryk et al., 2010; Cobb et al., 2003).
2. *Promote equity* through equitable recruitment and selection of participants, allocation of resources, and use of curriculum and instructional strategies that have demonstrated success with underrepresented populations (Kaser et al., 1999).
3. *Plan for the long term.* Recognizing that scale-up is not just an increase in number, but also of complexity, provide continuous, adaptive support over an extended time. Communicate clearly that change is not an event, but a process (Hall & Hord, 2001).
4. *Focus on instructional change that promotes depth of children's thinking, placing learning trajectories at the core* of the teacher/child/curriculum triad to ensure that curriculum, materials, instructional strategies, and assessments are aligned with national and state standards and a vision of high-quality math education, each other, (c) "best practice" as determined by research, especially formative assessment (Ball & Cohen, 1999; Bodilly, 1998; Bryk et al., 2010; Fullan, 2000; Kaser et al., 1999; National Mathematics Advisory Panel, 2008; Raudenbush, 2008; Sowder, 2007).
5. *Build expectations and camaraderie to support a consensus around adaptation.* Promote "buy-in" in multiple ways, such as dealing with all participants as equal partners and distributing resources to support the project. Establish and maintain cohort groups. Facilitate teachers visiting successful implementation sites. Build local leadership by involving principals and encouraging teachers to become teacher leaders (Berends et al., 2001; Borman, Hewes, Overman, & Brown, 2003; Elmore, 1996b; Fullan, 2000; Glennan, Bodilly, Galegher, & Kerr, 2004; Hall & Hord, 2001.)
6. *Provide professional development that is ongoing, intentional, reflective, focused on mathematics content knowledge and children's thinking, grounded in particular curriculum materials, situated in the classroom and the school.* Encourage

sharing, risk taking, and learning from and with peers. Aim at preparing to teach a specific curriculum and develop teachers' knowledge and beliefs that the curriculum is appropriate and its goals are valued and attainable. Situate work in the classroom, formatively evaluating teachers' fidelity of implementation and providing feedback and support from coaches in real time (Bodilly, 1998; Borman et al., 2003; Bryk et al., 2010; Cohen, 1996; Elmore, 1996b; Garet, Porter, Desimone, Birman, & Yoon, 2001; Guskey, 2000; Hall & Hord, 2001; Kaser et al., 1999; Klingner et al., 2003; Pellegrino, 2007; Schoen et al., 2003; Showers et al., 1987; Sowder, 2007). This also provides a common language for teachers in working with each other and other groups (Bryk et al., 2010).

7. *Ensure school leaders are a central force supporting the innovation* and provide *teachers continuous feedback that children are learning what they are taught and that these learnings* are valued. Leaders, especially principals, must show that the innovation is a high priority, through statements, resources, and continued commitment to permanency of the effort. An innovation champion leads the effort within each organization (Bodilly, 1998; Bryk et al., 2010; Glennan et al., 2004; Hall & Hord, 2001; Rogers, 2003, p. 434; Sarama et al., 1998).
8. *Give latitude for adaptation to teachers and schools, but maintain integrity.* Emphasize the similarities of the curriculum with sound practice and what teachers already are doing. Discourage uncoordinated innovations (Fullan, 2000; Huberman, 1992; Sarama et al., 1998; Snipes et al., 2002).
9. *Provide incentives for all participants, including intrinsic and extrinsic motivators linked to project work*, such as external expectations—from standards to validation from administrators. Show how the innovation is advantageous to and compatible with teachers' experiences and needs (Berends et al., 2001; Borman et al., 2003; Cohen, 1996; Darling-Hammond, 1996; Elmore, 1996a; Mohrman & Lawler III, 1996; Rogers, 2003).
10. *Maintain frequent, repeated communication, assessment (checking up), and follow-through efforts* emphasizing the purpose, expectations, and visions of the project, and involve key groups in continual improvement through cycles of data collection and problem solving (Fullan, 1992; Hall & Hord, 2001; Huberman, 1992; Kaser et al., 1999; Snipes et al., 2002). Throughout, connections with parents and community groups are especially important, to meet immediate and long-range (sustainability) goals.

2.3 How the TRIAD Implementation Was Designed to Militate Against Pollution and Dilution

This section addresses the TRIAD model regarding pollution and dilution. That is, for each of these 10 guidelines constituting the model, we describe the sources of possible pollution or dilution and then describe how the guideline was expected to minimize the threat to high-quality implementation.

1. Involve, and promote communication among, key groups around a shared vision of the innovation. This guideline jumpstarts and later institutionalizes the intervention; for example, by establishing a common understanding of the goals and how they are consistent with state or district standards and by planning for the ongoing socialization and training of new teachers (Elmore, 1996b; Fullan, 2000; Huberman, 1992; Kaser et al., 1999; Klingner et al., 2003; Sarama et al., 1998). To begin, the TRIAD model includes initial meetings with administration as well as teachers, including a presentation with a brochure and a question and answer session. Planning for this meeting includes preliminary discussions with key staff to ensure that the relationships between the district's goals and other instructional guidelines and those of the intervention can be explicated and defended. After the meeting, presenters emphasize that to move forward, all levels of administration must agree to participate. (Ideally, at least 80% of teachers, as well as parents, would agree to participate. However, for the purposes of generalizability, we did not ask teachers to sign consent forms in the largest study described here. This way we could generalize our findings to districts that used that model but had reluctant or hostile teachers.)

   Once agreements are obtained and the project started, ongoing planning with all key groups that minimizes dilution (see D1's in Fig. 1) includes specifics around the intervention, including clear, written specifications concerning the time and supports needed to implement (e.g., all teachers and paraprofessionals are scheduled to attend all trainings; sufficient resources are available in every classroom, from blocks to computers; admin). Furthermore, dilution may occur due to rapid turnover of preschool teachers, introducing an increasing proportion of teachers who are not involved in the initial implementation. Also, pollution (see P1's in Fig. 1) can occur if new (and also old) teachers introduce inconsistent elements into their practice. Therefore, plans are made for training and socializing new teachers, including a mentor system and including district supervisors and coaches in the project's training sessions so they would be well versed in introducing the intervention to any new teachers and to monitoring and supporting all teachers past the end of the project.

2. Promote equity. This ensures that every teacher has the support, both physical and personnel, necessary to teach and continue to teach the intervention properly. Without this, both pollution and dilution are likely (see D2's and P2 in Fig. 1), especially given the challenges present-day teachers face in implementing mathematics interventions. That is, without comprehensive curricular resources, training, and coaching, fidelity is unlikely to be achieved and, maintained; unfortunately, inequitable distribution of resources frequently plagues preschools, especially those serving children from lower SES communities. We worked with administrators to ensure all resources were provided to each teacher and coaches filled in for any missing materials throughout the implementation period.
3. Plan for the long term. The TRIAD professional development model includes regular professional development meetings, bi-weekly coaching visits, and peer coaching. Year 1 professional development is a "gentle introduction," in that child data are not collected and coaches have time to slowly get teachers comfortable with all aspects of the intervention. "One-shot" interventions may have a temporary effect, but frequently are diluted or even abandoned soon thereafter (Bodilly, 1998; Cuban & Usdan, 2003). TRIAD uses a dynamic, multilevel, feedback, and self-corrective strategy (Bryk et al., 2010; Coburn, 2003; Guskey, 2000). That is, all work with teachers is designed so that data about teachers' understandings and engagement is collected in real time and used to alter plans for the following sessions. Interviews with principals occur throughout the project, to ensure they not only remember the project (Foorman et al., 2007) but provide feedback on what is progressing successfully and what is not in their building (see guideline #1). In addition, administrators are provided with tools to offer additional support to teachers in their building and fulfill district requirements (to monitor instructional time and quality of instruction).
4. Focus on instructional change that promotes depth of children's thinking, placing learning trajectories at the core of the teacher/child/curriculum. This is the core of the TRIAD approach so ensuring this guideline is critical. Implemented well, this focus helps avoid pollution, mainly, by ensuring fidelity by helping teachers not only use the

effective instructional strategy of formative assessment (Penuel & Shepard, 2016) but also by helping teacher understand not only how to teach a given activity, but why—in other words, connecting their growing knowledge of how children think about and learning mathematics to the characteristics of a given activity that promote this learning. With an integrated understanding of the mathematics, children's developmental progressions, and instruction, teachers are less likely to introduce lethal mutations, changes to the core math content, or pedagogy that weakens the activity. This focus also prevents dilution, however, because teachers who observe and understand the depth of children's thinking and learning given these activities are more likely to continue to use the program with fidelity over years (Clements, Sarama, Wolfe, & Spitler, 2015).

5. Build expectations and camaraderie to support a consensus around adaptation. Dilution is less likely when teachers are part of a learning–teaching community. In TRIAD, teachers are intermingled (across schools) in small group work, but they also work in school clusters for other small groups. In this way, teachers are exposed to ideas from peers at other sites but had adequate time to address school specific concerns and address implementation as a team.
6. Provide professional development that is ongoing, intentional, reflective, focused on mathematics content knowledge and children's thinking, grounded in particular curriculum materials, situated in the classroom and the school. As guideline #4 is a main focus, this guideline is TRIAD's main leverage point—sine qua non of a high-quality implementation. That is, without well-trained and committed teachers, dilution (to the degree of no implementation) and/or pollution (low fidelity) is highly likely. TRAID's professional-development model includes both repeated (e.g., >10) full-day sessions of training in regular meetings and frequent coaching. Training includes all three components of each learning trajectory, the goal, the developmental progression, and the instructional activities and strategies (Fig. 2). All three are critical. To understand the goal, teachers study the mathematical content. Without knowledge of the mathematical content, pollution is almost assured. For example, teachers may believe that any series of colors in fabric is "a pattern" (Fox, 2005) or that "two triangles put together always make a square" (Thomas, 1982). In TRIAD, teachers learn core mathematics concepts and procedures for each topic. For example, they study the system of verbal counting based on cycling through 10 digits

and the concept of place value (i.e., content like that presented in National Research Council, 2009; Wu, 2011).

A key instructional use of learning trajectories is in formative assessment, an empirically supported pedagogical strategy (National Mathematics Advisory Panel, 2008; Penuel & Shepard, 2016). However, dilution is probable if teachers cannot accurately assess children's developmental level or select appropriate tasks and instructional strategies based on children's levels—the second and third components of a learning trajectory. To learn about the developmental progression, teachers analyze multiple video segments illustrating each level and discuss the mental "actions on objects" that constitute the defining cognitive components of each level; order tasks corresponding to those levels; and practice diagnosis in teams, with a couple of teachers exemplifying behaviors of children at different levels, and one teacher identifying the level of each. Furthermore, teachers need training in understanding, administering, and especially using data from new assessment strategies (Foorman et al., 2007). TRIAD training focuses mainly on the curriculum-embedded assessment of Building Blocks' "Small Group Record Sheets," but also provides teachers with tasks appropriate for clinical interviews.

Formative assessment requires more than identifying children's levels of thinking. Teachers must select and modify instructional activities and strategies that are appropriate and effective for each level. To learn about instructional tasks and strategies, teachers practice the curriculum's activities, but also analyze them to establish and justify their connection to a particular level of the developmental progression. They extend the team practice sessions to use their diagnoses to select and to modify activities to match instructional tasks to developmental levels of individuals.

Across all forms of professional development, the focus is on children's thinking and learning. Conversations about children learning serve as way to address implementation issues. Although early mathematics is often an uncomfortable topic for early childhood educators, the newness of the learning trajectories for all participants helps establish a sense of shared learning and community. Each session in the last third of professional development includes scheduled time to discuss "learning stories" (Perry, Perry, Dockett, & Harley, 2007). Teachers show their record keeping on small group record sheets, and sometimes videos, and discuss their use of learning trajectories in teaching children, including challenges, questions, and successes. These discussions promote peer learning and collaboration and also motivate peers to solve implementation difficulties.

Finally, we do not train the teachers and immediately begin evaluating the project. Instead, an entire year is spent in professional development and implementation without evaluations as a "gentle introduction" without the pressure of assessments.

7. Ensure school leaders are a central force supporting the innovation and provide teachers continuous feedback that children are learning what they are taught and that these learnings are valued. School leaders deal with myriad issues and problems and a single intervention can easily lose their attention (Foorman et al., 2007), vitiating the support teachers need from their leaders. In TRIAD, communication is initially established (see guideline #1) and is not maintained to ensure that school leaders provide support, encouragement, and feedback to teachers. Principals receive, and are offered training on, a "walk-through" fidelity instrument, which when used, heightened teachers' perception of the importance the leader assigned to the intervention. Leaders also include supervisors of early childhood education and mathematics education. Coaches from both these fields are invited to all trainings. All leaders are taught about the structure and research of the intervention, to avoid dilution and pollution that can occur when leaders tell teachers they are "professionals who can choose to include any activity from any source."
8. Give latitude for adaptation to teachers and schools, but maintain integrity. Adaptation to local contexts can be important, as can adaptation to children's interests and backgrounds. However, it is easy for teachers to mistakenly dilute or pollute instructional activities in these attempts. TRIAD training addresses this issue by explicitly comparing productive adaptations to lethal mutations (Brown & Campione, 1996). For example, teachers learn to change surface features (e.g., to fit a class "theme") without altering the core mathematical or pedagogical components of the instructional activities (Winter & Szulanski, 2001). They learn to modify an activity to fit children's interests and needs without (permanently) reducing the cognitive demand of instructional tasks after their initial introduction (Stein, Brover, & Hennigsen, 1996). The best productive adaptations are then shared among the sites and added to future training sessions.

From a school perspective, our fidelity form, including the "walk-through" version for principals, includes pacing guidelines. However, these forms are not based on compliance alone. That is, the emphasis is on integrity and pacing, but with latitude to make changes. For example, higher, not lower, scores are given to teachers who modify the schedule following children's development and learning. Similarly, higher scores are given for modifying activities, but only for productive adaptations.

9. Provide incentives for all participants, including intrinsic and extrinsic motivators linked to project work. TRIAD implementations connect the intervention to the standards and school reports teachers use. Early successes, even anecdotal, with teachers and children are shared with principals so they remain committed and pass on the complements to their teachers. Discussing the small group record sheets and sharing the learning stories not only motivate those implementing, but also subtly pressure nonimplementers to "up their game." Training sessions are designed to be happy and productive, providing opportunities to work with peers, to discuss education over good food, and generally to be treated as professionals. For example, they are asked to share their ideas for productive adaptations and trainers are serious about using feedback from them to trainings.
10. Maintain frequent, repeated communication, assessment (checking up), and follow-through efforts. Of course, assuming that simple introduction and training will lead to complete and continued implementation is invalid. Throughout the TRIAD implementation, project staff maintained connections with not only teachers and leaders, as described previously, but also with parents and local media.

3. RESEARCH EVALUATIONS: DID THE TRIAD DESIGN MITIGATE DILUTION AND POLLUTION?

Our colleagues and we have conducted several studies to evaluate instantiations of the TRIAD model and its components (Clements & Sarama, 2008a, 2008b; Clements, Sarama, Spitler, Lange, & Wolfe, 2011; Clements, Sarama, Wolfe, & Spitler, 2013; Sarama & Clements, 2009b; Sarama et al., 2008). Most of these studies used cluster randomized trials designs. Although financial and logistical constraints did not allow data collection that would directly address every source of dilution and pollution identified in Fig. 1 and in the previous section, successful outcomes do provide evidence that these challenges were addressed successfully.

3.1 Initial Instantiation and Evaluations of the TRIAD Model

The first evaluations were limited, designed to complete procedures and assessments and serve as a proof of concept that the model could be effective. In the first, we randomly assigned 25 preschool classrooms serving children at-risk for later school failure to either TRIAD or control groups (Sarama et al., 2008), including about half Head Start and half public preschool programs. This was successful so we implemented it more completely with 36 teachers, which we describe here (Clements & Sarama, 2008a).

3.1.1 Implementation

We attempted to implement most components of the TRIAD model (some, such as planning for the long term, guideline #3, and full realization of #10, were not possible in this limited scale-up). For example, we began by meeting with administrators and teachers to encourage them to have a shared understanding of the goals of the intervention (guideline #1). We provided incentives in the form of rationales for the intervention and the provision of curriculum materials and especially professional development (guideline #9).

To promote equity (guideline #2), the TRIAD intervention used the Building Blocks curriculum, which had been empirically validated in classrooms serving underrepresented populations (Clements & Sarama, 2007, 2008a). Every classroom received all components and, thanks to the support of administration, (guidelines #1 and #7) adequate auxiliary materials such as computers. All teachers were provided with the same professional development (see guideline #6).

We describe this professional development in detail because, like the measures, it shared most characteristics with similar work in all future studies of TRIAD (guidelines #4 and #6). The first year was a pilot/training year, because our previous experience and others' research suggested that teachers often need at least a year of experience before completely and effectively implementing a curriculum (Berends et al., 2001; Clements & Sarama, 2014; Cobb et al., 2003). Thus, teachers participated in 8 days of professional development during the school day in the first year (a "no stress, gentle introduction" to the curriculum with no data collection by researchers). Introductory discussions emphasized the "developmental appropriateness" of the intervention's mathematics education and its importance to the teachers and children (guideline #5), especially in promoting equity (guideline #2). This work focused on the learning trajectories for each mathematical topic, usually as woven into the *Building Blocks* curriculum (guideline #4). Training addressed each of the three components of the learning trajectories. To understand the *goals*, teachers learned core mathematics concepts and procedures for each topic. For example, they studied the system of verbal counting based on cycling through 10 digits and the concept of place value (based on content like that in National Research Council, 2009).

To understand the developmental progressions of levels of thinking, teachers studied multiple video segments illustrating each level and discussed the mental "actions on objects" that constitute the defining cognitive components of each level. To understand the instructional tasks, teachers studied the tasks, and they viewed, analyzed, and discussed video of the enactments of these tasks in classrooms. A central tool to study and to connect all three components was the internet-based software application, *Building Blocks Learning Trajectories* (BBLT, see Fig. 3). Throughout the 2 years, teachers study the BBLT and engaged in discussions of these videos and correlated text from the *Building Blocks* curriculum. *BBLT* provided scalable access to the learning trajectories via descriptions, videos, and commentaries. Two sequential aspects of the learning trajectories—the developmental progressions of children's thinking and connected instruction—are linked to the others (see Fig. 2). As an example, using these approaches, teachers studied the learning trajectory for counting. Initial presentations including viewing the levels of thinking that constitute the developmental progression, using the BBLT, and discussing the attributes, both mathematical and psychological, of each level. As an illustration, these included the cardinal principle (the last counting word "tells how many"), which is often underemphasized in curricula and teaching; and, at higher levels, competencies in counting on, and other counting strategies. Teachers used the "Test Yourself" feature of the BBLT to evaluate their abilities to diagnose children's level of counting (by identifying the level evinced by children in randomly selected videos). They also used the BBLT's links to view research-based instructional strategies to promote children's progress to the next level. Teachers worked in small groups to plan how activities from their curriculum might promote, or be modified to promote, learning for the relevant levels. The coaches joined the teachers in the participated in professional development, as well as several days of training on coaching, most of which focused on the unique aspects

BBLT provides scalable access to learning trajectories—*developmental progressions* of children's thinking and *instruction*—linked to each other. For example, teachers might choose the instruction (curriculum) view and see the screen on the left. Clicking a specific activity provides a description. Clicking more info slides the screen over to reveal descriptions, several videos of the activity "in action," notes on the video, and the level of thinking in the learning trajectory that activity is designed to develop, as shown below on the right.

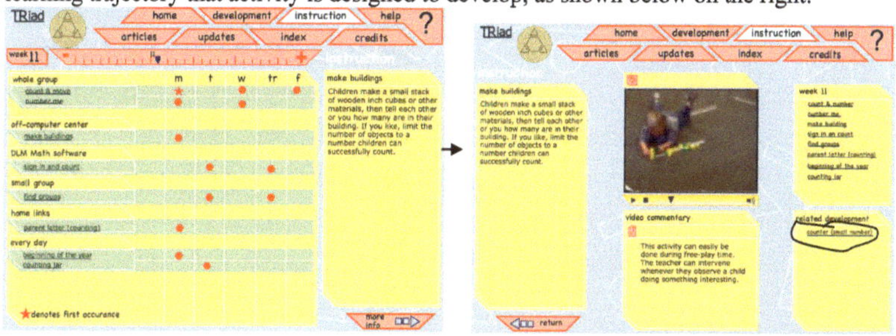

Clicking on the related developmental level (child's level of thinking), ringed above, switches to the development view *of that topic and that level of thinking*. Alternatively, the user may have been studying developmental sequences. After choosing development, teachers see a list of the mathematical topics and the developmental sequences. If they had chosen "Counting," then the

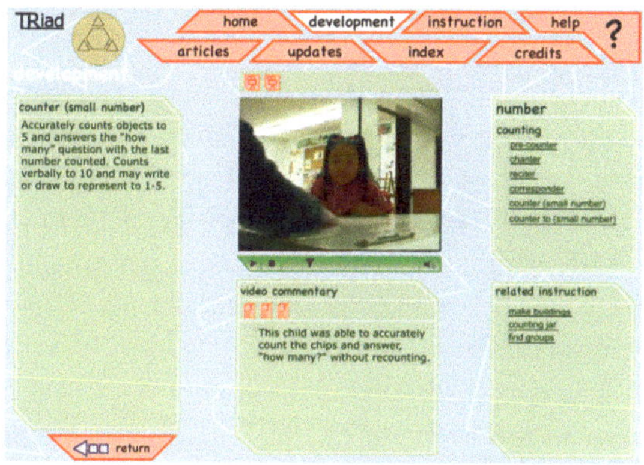

"Counter (Small Numbers)" level, and then "More Info," they would see the same screen as above. The level of thinking is illustrated by both video of clinical interview assessments *and* video of classroom activities in which children show that level thinking (the icons above the video allow the selection of alternative video), an approach that has received empirical support (Klingner et al., 2003).

Fig. 3 Building Blocks Learning Trajectory (BBLT) web application.

of coaching early mathematics education. Coaches worked with teachers during the year to avoid dilution of the intervention and to provide teachers with continual feedback and support.

Trainers and coaches also promoted productive adaptations and avoidance of lethal mutations (guideline #8). As an example, for board games, productive adaptations could include changing the die (from 1–6 dots to 1–3 dots for children who are at a lower level, or 5–10 dots or even two cubes for children at higher levels). Such adaptations could simply be changing the theme of the board and tokens to match a class theme (e.g., a trip through a farm and farm animals). In contrast, lethal mutations include changing the game to Candy Land or other games that use a spinner with colors where children just advance to the next instance of that color, eliminating number from the activity. In subsequent sessions, teachers shared and kept lists of their own productive adaptations and discussed how they avoided lethal mutations.

In year 2, teachers and coaches participated in an additional 5 days of professional development (guidelines #6 and #10). They continued to study the learning trajectories, including discussions of how they conducted various curricular activities the previous year. As part of this work, teachers brought case studies of particular situations that occurred in their classrooms to the group to facilitate these discussions; thus, this work included elements of lesson study.

3.1.2 Findings

TRIAD children made significantly and substantially greater gains in mathematics achievement than the control children (effect size = 1.07). There was no evidence that contextual variables influenced these positive effects, such as the Head Start or public programs. Similarly, there was no evidence that effects differed by children's ethnicity or gender. This provides evidence that TRIAD's guidelines worked as a whole, and those addressing equity were successful in particular.

The two observational measures provide additional evidence that greater scores in achievement by the TRIAD group resulted from the implementation of the guidelines. Findings on the fidelity instrument indicate that the teachers implemented the curriculum with "good" fidelity (that is, on a 5-point Likert scales from "strongly disagree" to from "strongly agree" for each item evaluating the quality of teaching, the average was "agree"). This provides support for guidelines especially #4, #5, #6, and #8 as teachers as a group implemented the intervention with high-quality fidelity. In addition, the higher the teacher's fidelity score, the higher the average child gain in achievement in her classroom (although small variance resulted in lack of statistical significance). Furthermore, the TRIAD intervention resulted in consistently greater scores on the quality and quantity of the mathematics environments and teaching in the TRIAD, compared to control, classes. TRIAD classrooms had higher average scores on general classroom behaviors such as teaching more mathematics, showing knowledge of, enjoyment in, and enthusiasm for mathematics, and using effective management and instructional strategies. In summary, the intervention increased the quantity and quality of the mathematics environment and teaching in preschool classrooms, providing empirical support for the guidelines, especially #4, #5, #6, and #9.

3.2 Full-Scale Implementation and Evaluation of TRIAD

Our largest implementation and evaluation to date involved 1375 preschoolers in 106 classrooms serving low-resource communities in two states. Schools in two large urban cities in two states were randomly assigned to three treatment groups: TRIAD, TRIAD with Follow Through (TRIAD-FT), and control (business as usual—in both cities, this involved standards, curriculum, and expectations in preschool mathematics). The TRIAD-FT differed from the TRIAD implementation only after the preschool year, so they are combined for analyses reported here.

3.2.1 Implementation

This implementation was also the most valid and therefore generalizable because the administration of both school systems agreed to the implementation across their districts. Therefore, we could both randomly *select* and randomly assign schools (with small exceptions—schools that had worked with us in previous studies were not included in the selection pool). Further, teachers were not volunteers—a common limitation in many such studies, especially as math is often not the favorite subject of preschool teachers (Copley, 2004; Sarama & DiBiase, 2004).

We attempted a complete implementation of the TRIAD-based intervention in two school districts. We again began with extensive explanations of the intervention, including the rationale for it and the particulars of it, to broad audiences of administrators, teachers, and parents (guideline #1). In this effort, we include planning for the long term, (guideline #3); for example, by planning for training both teachers who were randomly assigned to the control group and those who are new to the district. We also more fully realized guideline #10. For example, we scheduled meetings with district and school leaders more frequently and, when possible, school boards to update them on our results and progress. As another example, parents received letters explaining what their children were doing that week and how they could support their children's learning at home. Beyond this, they were asked to give the project staff advice on the usefulness of these letters and on the way the intervention was or was not helping their children. Such efforts are important to avoid dilution of implementation over years. Leaders did respond well, buying supplemental materials at one site, and actively working on integrating the new curriculum with other curricula at the other. They released and encouraged early childhood coaches to attend the trainings. Several early childhood center leaders asked for additional support after the trainings proper.

The professional development followed the same model as previously described; here we note only the differences. We added a distinction between "mentors" (coaches who were trained by and worked closely with project staff) and in-school teacher coaches within each building. Assigning schools instead of classrooms allowed us to build consensus, support, and camaraderie within each building. Both changes were hypothesized to better implement guideline #5. As much as possible, we included each districts' early childhood and mathematics specialists in these meetings and our professional development to plan for long-term implementation. We encouraged them to rely on teachers in our sessions as teacher leaders to support wider implementation after the conclusion of the data collection (guidelines #3, #7, and #10).

3.2.2 Findings

Teachers implemented the intervention with adequate fidelity (Clements et al., 2011). Again, the modal category for mentor-rated Likert items was "agree." That is, most teachers implemented all aspects of the *Building Blocks* curriculum. Along with previous studies, this evidence indicates that interventions in preschool mathematics education, such as this one, can be successfully implemented on a large scale. Similarly, the TRIAD intervention enabled teachers to develop richer classroom environments for mathematics than those of the control classrooms as measured by the COEMET (the Classroom Observation of Early Mathematics Environment and Teaching measures the quality of the mathematics environment and activities with a half-day observation and can be used with different curricula, see Clements et al., 2011). The TRIAD classes scored significantly higher than the control classes on four components of this measure: the classroom culture subscore, the specific math activities (SMAs) subscore, the number of SMAs, and the number of computers on and working for students to use. The classroom culture subscore assesses teachers' general approach to mathematics education, indicated by "environment and interaction" variables such as responsiveness to children, use of "teachable moments" as well as "personal attributes of the teacher" variables, including appearing knowledgeable and confident about mathematics as well as showing enjoyment in, curiosity about, and enthusiasm for, teaching mathematics. These suggest that the TRIAD intervention successfully altered teachers' beliefs and dispositions, beyond specific curricular practices (guideline #6). Such practices were also positively affected, given that teachers used the computer component of the TRIAD curriculum and engaged their children in a greater number of explicit, targeted, and mathematics activities. The higher scores for the SMA subscale suggest that the SMAs that teachers in TRIAD classrooms conducted were of a higher quality than those in the control classrooms. We will return to the mediational role of these practices after examining the effects on mathematics achievement. Again, these data support the efficacy of implementing guidelines #4, #5, #6, and #8 to avoid dilution and pollution and instead implement with high fidelity, achieving high-quality mathematics teaching and environments.

Thus, children in the Building Blocks group learned more mathematics than the children in the control group (effect size, $g = 0.72$). This is significant, because, unlike most previous research, the counterfactual condition was not the "practice-as-usual" control condition involving no published mathematics curriculum and little district-wide emphasis on mathematics. Both districts had placed new emphasis on pre-K mathematics and adopted new literacy curricula that included specific mathematics components. In addition, the TRIAD intervention itself, especially the first year's "gentle introduction," generated considerable spillover of early mathematics pedagogical practices into control classrooms.

The use of curriculum demonstrated as effective with underrepresented populations, with adequate support for teachers, is supported (guideline #2). Consistent with other research, analyses support the use of curricula that have been empirically supported. Analyses also indicate that the computer software component of the *Building Blocks* curriculum makes a unique contribution to children's learning.

Evidence suggested that the *Building Blocks* curriculum was equally effective in classes serving low- or mixed-income families or with schools that have a higher or lower percentage of children with limited English proficiency. Also, the intervention was equally effective for girls and boys, and for children with or without IEPs. There was evidence that the intervention was differentially effective for only a single comparison: African-American children learned less than other children in the same control classrooms and African-American children learned more than other children in the same Building Blocks classrooms. It may be that this TRIAD intervention is particularly effective in ameliorating the negative effects of low expectations for learning of mathematics (National Mathematics Advisory Panel, 2008).

Thus, there is some support for the implementation avoiding dilution and pollutions from inequitable support and thereby promoting equity (guideline #2, although future research may wish to alter the intervention to close or narrow other gaps as it did for African-American children here).

3.3 Fighting Dilution Over Time: TRIAD and Sustainability

Avoiding dilution during the intervention is necessary, but it may be more important to investigate the dilution that most interventions suffer once the "push" for the intervention has passed—especially for those like TRIAD, which is introduced from outside the school district. A full concept of successful scale-up requires not only consequential implementation and diffusion but also endurance over long periods of time and a transfer of responsibility from any external organization to the internal resources of a school district (Coburn, 2003; Dearing & Meyer, 2006). Does implementation become diluted or polluted over time? Or is it *sustainable*—continued professional practice over time, with a focus on the maintenance of core beliefs and values, and the use of these core beliefs to guide adaptations (Century, Rudnick, & Freeman, 2012; Scheirer, 2005; Scheirer & Dearing, 2011; Timperley, 2011). (Persistence of effects, continuation of the effects of an intervention on individual children's trajectories of learning, is a distinct issue, discussed in other publications, Clements et al., 2017; Clements et al., 2013; Sarama, Clements, Wolfe, & Spitler, 2012; Sarama, Lange, Clements, & Wolfe, 2012.)

We evaluated the fidelity of implementation and the sustainability of TRIAD years after any support from the funded project ended in two studies. The first measured teachers' implementation 2 years later (Clements et al., 2015). Although a logical expectation would be that, after the cessation of external support and professional development provided by the intervention, teachers would show a pattern of decreasing levels of fidelity, these teachers demonstrated increasing levels of fidelity, continuing to demonstrate high levels of sustained fidelity in their implementation of the curriculum and underlying pedagogy. Different profiles of variables predicted separate aspects of sustainability, but by far the most consistent factor was teachers' recognition of children's learning. Teachers who noticed children's substantial development increased their efforts to implement all the components of the intervention with greater fidelity (Clements et al., 2015).

In our latest study, teachers again demonstrated sustained or increasing levels of fidelity as long as 6 years after the end of the intervention without support from the project (Sarama, Clements, Wolfe, & Spitler, 2016). Notable is these teachers' encouragement and support for discussions of mathematics and their use of formative assessment.

We believe there is considerable evidence that dilution especially was mitigated by implementing guidelines #1, #7, and #10: Communication among key groups around a shared vision of the innovation, ensuring school leaders are a central force, and maintaining communication, assessment, and follow-through efforts. Most project activities would not be achieved consistently throughout the districts without such communication and continued collaboration. We found it necessary to repeatedly provide higher-level administrators with updates and reminders of the projects' goals and activities. Similarly, every change in administration had to be monitored and new people introduced to the project and its successes quickly. The early introduction of the project was facilitated by prior awareness of the researchers' work and a strong commitment to raising mathematics achievement.

At the classroom level, both dilution and pollution were minimized because of a strong and consistent implementation of guidelines #6 and #7, addressing professional development and support from leaders, especially coaches. Effects of TRIAD depend on the development of teachers' skills and knowledge. The classroom culture so heavily dependent on the teacher is a consistent mediator of those positive effects. Note that we provided approximately 75 h of out-of-class teacher training as well as hours of mentoring in the classroom, which is substantially more than offered to most teachers, only 6% of whom participate in mathematics professional development for more than 24 h over a year (Birman et al., 2007). A total of 50–70 h of professional development is consistent with previous research documenting what is necessary to achieve measurable effectiveness (Yoon, Duncan, Lee, Scarloss, & Shapley, 2007). Similarly, findings indicate that learning trajectories played a critical role in both teachers' and children's learning, supporting guideline #4: Focus on instructional change that promotes depth of children's thinking, placing learning trajectories at the core.

Finally, teachers repeatedly commented that the project staff respected and collaborated with them as professionals, important for generating a self-sustaining learning community (avoiding dilution via camaraderie support adaptation, guideline #5). After some initial trepidation, teachers came to view peer coaching within schools as a positive characteristic of the implementation. In-class support from mentors and technology support staff (guideline #6) were also appreciated and viewed as important by the teachers after they were sure they were there to help and not "police" implementation. Mentors needed to have the knowledge and skill to demonstrate the implementation of all components of the curriculum, doing activities right in teachers' classroom on many occasions and inviting discussion centered on students' learning.

4. FINAL WORDS

The goal of the TRIAD project was to increase knowledge of scaling up by conducting research that investigates the effectiveness of a research-based mathematics education intervention implemented in varied pre-K settings with diverse student populations. Perhaps most unique is TRIAD's consistent emphasis on teaching for understanding following developmental guidelines, or learning trajectories, as well as its use of technology at multiple levels. TRIAD had a moderate to strong positive effect on children's mathematics achievement and did not harm literacy skills while *increasing* most language competencies measured. There was no evidence the approach is differentially effective for participants in different states or types of programs, or for children of different SES, ethnic group, or gender. The moderate the large effect sizes we obtained is notable considering that other comprehensive reform programs, including multitiered teacher support, sustained professional development, and in-class coaching, achieve effect sizes such as 0.24 only with great effort (Balfanz, Mac Iver, & Byrnes, 2006).

Most people who discuss TRIAD focus on these findings regarding children's learning, and of course without this evidence, little else matters. However, to evaluate the ultimate effects of the project, these longitudinal results are arguably more important—if the intervention is diluted and polluted over time, is it worth the cost and effort? On the other hand, if it *grows, the number of children positively affected is potentially an order of magnitude greater.*

Teachers seemed to have internalized the program (Timperley, Wilson, Barrar, & Fung, 2007). Through the professional development and support, and then, becoming empowered by their own knowledge of the trajectories and the practices to support learners through the trajectories, they became progressively more faithful to the intended program, instead of drifting from it as time elapsed and support disappeared, dilution found in other studies (Datnow, 2005; Hargreaves, 2002).

The TRIAD model is not simply about a new curriculum or training teachers to use it. Success required complex changes, including a change in instructional structures, pedagogical strategies, and classroom communication and culture (Grubb, 2008). Given the importance of early competence in mathematics (e.g., Duncan et al., 2007; Paris, Morrison, & Miller, 2006), the TRIAD implementation described here has implications for practice and policy, as well as research. TRIAD's guidelines should be considered when planning to increase the quality and quantity of not only early mathematics but also to mitigate dilution and pollution threatening any intervention.

ACKNOWLEDGMENTS

This research was supported by the Institute of Education Sciences, US Department of Education through Grants R305A120813, R305K05157, and R305A110188. The opinions expressed are those of the authors and do not represent views of the US Department of Education. Although the research is concerned with theoretical issues, not particular curricula, a small component of the intervention used in this research have been published by the authors and their collaborators on the project, who thus could have a vested interest in the results. Researchers from an independent institution oversaw the research design, data collection, and analysis and confirmed findings and procedures. The authors wish to express appreciation to the school districts, teachers, and students who participated in this research.

REFERENCES

Baker, E. L. (2007). Principles for scaling up: Choosing, measuring effects, and promoting the widespread use of educational innovation. In B. Schneider & S.-K. McDonald (Eds.), *Scale up in education: Ideas in principle: Vol. 1.* (pp. 37–54). Lanham, MD: Rowman & Littlefield.

Balfanz, R., Mac Iver, D. J., & Byrnes, V. (2006). The implementation and impact of evidence-based mathematics reforms in high-poverty middle schools: A multi-site, multi-year study. *Journal for Research in Mathematics Education, 37*, 33–64.

Ball, D. L., & Cohen, D. K. (1999). *Instruction, capacity, and improvement.* Philadelphia, PA: Consortium for Policy Research in Education, University of Pennsylvania.

Baroody, A. J., & Coslick, R. T. (1998). *Fostering children's mathematical power: An investigative approach to K-8 mathematics instruction.* Mahwah, NJ: Erlbaum.

Berends, M., Kirby, S. N., Naftel, S., & McKelvey, C. (2001). *Implementation and performance in new American schools: Three years into scale-up.* Santa Monica, CA: Rand Education.

Birman, B. F., LeFloch, K. C., Klekotka, A., Ludwig, M., Taylor, J., Walters, K., … Yoon, K. S. (2007). *State and local implementation of the No Child Left Behind Act, Volume II—Teacher quality under NCLB: Interim report.* Washington, DC: U.S. Department of Education Office of Planning Evaluation and Policy Development, Policy and Program Studies Service.

Bodilly, S. J. (1998). *Lessons from new American Schools' scale-up phase.* Santa Monica, CA: RAND Education.

Borman, G. D., Hewes, G. M., Overman, L. T., & Brown, S. (2003). Comprehensive school reform and achievement: A meta-analysis. *Review of Educational Research, 73*, 125–230. http://dx.doi.org/10.3102/00346543073002125.

Bredekamp, S. (2004). Standards for preschool and kindergarten mathematics education. In D. H. Clements, J. Sarama, & A.-M. DiBiase (Eds.), *Engaging young children in mathematics: Standards for early childhood mathematics education* (pp. 77–82). Mahwah, NJ: Erlbaum.

Brown, A. L., & Campione, J. C. (1996). Psychological theory and the design of innovative learning environments: On procedures, principles, and systems. In R. Glaser (Ed.), *Innovations in learning: New environments for education* (pp. 289–325). Mahwah, NJ: Erlbaum.

Bryk, A. S., Sebring, P. B., Allensworth, E., Suppescu, S., & Easton, J. Q. (2010). *Organizing schools for improvement: Lessons from Chicago*. Chicago, IL: University of Chicago Press.

Carpenter, T. P., & Franke, M. L. (2004). Cognitively guided instruction: Challenging the core of educational practice. In J. Glennan, K. Thomas, S. J. Bodilly, J. R. Galegher, & K. A. Kerr (Eds.), *Expanding the reach of education reforms: Perspectives from leaders in the scale-up of educational interventions* (pp. 41–80). Santa Monica, CA: RAND Corporation.

Century, J., Rudnick, M., & Freeman, C. (2012). In *Defining and measuring sustainability of reform: Factors that affect our abilities to generate enduring change. Paper presented at the annual meeting of the American Educational Research Association, Vancouver, British, Columbia, Canada.*

Clements, D. H., & Sarama, J. (2004). Learning trajectories in mathematics education. *Mathematical Thinking and Learning*, *6*, 81–89. http://dx.doi.org/10.1207/s15327833mtl0602_1.

Clements, D. H., & Sarama, J. (2007). Effects of a preschool mathematics curriculum: Summative research on the *Building Blocks* project. *Journal for Research in Mathematics Education*, *38*(2), 136–163.

Clements, D. H., & Sarama, J. (2008a). Experimental evaluation of the effects of a research-based preschool mathematics curriculum. *American Educational Research Journal*, *45*, 443–494. http://dx.doi.org/10.3102/0002831207312908.

Clements, D. H., & Sarama, J. (2008b). In *Scaling-up interventions: The case of mathematics. Paper presented at the American Educational Research Association, New York, NY.*

Clements, D. H., & Sarama, J. (2011). Early childhood mathematics intervention. *Science*, *333*(6045), 968–970. http://dx.doi.org/10.1126/science.1204537.

Clements, D. H., & Sarama, J. (2014). *Learning and teaching early math: The learning trajectories approach*. New York, NY: Routledge.

Clements, D. H., Sarama, J., & DiBiase, A.-M. (2003). Preschool and kindergarten mathematics: A national conference. *Teaching Children Mathematics*, *8*, 510–514.

Clements, D. H., Sarama, J., Spitler, M. E., Lange, A. A., & Wolfe, C. B. (2011). Mathematics learned by young children in an intervention based on learning trajectories: A large-scale cluster randomized trial. *Journal for Research in Mathematics Education*, *42*(2), 127–166.

Clements, D. H., Sarama, J., Wolfe, C. B., & Spitler, M. E. (2013). Longitudinal evaluation of a scale-up model for teaching mathematics with trajectories and technologies: Persistence of effects in the third year. *American Educational Research Journal*, *50*(4), 812–850. http://dx.doi.org/10.3102/0002831212469270.

Clements, D. H., Sarama, J., Wolfe, C. B., & Spitler, M. E. (2015). Sustainability of a scale-up intervention in early mathematics: Longitudinal evaluation of implementation fidelity. *Early Education and Development*, *26*(3), 427–449. http://dx.doi.org/10.1080/10409289.2015.968242.

Clements, D. H., Sarama, J., Layzer, C., Unlu, F., Wolfe, C. B., Fesler, L., … Spitler, M. E. (2017). *Effects of TRIAD on mathematics achievement: Long-term impacts*. Submitted for publication.

Cobb, P., McClain, K., de Silva, T., & Dean, C. (2003). Situating teachers' instructional practices in the institutional setting of the school and district. *Educational Researcher*, *32*(6), 13–24.

Coburn, C. E. (2003). Rethinking scale: Moving beyond numbers to deep and lasting change. *Educational Researcher*, *32*(6), 3–12.

Cohen, D. K. (1996). Rewarding teachers for student performance. In S. H. Fuhrman & J. A. O'Day (Eds.), *Rewards and reforms: Creating educational incentives that work* (pp. 61–112). San Francisco, CA: Jossey Bass.

Copley, J. V. (2004). The early childhood collaborative: A professional development model to communicate and implement the standards. In D. H. Clements, J. Sarama, & A.-M. DiBiase (Eds.), *Engaging young children in mathematics: Standards for early childhood mathematics education* (pp. 401–414). Mahwah, NJ: Erlbaum.

Corcoran, T. B. (2007). *Teaching matters: How state and local policymakers can improve the quality of teachers and teaching*. CPRE Policy Briefs (pp. 1–15).

Cuban, L., & Usdan, M. (Eds.), (2003). *Powerful reforms with shallow roots: Improving America's urban schools*. New York, NY: Teachers College.

Darling-Hammond, L. (1996). Restructuring schools for high performance. In S. H. Fuhrman & J. A. O'Day (Eds.), *Rewards and reform: Creating educational incentives that work* (pp. 144–192). San Francisco: Jossey-Bass.

Darling-Hammond, L. (1997). *The right to learn: A blueprint for creating schools that work*. San Francisco: Jossey-Bass.

Datnow, A. (2005). The sustainability of comprehensive school reform models in changing district and state contexts. *Educational Administrative Quarterly*, *41*(1), 121–153.

Dearing, J. W., & Meyer, G. (2006). Revisiting diffusion theory. In A. Singhal & J. W. Dearing (Eds.), *Communication of innovations: A journey with Ev Rogers* (pp. 29–60). New Delhi: Sage.

Duncan, G. J., Dowsett, C. J., Claessens, A., Magnuson, K., Huston, A. C., Klebanov, P., ... Japel, C. (2007). School readiness and later achievement. *Developmental Psychology*, *43*(6), 1428–1446. http://dx.doi.org/10.1037/0012-1649.43.6.1428.

Elmore, R. F. (1996a). Getting to scale with good educational practices. In S. H. Fuhrman & J. A. O'Day (Eds.), *Rewards and reform: Creating educational incentives that work* (pp. 294–329). San Francisco: Jossey-Bass.

Elmore, R. F. (1996b). Getting to scale with good educational practices. *Harvard Educational Review*, *66*, 1–25.

Ferguson, R. F. (1991). Paying for public education: New evidence on how and why money matters. *Harvard Journal on Legislation*, *28*(2), 465–498.

Fox, J. (2005). Child-initiated mathematical patterning in the pre-compulsory years. In H. L. Chick & J. L. Vincent (Eds.), *Proceedings of the 29th conference of the international group for the psychology in mathematics education (Vol. 2, pp. 313–320). Melbourne, AU: PME.*

Foorman, B. R., Santi, K. L., & Berger, L. (2007). Scaling assessment-driven instruction using the internet and handheld computers, Volume 2. In B. Schneider & S.-K. McDonald (Eds.), *Scale up in practice* (pp. 69–89). Lanham, MD: Rowman & Littlefield.

Fullan, M. G. (1992). *Successful school improvement*. Philadelphia, PA: Open University Press.

Fullan, M. G. (2000). The return of large-scale reform. *Journal of Educational Change*, *1*, 5–28.

Galindo, C., & Sonnenschein, S. (2015). Decreasing the SES math achievement gap: Initial math proficiency and home learning environments. *Contemporary Educational Psychology*, *43*, 25–38. http://dx.doi.org/10.1016/j.cedpsych.2015.08.003.

Garet, M. S., Porter, A. C., Desimone, L., Birman, B. F., & Yoon, K. S. (2001). What makes professional development effective? Results from a national sample of teachers. *American Educational Research Journal*, *38*, 915–945.

Glennan, T. K., Jr. (2004). Expanding the reach of education reforms. In S. J. Bodilly, J. R. Galegher, & K. A. Kerr (Eds.), *Perspectives from leaders in the scale-up of educational interventions*. Santa Monica, CA: RAND Corporation.

Grubb, W. N. (2008). Multiple resource, multiple outcomes: Testing the "improved" school finance with NELS88. *American Educational Research Journal*, *45*, 104–144.

Guskey, T. R. (Ed.), (2000). *Evaluating professional development*. Thousand Oaks, CA: Corwin Press.

Hall, G. E., & Hord, S. M. (2001). *Implementing change: Patterns, principles, and potholes.* Boston, MA: Allyn and Bacon.

Halpern, R. (2004). Parent support and education: Past history, future prospects. *Applied Research in Child Development, 6*(1), 4–11.

Hargreaves, A. (2002). Sustainability of educational change: The role of social geographies. *Journal of Educational Change, 3*(3–4), 189–214.

Heck, D. J., Weiss, I. R., Boyd, S., & Howard, M. (2002). *Lessons learned about planning and implementing statewide systemic initiatives in mathematics and science education. In Paper presented at the annual meeting of the American Educational Research Association.* New Orleans, LA. http://www.horizon-research.com/public.htm.

Hiebert, J. C. (1999). Relationships between research and the NCTM standards. *Journal for Research in Mathematics Education, 30,* 3–19.

Huberman, M. (1992). Critical introduction. In M. G. Fullan (Ed.), *Successful school improvement* (pp. 1–20). Philadelphia, PA: Open University Press.

Kaser, J. S., Bourexis, P. S., Loucks-Horsley, S., & Raizen, S. A. (1999). *Enhancing program quality in science and mathematics.* Thousand Oaks, CA: Corwin Press.

Klingner, J. K., Ahwee, S., Pilonieta, P., & Menendez, R. (2003). Barriers and facilitators in scaling up research-based practices. *Exceptional Children, 69,* 411–429.

Lawless, K. A., & Pellegrino, J. W. (2007). Professional development in integrating technology into teaching and learning knowns, unknowns, and ways to pursue better questions and answers. *Review of Educational Research,* 77(4), 575–614.

Mohrman, S. A., & Lawler, E. E., III. (1996). Motivation for school reform. In S. H. Fuhrman & J. A. O'Day (Eds.), *Rewards and reform: Creating educational incentives that work* (pp. 115–143). San Francisco: Jossey-Bass.

NAEYC. (2002). *NAEYC standards for early childhood professional preparation.* Washington, DC: National Association for the Education of Young Children.

National Mathematics Advisory Panel. (2008). *Foundations for success: The final report of the National Mathematics Advisory Panel.* Washington, DC: U.S. Department of Education, Office of Planning, Evaluation and Policy Development.

National Research Council. (2007). *Taking science to school: Learning and teaching sciences in grades K-8.* Washington, DC: National Academies Press.

National Research Council. (2009). *Mathematics learning in early childhood: Paths toward excellence and equity.* Washington, DC: National Academy Press.

Paris, S. G., Morrison, F. J., & Miller, K. F. (2006). Academic pathways from preschool through elementary school. In P. Alexander & P. Winne (Eds.), *Handbook of research in educational psychology* (pp. 61–85). Mahwah, NJ: Erlbaum.

Peisner-Feinberg, E. S., Burchinal, M. R., Clifford, R. M., Culkin, M. L., Howes, C., Kagan, S. L., … Zelazo, J. (1999). *The children of the cost, quality, and outcomes study go to school.* Chapel Hill, NC: Frank Porter Graham Child Development Center, University of North Carolina at Chapel Hill.

Pellegrino, J. W. (2007). From early reading to high school mathematics: Matching case studies of four educational innovations against principles for effective scale up. In B. Schneider & S.-K. McDonald (Eds.), *Scale up in practice* (pp. 131–139). Lanham, MD: Rowman and Littlefield.

Penuel, W. R., & Shepard, L. A. (2016). Assessment and teaching. In D. H. Gitomer & C. A. Bell (Eds.), *Handbook of research on teaching* (5th ed., pp. 787–850). Washington, DC: American Educational Research Association.

Perry, B., Perry, B., Dockett, S., & Harley, E. (2007). Learning stories and children's powerful mathematics. *Early Childhood Research & Practice, 9*(2), 117–134.

Ramey, C. T., & Ramey, S. L. (1998). Early intervention and early experience. *American Psychologist, 53,* 109–120.

Raudenbush, S. W. (2008). Advancing educational policy by advancing research on instruction. *American Educational Research Journal, 45*, 206–230.
Rogers, E. M. (2003). *Diffusion of innovations* (5th ed.). New York, NY: The Free Press.
Sarama, J. (2002). Listening to teachers: Planning for professional development. *Teaching Children Mathematics, 9*, 36–39.
Sarama, J., & Clements, D. H. (2009a). *Early childhood mathematics education research: Learning trajectories for young children*. New York, NY: Routledge.
Sarama, J., & Clements, D. H. (2009b). In *Scaling up successful interventions: Multidisciplinary perspectives. Paper presented at the American Educational Research Association, San Diego, CA.*
Sarama, J., & Clements, D. H. (2013). Lessons learned in the implementation of the TRIAD scale-up model: Teaching early mathematics with trajectories and technologies. In T. G. Halle, A. J. Metz, & I. Martinez-Beck (Eds.), *Applying implementation science in early childhood programs and systems* (pp. 173–191). Baltimore, MD: Brookes.
Sarama, J., & DiBiase, A.-M. (2004). The professional development challenge in preschool mathematics. In D. H. Clements, J. Sarama, & A.-M. DiBiase (Eds.), *Engaging young children in mathematics: Standards for early childhood mathematics education* (pp. 415–446). Mahwah, NJ: Erlbaum.
Sarama, J., Clements, D. H., & Henry, J. J. (1998). Network of influences in an implementation of a mathematics curriculum innovation. *International Journal of Computers for Mathematical Learning, 3*, 113–148.
Sarama, J., Clements, D. H., Starkey, P., Klein, A., & Wakeley, A. (2008). Scaling up the implementation of a pre-kindergarten mathematics curriculum: Teaching for understanding with trajectories and technologies. *Journal of Research on Educational Effectiveness, 1*, 89–119. http://dx.doi.org/10.1080/19345740801941332.
Sarama, J., Clements, D. H., Wolfe, C. B., & Spitler, M. E. (2012a). Longitudinal evaluation of a scale-up model for teaching mathematics with trajectories and technologies. *Journal of Research on Educational Effectiveness, 5*(2), 105–135. http://dx.doi.org/10.1080/119345747.2011, 627980.
Sarama, J., Lange, A., Clements, D. H., & Wolfe, C. B. (2012b). The impacts of an early mathematics curriculum on emerging literacy and language. *Early Childhood Research Quarterly, 27*, 489–502. http://dx.doi.org/10.1016/j.ecresq.2011.12.002.
Sarama, J., Clements, D. H., Wolfe, C. B., & Spitler, M. E. (2016). Professional development in early mathematics: Effects of an intervention based on learning trajectories on teachers' practices. *Nordic Studies in Mathematics Education, 21*(4), 29–55.
Scheirer, M. A. (2005). Is sustainability possible? A review and commentary on empirical studies of program sustainability. *The American Journal of Evaluation, 26*(3), 320–347.
Scheirer, M. A., & Dearing, J. W. (2011). An agenda for research on the sustainability of public health programs. *American Journal of Public Health, 101*(11), 2059–2067.
Schoen, H. L., Cebulla, K. J., Finn, K. F., & Fi, C. (2003). Teacher variables that relate to student achievement when using a standards-based curriculum. *Journal for Research in Mathematics Education, 34*(3), 228–259.
Showers, B., Joyce, B., & Bennett, B. (1987). Synthesis of research on staff development: A framework for future study and a state-of-the-art analysis. *Educational Leadership, 45*(3), 77–87.
Snipes, J., Doolittle, F., & Herlihy, C. (2002). *Foundations for success: Case studies of how urban school systems improve student achievement.* Washington, DC: Council of the Great City Schools.
Sowder, J. T. (2007). The mathematical education and development of teachers. In F. K. Lester, Jr. (Ed.), *Second handbook of research on mathematics teaching and learning: Vol. 1* (pp. 157–223). New York, NY: Information Age Publishing.
Spillane, J. P. (2000). Cognition and policy implementation: District policy-makers and the reform of mathematics education. *Cognition and Instruction, 18*, 141–179.

Starkey, P., Klein, A., Chang, I., Qi, D., Lijuan, P., & Yang, Z. (1999, April). In *Environmental supports for young children's mathematical development in China and the United States. Paper presented at the society for research in child development, Albuquerque, NM.*

Stein, M. K., Brover, B. W., & Hennigsen, M. (1996). Building student capacity for mathematical thinking and reasoning: An analysis of mathematical tasks used in reform classroom. *American Educational Research Journal, 33*, 455–488.

Teddlie, C., & Stringfield, S. (1993). *Schools make a difference: Lessons learned from a 10-year study of school effects.* New York: Teachers College Press.

Thomas, B. (1982). *An abstract of kindergarten teachers' elicitation and utilization of children's prior knowledge in the teaching of shape concepts.* Unpublished manuscript. New York, NY: School of Education, Health, Nursing, and Arts Professions, New York University.

Timperley, H. (2011). *Realizing the power of professional learning.* London: Open University Press.

Timperley, H., Wilson, A., Barrar, H., & Fung, I. (2007). *Teacher professional development: Best evidence synthesis iteration.* Wellington, New Zealand: Ministry of Education.

U. S. Department of Education. (1999). *New teachers for a new century: The future of early childhood professional preparation.* Washington, DC: Author.

Weiss, I. R. (2002). Systemic reform in mathematics education: What have we learned? In *Research presession of the 80th annual meeting of the National Council of Teachers of Mathematics.* Las Vegas, NV.

Winter, S. G., & Szulanski, G. (2001). Replication as strategy. *Organization Science, 12*, 730–743.

Wu, H.-H. (2011). *Understanding numbers in elementary school mathematics.* Providence, RI: American Mathematical Society.

Yoon, K. S., Duncan, T., Lee, S. W.-Y., Scarloss, B., & Shapley, K. L. (2007). *Reviewing the evidence on how teacher professional development affects student achievement (Issues & Answers Report, REL 2007–No. 033).* Retrieved from Washington, DC. http://ies.ed.gov/ncee/edlabs.

CHAPTER FOUR

Coaching in Early Mathematics

Carrie Germeroth[1], Julie Sarama[1]
University of Denver, Denver, CO, United States
[1]Corresponding author: e-mail address: carrie.germeroth@du.edu; Julie.Sarama@du.edu

Contents

Abstract

Falling scores in math have prompted a renewed interest in math instruction at early ages. By their own admission, early childhood educators are generally underprepared and not always comfortable teaching math. Professional development (PD) in early

Advances in Child Development and Behavior, Volume 53
ISSN 0065-2407
http://dx.doi.org/10.1016/bs.acdb.2017.04.003

mathematics is widely considered a main way to increase teachers' skills and efficacy (e.g., Guskey, 2000; Hyson & Woods, 2014; Munby, Russell, & Martin, 2001; Piasta, Logan, Pelatti, Capps, & Petrill, 2015; Richardson & Placier, 2001; Sarama, Clements, Wolfe, & Spitler, 2016; Sarama & DiBiase, 2004; Zaslow, 2014). However, it has been documented that stand-alone PD is not as effective in changing practice (e.g., Biancarosa & Bryk, 2011; Garet et al., 2008; Guskey, 2000; Hyson & Woods, 2014; Institute of Medicine and National Research Council, 2015; Joyce & Showers, 2002; Zaslow, 2014). Site-embedded ongoing support in the form of coaching or mentoring has been shown to be critical for successful implementation (Neuman & Cunningham, 2009; Powell, Diamond, Burchinal, & Koehler, 2010). In this chapter, we describe coaching models and abstract characteristics of effective coaching from the research. With this background, we provide an in-depth view of the coaching aspect of two large empirical studies in early mathematics. We introduce the theoretical framework from which the coaching models for these projects were developed and describe the research on which they were based. We then summarize how the planned models were instantiated and challenges to their implementation within each project. In the final section, we summarize what we have learned and described implications and challenges for the field.

1. OVERVIEW OF COACHING RESEARCH

The transforming the workforce report (Institute of Medicine and National Research Council, 2015) outlines several factors that impact the quality and development of the early childhood workforce. The report suggests that to develop and sustain professional competencies direct mechanisms including preparation programs, training, mentoring and coaching, and in-service professional development (PD) are all critical. For early childhood educators, PD can serve to equalize knowledge and skills across a workforce that has entered teaching without a common foundation of preservice preparation. Teaching is a dynamic profession, and ongoing PD is a necessary educational component if teachers are to continually refine practices for student growth (Darling-Hammond & Richardson, 2009). To continue their growth as educators, PD is a powerful resource to help augment the teachers' skills and knowledge so that they can best serve their students (Neuman & Cunningham, 2009).

Research on effective PD suggests that the development of effective teaching needs to occur in the context of teachers' own practices (Ball & Cohen, 1999; Pianta, Mashburn, Downer, Hamre, & Justice, 2008). Thus, a reasonable hypothesis is that coaching may be the most effective method of facilitating teacher's expansion of skills. However, the empirical evidence on coaching is limited and not conclusive. One review of studies conducted in a variety of settings including childcare centers, Head Start programs, family childcare programs, prekindergarten/preschool programs, and early elementary grades concluded that there was at least limited evidence that coaching contributed to improved quality and practices (Isner et al., 2011). However, the review could attribute the improvements that did occur to specific aspects of the coaching. The review also reported that coaching positively affected children's language and literacy when the coaching model focused on language and literacy teaching methods (Isner et al., 2011).

A metaanalysis on early childhood PD reported that the combination of training and coaching increased teachers' use of new skills (Sheridan, Edwards, Marvin, & Knoche, 2009). A separate review agreed with this conclusion, although again, could not attribute the improvement to any specific aspects of the coaching (Gupta & Daniels, 2012). However, coaching does seem to be a critical component of the combination of training with in-class support. For example, another metaanalysis found that when teachers were provided with training that consisted of only theory and discussion, a mere 10% demonstrated knowledge gained and none used the skills that were discussed within the classroom. However, when coaching was added to the training experience, 95% of the participants demonstrated knowledge gained, demonstrated the new skills, and used the new skills in the classroom, respectively (Joyce & Showers, 2002). Other studies support the hypothesis that coaching makes a unique contribution (Kretlow, Wood, & Cooke, 2011; Neuman & Cunningham, 2009). Other studies similarly found more improvements in literacy instruction when training is complemented by strong in-class coaching or mentoring (Biancarosa & Bryk, 2011; Garet et al., 2008; Gersten, Dimino, Jayanthi, Kim, & Santoro, 2010). In summary, PD that includes layered and combined training and on-site coaching can increase the effectiveness of instructional practices (Biancarosa & Bryk, 2011; Buysse, Castro, & Peisner-Feinberg, 2010; Carlisle, Kelcey, Rowan, & Phelps, 2011). Such coaching may not have to require in-person visits. Online coaching to be promising to improve practice in early childhood settings (Hemmeter, Snyder, Kinder, & Artman, 2011; Powell & Diamond, 2011).

1.1 Coaching on Math

Several studies have examined the effect of coaching to improve teacher practices and student achievement in language and literacy (Rudd,

Lambert, Satterwhite, & Smith, 2009; Shidler, 2009), suggesting that coaching might be more successful when focused on one domain (Isner et al., 2011). Unfortunately, few studies have focused on coaching mathematics, especially in early childhood education (although there is recent interest, Bay-Williams & McGatha, 2014). Thus, there is a need to investigate if coaching in the domain of math will have the same efficacy as it does for other domains more typically coached in early childhood, such as language, literacy, and social–emotional development. There is some evidence conducted at the middle school level that found student achievement increased when teachers received math coaching (Obara & Sloan, 2009). Another study found math coaching for teachers of grades 3, 4, and 5 over a 3-year period resulted in greater student achievement—given expert coaches (Campbell & Malkus, 2011; Erchick et al., 2007). These results, though not in early childhood, still hold promise for the usefulness of math coaching.

However, any such effort will confront that the learning and teaching of mathematics present special challenges. The importance of mathematics is well known, especially in a global economy with the vast majority of jobs requiring more sophisticated skills than in the past (Doig, McCrae, & Rowe, 2003). However (Sarama & Clements, 2013), US students do not perform well compared to students from many other nations, as early as kindergarten (Stigler, Lee, & Stevenson, 1990) and even the preschool years (Yuzawa, Bart, Kinne, Sukemune, & Kataoka, 1999). Such achievement gaps are most pronounced in US children from low-resource communities (Siegler, 1993). This is especially worrisome because early knowledge strongly affects later success in mathematics (National Mathematics Advisory Panel, 2008).

Thus, we need to improve the basic academic mathematical concepts, skills, and motivation of low-achieving children early in their school career (Claessens, Duncan, & Engel, 2007). Although this can be achieved by providing high-quality experiences in early mathematics (Doig et al., 2003; Thomson, Rowe, Underwood, & Peck, 2005), few early childhood programs or settings do so (National Research Council, 2009), and few teachers of early childhood are prepared to do so (Sarama, 2002; Sarama & DiBiase, 2004).

In summary, improving mathematics education within early childhood settings holds challenges that range from the logistical (e.g., all-day PD may take teachers away from emotionally dependent children) to the philosophical and motivational (why "push down" math onto young children?)

to the practical (many teachers lack knowledge of the content of mathematics, as well as its learning and teaching). Any effort to support a high-quality early math education must meet these challenges (Sarama & Clements, 2013).

Research on effective math PD has also shown intensive coaching and mentoring to be a critical component (Clements & Sarama, 2008, 2012). Teachers participating in one project with the goal of implementing a preschool math curriculum received offsite intensive training that was focused on and connected to classroom practice and was complemented by classroom-based coaching. The training components were intensive involving 2 years of work with teachers, including 12 full-day PD sessions and twice-per month visits from coaches (Clements & Sarama, 2009). In general, successful models for professional learning in math seem to require intensive PD sessions, are conducted school-wide, and are followed by some form of ongoing on-site coaching support (Bobis et al., 2005; Bright, Bowman, & Vacc, 1997; Clements & Sarama, 2008, 2012; Clements, Sarama, Spitler, Lange, & Wolfe, 2011; Sarama, Clements, Starkey, Klein, & Wakely, 2008; Wright, Martland, Stafford, & Stanger, 2002).

1.2 Coaching in Education

Coaching typically involves a collaboration in which the coach uses various strategies to transfer or impart knowledge and skills to a teacher. Coaching is often employed to help teachers improve their practice generally or provide targeted assistance in supporting the implementation of instruction for a content area (Isner et al., 2011; Whitebook & Bellm, 2013). There are many forms and practices of coaching; however, the consistent theme among most is that coaching is a form of PD that involves ongoing classroom modeling, observations, reflection, and evaluation (Sheridan et al., 2009; Whitebook & Bellm, 2013). The coaching relationship may last from as little as a few months to 1 year or more (Isner et al., 2011; Whitebook & Bellm, 2013). The terms mentoring and coaching have often been used interchangeably, and we will do so in this chapter. Some make distinctions between the two, assigning mentors as more of an expert paired with a novice and coaching more about the relationship between the coach and the teacher who may have been in the classroom for years. Typically, coaches/mentors of early childhood educators are selected based on their education credentials and work experience in an early childhood care and education setting.

1.3 Coaching Models

There are multiple models for coaching/mentoring that all share similar characteristics. We focus on four that have been identified as those used by math coaches (Barlow, Burroughs, Harmon, Sutton, & Yopp, 2014). For this chapter, we briefly describe cognitive coaching (Costa & Garmston, 2002), instructional coaching (Knight, 2007), content-focused coaching (West & Staub, 2003), and a guide to mathematics coaching (Hull, Balka, & Miles, 2009). Though these models differ somewhat in approach, they all address coaching roles, interactions with the teacher, reflection as a core component, and managing the teacher–coach relationship.

1.3.1 Cognitive Coaching

Cognitive coaching (Costa & Garmston, 2002) utilizes reflective questioning strategies to support teachers in honing their professional practices and knowledge. This approach aims for the teacher to arrive at self-directed growth with the coach as a mediator. The theory is that teachers' thoughts and perceptions drive their behavior. This model follows a three-phase process of prelesson conference, lesson observation, and a postlesson debrief. Coaches must meet with the teacher beforehand to know what to look for in an observation. During these conversations, coaches use reflective listening, questioning, and paraphrasing to mediate teacher growth. For example, to mediate a teacher in their reflective process, a coach might paraphrase what a teacher said to focus on a personal goal by prompting "Tell me more about that ..." This action is designed to support the teacher in clarifying their thinking.

1.3.2 Content-Focused Coaching

Content-focused coaching (West & Staub, 2003) is more direct in focusing on the content of what is being coached. In this model, the coach is assumed to have a high level of content knowledge and more experience than the teacher. This assumption drives much of how the coaching process unfolds. Unlike cognitive coaching, the emphasis is more on student learning of a content area and teacher's practices within that content area. Coaching interactions may focus on designing lessons for that content area. Less likely to use teacher self-direction to arrive at goals, the content-focused coach will use direct evidence of student learning, student data, or work samples to guide a coaching conversation. However, similar to cognitive coaching, a three-phase process is generally employed with the pre- and post-conversation and observation.

1.3.3 Instructional Coaching

Instructional coaching (Knight, 2007) treats coaching as a partnership. The coach actively listens so as to understand the teacher's point of view, as in cognitive coaching. The theoretical foundations emphasize seven principles to ground the coaching relationship: equality, choice, voice, dialog, reflection, praxis, and reciprocity. This model is heavily based on relationship building, viewing personal connections between the coach and teacher as having a profound impact on the quality of the experience. The coach focuses on the positive, using direct observations of student behavior, content, instruction, and formative assessment. Again similar to cognitive coaching, a three-phase process is generally employed with the pre- and postconversation and observation.

1.3.4 Mathematics Coaching

Mathematics coaching (Hull et al., 2009) interweaves information about instructional strategies and content knowledge. This approach views the coach as having a solid background in mathematics content and pedagogy. In their role, coaches must also have knowledge of adult learning styles and strategies, group dynamics, and social norms. Developing trust and a positive relationship is central to the coach's role in this model. Mathematics coaches use data, analysis, interpretation, and application in their conversations with teachers. The focus of coaching is directly working with classroom teachers to improve student learning. Coaches similarly follow a pre- and post-observation and conversation process.

The four models briefly reviewed share similarities in process with a three-phase cycle. Additionally, all emphasize the importance of the professional relationship. The primary differences are more philosophical. Cognitive coaches mediate self-directed growth. Content and mathematics coaches focus on teacher knowledge and student evidence, while partnership and relationships are central to instructional coaches. Finally, a particular model may provide a framework for coaching, but often it is the case that characteristics unique to a school or district determine how that model is enacted.

1.4 Characteristics of Effective Coaches

Of course, coaching expertise is not equal across coaches using any model, and the effectiveness of the coach is sure to have an impact on the gains experienced by the teacher (Campbell & Malkus, 2011; Poglinco et al., 2003). The Institute of Medicine Transforming the Workforce report suggests that the knowledge and skill of coaches are critical to effective professional learning. They need to have the content, general pedagogical knowledge, and pedagogical content knowledge, as well as competence in effective coaching. Coaching is crucial, but simply having access to coaching does not ensure improvements (Garet et al., 2008; Hsieh, Hemmeter, McCollum, & Ostrosky, 2009). The coaching should also be ongoing and conducted by knowledgeable coaches (Biancarosa & Bryk, 2011; Carlisle & Berebitsky, 2010; Powell et al., 2010). Several concrete coaching practices that facilitate a positive experience for the teacher have been highlighted in the literature.

1. Coaches should have both content and pedagogical knowledge for the domain being coached as well as the ability to cultivate a positive and trusting relationship with teachers (e.g., Campbell & Malkus, 2011; Garet et al., 2008). Coaching is multifaceted in that coaches need to be able to gage both the complexity of what teachers are being instructed to implement, and the newness of the content in terms of previous knowledge and practice of the teachers (Joyce & Showers, 2002).
2. Coaches should assist teachers with lesson planning, providing resources, modeling, and coteaching. Technique modeling, coteaching, and acting as a resource provider are also essential (Fox, Hemmeter, Snyder, Binder, & Clarke, 2011; Neuman & Cunningham, 2009). These important, though often overlooked, aspects of the coach–teacher relationship are essential to effective coaching. For one, they help break the misconception that coaches are not (or should not be) an active part of the classroom. In fact, teachers and students see much more substantial gains when coaches move away from simple observation and feedback and move toward collaboration and active involvement (Neuman & Cunningham, 2009).
3. Coaches should also be continually working with teachers to set goals and adjust those goals to meet the teacher's current needs. For instance, as coaches and teachers enter into a professional relationship, curriculum and lesson planning, as well as goal setting, should be made priorities (Neuman & Cunningham, 2009; Skiffington, Washburn, & Elliott, 2011).

4. Coaches should facilitate reflection. Effective coaches provide descriptive, not evaluative feedback using concrete observations from the classroom. Engaging the teacher in collaborative problem solving to improve practice (Guiney, 2001; Harwell-Kee, 1999).

2. COACHING/MENTORING MODEL FOR TWO LARGE-SCALE PROJECTS IN EARLY MATH

In the remainder of this chapter, we describe two large-scale projects in early mathematics that were designed to rely heavily on coaching. First we introduce the theoretical framework from which the coaching models were developed, and the research on which they were based. We then summarize how the planned models were instantiated and challenges to their implementation.

Our mentoring/coaching model is dynamic. Initial questionnaire data are not sufficient for knowing who will need additional assistance from coaches and mentors. For example, some teachers present themselves well and even misrepresent how much of the curriculum they are teaching; only coaching/mentoring visits reveal and address such problems. Therefore, we created a dynamic model in which additional attention by both mentors and peer coaches is given until adequate fidelity of implementation is achieved. Several additional features of our training plus in-class coaching model also encourage sensitivity to individual needs of teachers, including (a) time to learn and work with cohort groups; (b) job-embeddedness—addressing concrete, immediate concern with practical problems of implementation; (c) opportunities for practice, receiving individual, nonthreatening feedback from mentors/coaches; and (d) software and web resources that facilitate individualization. In summary, our PD follows research-based guidelines and provides a combination of experiences (Guskey, 2000; Showers, Joyce, & Bennett, 1987; Hyson & Woods, 2014; Munby, Russell, & Martin, 2001; Piasta, Logan, Pelatti, Capps, & Petrill, 2015; Richardson & Placier, 2001; Sarama, Clements, Wolfe, & Spitler, 2016; Zaslow, 2014). Research indicates that when staff development includes ongoing coaching, classroom innovations continue at a 90% level after external funding ceases (Copley, 2004; Costa & Garmston, 1994; Nettles, 1993).

2.1 TRIAD (Technology-Enhanced, Research-Based, Instruction, Assessment, and PD)

We created a research-based model for scale-up called TRIAD (see Sarama et al., 2008; Sarama, Clements, Wolfe, & Spitler, 2012), with the intent to generalize to both other subject matter areas and other grade levels. We implemented in the early childhood years because long-range benefits to children are greatest for interventions in that period (Carneiro & Heckman, 2003; Clements & Sarama, 2009) and because there is conflicting evidence as to whether the effects of early interventions persist or fade.

The rationale for situating the intervention in mathematics is that children from low-resource communities and who are members of linguistic and ethnic minority groups demonstrate significantly lower levels of mathematics achievement than children from higher resource, nonminority communities (Denton & West, 2002; Duncan & Magnuson, 2013; National Mathematics Advisory Panel, 2008). Such differences are evident from the earliest years (National Research Council, 2001, 2009; Sarama & Clements, 2009). This is important because early mathematics competence predicts later achievement even into high school (National Mathematics Advisory Panel, 2008; National Research Council, 2009; Stevenson & Newman, 1986), with persistent difficulties with mathematics one of the strongest predictors of failure to graduate high school and enter college (Duncan & Magnuson, 2011). Interventions to address these early differences may benefit low-resource and minority children more because they have fewer educational opportunities in their homes and communities (Brooks-Gunn, 2003; Carneiro & Heckman, 2003; Natriello, McDill, & Pallas, 1990; Raudenbush, 2009). Such interventions have been shown to be effective; however, most have not been taken to scale and often used the individual child as the unit of analysis, despite their assignment to treatments by class or school, which can inflate findings (Case, Griffin, & Kelly, 1999; for a review, see Clements & Sarama, 2011; Griffin & Case, 1997; Griffin, Case, & Siegler, 1994; Klein, Starkey, Clements, Sarama, & Iyer, 2008; National Mathematics Advisory Panel, 2008; Starkey, Klein, & Wakeley, 2004). Thus, an important practical and theoretical challenge is scaling up successful mathematics interventions in the early and primary grades in the United States and especially ensuring the persistence of the effects of such interventions.

The TRIAD model was created to address this challenge. In this study, we evaluated the persistence of effects of a TRIAD implementation, with and without the follow-through component, in its third and final year. That is, two experimental groups received the TRIAD implementation in prekindergarten (pre-K), and one of these groups also received the follow-through intervention in their kindergarten and first-grade year. Here we report the effects at the end of the students' first-grade year and whether these effects were equivalent for various subpopulations, as well as possible indirect, or mediational, effects through particular pedagogical practices (Clements et al., 2011).

2.2 TRIAD Theoretical Framework

TRIAD's theoretical framework (Sarama et al., 2008) is an elaboration of the network of influences theory (Sarama, Clements, & Henry, 1998). Scale-up is considered to involve multiple coordinated efforts to maintain the integrity of the vision and practices of an innovation through increasingly numerous and complex socially mediated filters, through phases of introduction, initial adoption, implementation, and institutionalization.

The TRIAD model scales up the support of "interactions among teachers and children around educational material" (Ball & Cohen, 1999, p. 3). This strategy creates extensive opportunities for teachers to focus on mathematics, goals, and students' thinking and learning, which improves teachers' knowledge of subject matter, teaching, and learning and increases student achievement (Ball & Cohen, 1999; Cohen, 1996; Schoen, Cebulla, Finn, & Fi, 2003; Sowder, 2007). Research suggests that the most important feature of a high-quality educational environment is a knowledgeable and responsive adult and that PD can foster these characteristics (Darling-Hammond, 1997; National Research Council, 2001; Sarama & DiBiase, 2004; Schoen et al., 2003; Sowder, 2007). Use of demonstrations, practice, and feedback, especially from coaches, increases the positive effects of information-only training (Fixsen, Naoom, Blase, Friedman, & Wallace, 2005; Pellegrino, 2007; Showers et al., 1987). The PD in TRIAD provides a promising path for developing teachers' understanding of learning, teaching, curriculum, and assessment by focusing on research-based models of students' thinking and learning (cf. Bredekamp, 2004; Carpenter & Franke, 2004; Hiebert, 1999; Klingner, Ahwee, Pilonieta, & Menendez, 2003). Research-based learning trajectories are TRIAD's core (Clements & Sarama, 2004a, 2004b; Clements, Sarama, & DiBiase, 2003). Learning trajectories have three components: a goal (that is, an aspect of a mathematical domain students should learn), a developmental progression of levels of thinking, and instruction that helps them move along that trajectory. Thus, they facilitate teachers' learning about mathematics, how students think about and learn this mathematics, and how such learning is supported by the curriculum and its teaching strategies.

They address domain-specific components of learning and teaching that have the strongest impact on cognitive outcomes (Lawless & Pellegrino, 2007). By illuminating potential developmental progressions, they bring coherence and consistency to goals, curricula, and assessments (Clements & Sarama, 2009; Sarama & Clements, 2009) and help teachers focus on the "conceptual storyline" of the curriculum, a critical element that is often missed (Heck, Weiss, Boyd, & Howard, 2002; Weiss, 2002).

2.3 Research-Based Guidelines for Scaling Up

The wider social context in which these groups are embedded, the American educational system, has a persistent "grammar of schooling" (Tyack & Tobin, 1992) composed of rigid cultural beliefs about proper teaching, learning, and knowledge. A historical analysis of innovations reveals that those challenging this grammar tend to be short lived (Tyack & Tobin, 1992). Especially because we were attempting change close to the "core" of educational practice (Elmore, 1996), based on our innovation on suggestions and cautions from the literature.

TRIAD follows the 10 guidelines we abstracted.

1. Involve, and promote communication among, key groups concerned with young children (Huberman, 1992; Kaser, Bourexis, Loucks-Horsley, & Raizen, 1999), emphasizing a shared understanding of, and connections between, the project's goals, national and state standards, and greater societal need (Elmore, 1996; Fullan, 2000; Sarama et al., 1998). Promote clarity of these goals, of leadership, and of all participants' responsibilities and accountability (Kaser et al., 1999). Begin creating (and eventually institutionalize) a support infrastructure (Kaser et al., 1999).
2. Promote equity through equitable recruitment and selection of participants, allocation of resources, and use of curriculum and instructional strategies that have demonstrated success with minorities and females (Kaser et al., 1999).
3. Plan for the long term, encouraging active participation with a "start small and build" strategy (Fullan, 1992).
4. Focus on instructional change that promotes depth and quality of children's thinking (Ball & Cohen, 1999; Fullan, 2000), placing

standards- and research-based learning trajectories at the core of the teacher/child/curriculum triad to ensure that curriculum, materials, instructional strategies, and assessments are aligned with (a) national and state standards and a vision of high-quality mathematics education, (b) each other, and (c) "best practice" as determined by research and the wisdom of expert practice (Bodilly, 1998; Clements, 2002b; Clements & Sarama, 2004a, 2004b, 2004c; Kaser et al., 1999).

5. Build expectation and camaraderie to support a consensus around adaptation. Reform strategies often perversely isolate teachers, either gathering them together in isolation or nurturing "special" teachers in a building (Elmore, 1996). Instead, (a) promote "buy-in" in multiple ways, such as dealing with all participants as equal partners, and distributing resources to support the project (Berends, Kirby, Naftel, & McKelvey, 2001); (b) establish and maintain cohort groups (Jacobson, Emihovich, Petrie, Helfrich, & Stevenson, 1998); (c) facilitate teachers visiting successful implementation sites and talking with other teachers there; and (d) build local leadership by involving principals and encouraging teachers to become teacher leaders (Fullan, 1992, 2000).
6. Provide PD that is multifaceted; extensive; ongoing; reflective; focused on common actions and problems of practice, and especially on children's thinking; grounded in particular curriculum material; and situated partially in the classroom. Encourage sharing, risk taking, and learning from and with peers (Bodilly, 1998; Cohen, 1996; Fullan, 1992; Kaser et al., 1999). Develop teachers' knowledge and beliefs that the curriculum is appropriate and its goals are valued and attainable (Elmore, 1996), keeping all PD activities targeted toward those goals (Fullan, 2000).
7. Give latitude for adaptation to teachers and schools, but maintain follow through and integrity (Fullan, 1992, 2000; Huberman, 1992). Emphasize the similarities of the curriculum with sound early childhood practice and what teachers already are doing. Do not allow dilution due to uncoordinated innovations (Fullan, 2000; Sarama et al., 1998).
8. Maintain frequent, repeated communication, and follow-through efforts emphasizing the purpose, expectations, and visions of the project (Fullan, 1992; Kaser et al., 1999). Conduct formative and summative

evaluations connected to specific benchmarks for expected outcomes and methods for improvement (Huberman, 1992; Kaser et al., 1999).

9. Give teachers continuous feedback from sources they trust that children are learning what they are taught and that these learnings are valued (Bodilly, 1998; Elmore, 1996; Sarama et al., 1998).
10. Provide incentives for all participants, including intrinsic and extrinsic motivators linked to project work (e.g., external expectations—from standards listed in the first guideline to pressures from administration; Berends et al., 2001; Cohen, 1996; Darling-Hammond, 1996; Elmore, 1996; Jacobson & Battaglia, 2001; Mohrman & Lawler, 1996).

TRIAD includes collaboration of key groups to establish and maintain (a) a pre-K mathematics curriculum, with all components of the curriculum—a teacher's manual, demonstration videotapes, manipulatives, software, teaching strategies, assessments, and PD—based on a common core of understanding the learning trajectories through which children develop mathematically; (b) PD for teachers; (c) on-site support for teacher by facilitators during the school year; and (d) supportive roles and materials for parents. TRIAD's collaboration with key groups is dedicated to developing organizational structures that intensify and focus, rather than dissipate and scatter, teachers' motivation to engage in and maintain this challenging practice (Elmore, 1996).

2.4 TRIAD Intervention

The TRIAD implementation occurred in 42 schools in two urban districts serving low-income communities, randomly assigned to three conditions. Following the research-based guidelines previously listed, the TRIAD model emphasizes PD and curriculum implementation.

The TRIAD model provides multiple forms of support for teachers. The main two supports were the sequence of PD sessions and in-class coaching. The initial PD sessions included a brief description of the study and an overview of the integrated curriculum and its goals. From that point forward, most sessions included a short introduction to the learning trajectory for a topic and hands-on experience and interactions with peers and staff around the curriculum's activities. We emphasized using materials as intended by the developers rather than on teachers adapting them, an approach supported by research (Weiss, 2002).

In a typical PD session, one of the coprinciple investigators presented upcoming mathematics content and research on the teaching and learning of that content, the group discussed questions and issues arising from the presentation, and teachers presented and discussed their own classroom experiences with the curriculum. This was followed by 2 h of hands-on experience with the curriculum for children, in which teachers rotated through practicing small group, computer, whole group, and "math throughout the year" activities. These experiences emphasized the understanding and use of learning trajectories, assessment with the small-group record sheets, and questions from teachers.

Part of each session was spent on teachers watching videos of the activities. Discussion of the videos centered on student learning and instructional strategies. The videos were designed to communicate the vision of the curriculum in action and to make the ideas and processes accessible, memorable, engaging, and therefore usable. This type of intense, focused work with video examples, interactions with peers around the curriculum's activities, and support throughout the year was designed to positively change teachers' perceptions, because "most teachers report that they believe their children are capable of fine work, but what they think they know from daily experience often hedges that belief with limited expectations" (Ball & Cohen, 1999, p. 8).

Coaching was the second main component. We collaborated with peer coaches (teachers in the study who served the role of "leaders" within their schools) and expert coaches, whom we called "mentors" to distinguish them from the peer coaches. Coaching by these mentors was conducted throughout the year. Mentors included coordinators/specialists from within the study schools as well as several project staffs. Mentors were responsible for supporting both teachers and the peer coaches. This type of real time, classroom-based observation, demonstration, coaching is viewed in the TRIAD model as indispensable. Section 2.5 describes the role of the mentors.

2.5 Mentor Role

This TRIAD coaching included shared decision making, one-to-one consultation, monitoring, and reinforcement. Coaches worked with teachers in their classrooms throughout the year, visiting teachers in their classrooms no less than once per month, discussing students' learning and teaching based on the LTs. Mentors' observations and reflection sessions were guided by the project's "I-Fidelity" tool, which was a teacher (thus the "I") version of the project's fidelity of implementation measure with the Likert ratings removed. That is, descriptions of high-fidelity practice were maintained, but no evaluative structures were included. After the observation, the teacher and mentor read and discussed the descriptions and reflected on the teacher's practice, making notes for improvement that stayed with the teacher. (Mentors also completed the fidelity of implementation evaluations several times per year, which informed their work with teachers, but were not shared with anyone outside of the researchers.) When visiting peer coaches, mentors also provided guidance on the peer coaches' role and relationships to ensure the coaching relationships were working at all levels.

Mentors reported to site coordinators after each visit with teachers to monitor the facilitation process. Any concerns were immediately communicated to the Implementation Director and Project Directors who then collaborated to increase support. Finally, mentors participated in, and supported professional learning, during all project PD activities for teachers (e.g., every training session).

2.6 Mentor Training

In addition to attending all PD sessions with teachers, mentors participated in a 5-day institute developed and piloted in a previous instantiation of the TRIAD project. The institute included, 1/2-day introduction, 3 days on learning the mathematics intervention, and one 1/2 day on mentoring per se, all conducted by project staff. All mentors met once per month via distance learning.

2.7 Implementation of Mentoring in TRIAD

Mentors supported teachers implementing the Building Blocks (BB) mathematics curriculum. On average, each mentor visited eight teachers. In general, most teachers had only one mentor. In a few instances, teachers may have had 2 or 3 mentors due to staff changes or the need for additional support (Fig. 1).

Mentor logs were coded for the second year of implementation to evaluate the extent to which mentoring occurred as intended. Mentor logs provide information on the date, teacher, duration, and type of visit, as well as qualitative notes about the visit. Two research assistants independently coded the qualitative notes from mentor logs. Researchers were trained on a coding rubric until 100% agreement was achieved across the coding rubric. Each researcher coded all the logs and any discrepancies in rating

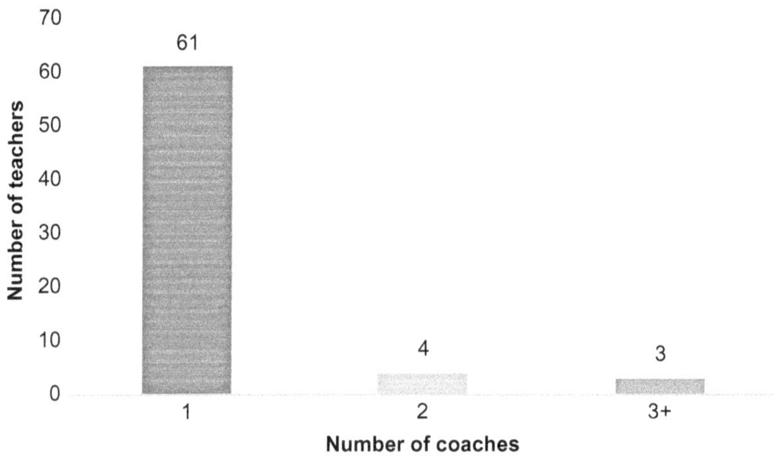

Fig. 1 Number of coaches by teacher.

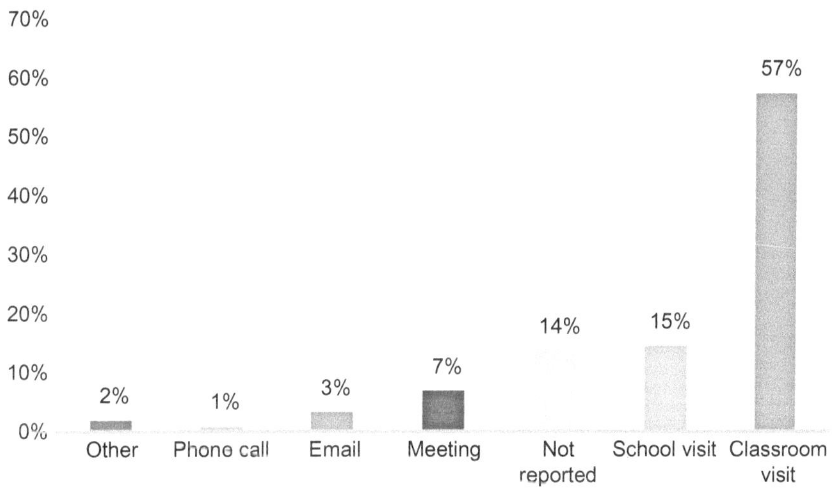

Fig. 2 Frequencies of method of contact by research group.

were resolved to reach consensus. Researcher codes were compared and collapsed to determine larger themes in the data.

Mentors recorded their method of contact on logs. Primarily mentors were visiting teachers' classrooms (Fig. 2). Meetings represent when a mentor conducted a small group or grade level meeting with teachers to provide training on a specific implementation topic. School visits may have been conducted to collected data from the school such as attendance or other school level information.

Often coaches/mentors are pulled in multiple directions unable to focus on their role. The mentor notes revealed six prominent roles of TRIAD mentors (Fig. 3). TRIAD mentors primarily spent their time conducting classroom observations and provide interventions, central duties for their role. As stated previously, fidelity of implementation checks were an important component of the mentor role. The fidelity checks provided real-time information to the project directors regarding implementation. During the second year of implementation which is the period of analysis for the mentor logs, a total of 96 fidelity check visits were reported on the coach logs. On average, each teacher received at least two fidelity checks during the school year. Descriptions of each mentor role observed are further described below.

- *Observations*: Mentor takes a passive role in the classroom. Comments describe what was seen.
    - *Teacher observation*: Mentor describes teacher, and only teacher, actions.
    - *Student observation*: Mentor describes student(s), and only student(s), actions.
    - *Classroom observation*: Mentor describes general classroom environment possibly in conjunction with student(s) and/or teacher.
- *Interventions*: Mentor takes an active role assisting with curriculum access.
    - *Teacher-initiated intervention*: Mentor intervenes/provides support at teacher's behest.

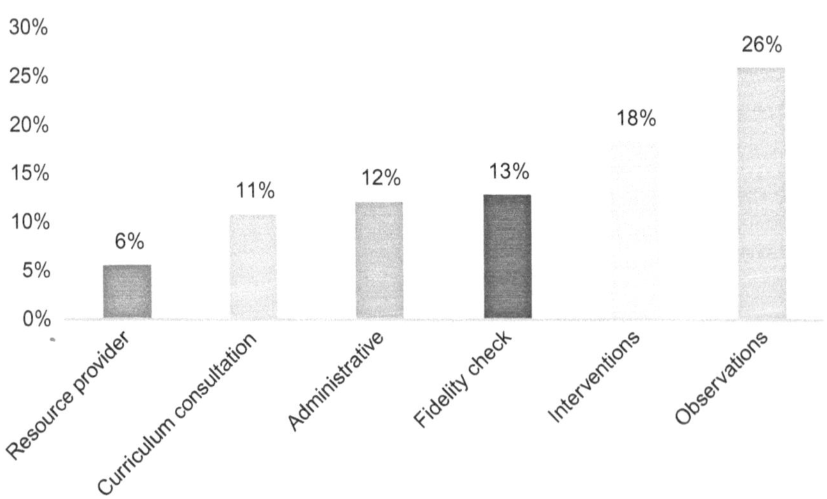

Fig. 3 Frequencies of coaching roles by research group.

  - *Mentor works directly with students*: Mentor provides direct instruction/interaction to students in classroom.
  - *Mentor models*: Mentor demonstrates appropriate curriculum implementation.
  - *Mentor corrects or prompts*: Mentor provides immediate teacher feedback on incorrect implementation.
  - *Mentor-initiated intervention*: Mentor intervenes/provides support of own volition.
  - *Mentor works directly with students*: Mentor provides direct instruction/interaction to students in classroom.
- *Fidelity check*: Complete fidelity checklist during observation report any concerns to the Implementation Director and Principal Investigators.
- *Administrative*: Any necessary tasks related to the mentoring role but not directly involving students or teachers such as making copies, picking up consent forms, etc.
- *Curriculum consultation*: Informal discussion between teacher(s) and mentor regarding aspects of curriculum.
- *Resource provider*: Bring curriculum-related materials into classroom.

2.8 Effectiveness of TRIAD as a PD Model

In summary, the mentor logs demonstrate that the TRIAD mentors performed their duties in alignment with the overall TRIAD PD model. Moreover, the mentor logs from the second implementation year also provide a glimpse at a mentoring model that meets effectiveness criteria from the greater literature on coaching. The coaching component is but one piece of the overall PD model, and although we hypothesize it to be a critical piece, we are not making causal claims about the specific contribution of coaching to the effectiveness of the intervention. Instead we posit that when coaching/mentoring are delivered in a manner that meets research-based criteria coupled with the other components critical to the TRIAD model, then we will see positive results. This is what we found. Children in the TRIAD group learned more mathematics than the children in the control group, with an effect size of 0.72 (Clements & Sarama, 2007). This effect size decreased over the kindergarten and first-grade years, but did not disappear. When examining only those students who experienced the assigned treatments (treatment-on-the-treated, TOT) from pre-K through first-grade analyses that included at the end of pre-K, the TRIAD groups (TRIAD-FT and TRIAD-NFT, whose treatments were identical in pre-K) were statistically significantly higher in mathematics achievement than the control group (Clements et al., 2011). By the end of kindergarten, both groups similarly outperformed the control group. At the end of first-grade results continue to support the benefits of high-quality, research-based mathematics instruction during pre-K as both groups continue to outperform the control conditions in TOT analyses (Clements, Sarama, Wolfe, & Spitler, 2013).

We conducted several analyses to measure the specific contribution of coaching to these positive effects. Unfortunately, these yielded no useful data. Using coaching hours as a positive mediator failed due to our dynamic model. That is, because coaches were asked to spend more hours with teachers who were having more difficulties, one could not hypothesize that more hours would lead to better results for children. This leads to the possibility that coaching hours could be a "negative mediator"—the higher the coaching hours, the lower the effects. This did not work either because coaches spent at least the minimum amount with all teachers, and, despite the request to spend more hours teacher scoring lower fidelity, a common pattern was that coaches also spent more time than average with their favorite teachers those who were doing well. Therefore, future research projects may wish to investigate alternative models to design the coaching and data collection to allow the use of proper mediational models. It may have been better, at least for our research purposes, to have a fixed schedules for coaching. Or, perhaps we should have earlier recognized and minimized the natural human tendency to spend more time those teachers whose classrooms were pleasant and easy to work within. Future research should consider these issues to better contribute to the research literature on coaching.

2.9 EMERGE (Early Math, Early Regulation: Good for Everyone)

The EMERGE project was designed to investigate whether adding a self-regulation component to an early mathematics curriculum provides enhanced child outcomes in multiple domains, above and beyond those yielded by the mathematics curriculum alone. The main rationale for such a study was that early childhood schools and centers are searching for solutions to the dilemma of child-centered or content-centered strategies. Child-centered programs have a long history. However, concerns about children's achievement and the pressure of content-specific standards have set up a perceived conflict, in which teachers often believe they are being asked to abandon child-centered approaches, or, at least, to compromise and squeeze in, as one teachers put it, "Literacy on Monday–Wednesday–Friday, math on Tuesday–Thursday, and social–emotional during our shortened play periods." The PD approach developed under EMERGE was designed to view both approaches in a positive light, in which teaching and learning content was an appropriate, important, joyful, and equitable component of early childhood education within a child-centered context. If empirically supported, the EMERGE approach could serve as a model that others could use to successfully and *synergistically* combine these strategies so the whole was *more* than the sum of its integrated, not conflicting, parts.

2.10 EMERGE Theoretical Framework

EMERGE was designed to follow the same research-based guidelines as those in TRIAD to provide a combination of PD experiences (Guskey, 2000; Showers et al., 1987). The two interventions synthesized were both theoretically and empirically grounded. The BB curriculum was previously discussed. The second intervention was the use of scaffolding designed to promote self-regulation (SSR) through make-believe play from the Tools of the Mind curriculum (TotM, Bodrova & Leong, 2007), an approach at the core of the child-centered approach. Self-regulation is important in facilitating learning and retention, especially for children at risk for later school failure. When explicit content-oriented instruction is mistakenly implemented as (only) teacher-led activities at the expense of engaging children in activities of their choice, children practice being "teacher-regulated" and are not given an opportunity to develop self-regulated behaviors which affects their ability to later engage in learning behaviors of their own accord.

Our hypothesis was that the dichotomy between explicit instruction and child-centered approach is a false one and that it is possible to design a program that would combine an explicit focus on content with equally explicit focus on promoting self-regulatory behaviors. The use of specific pedagogical strategies that optimize make-believe play has been proven successful in improving young children's self-regulation competencies and academic achievement (e.g., Bodrova & Leong, 2005). Used as a part of a comprehensive preschool curriculum as well as a part of an early literacy intervention, these strategies had been proven successful in improving young children's self-regulation and academic achievement (Barnett, Yarosz, Thomas, & Hornbeck, 2006; Bodrova, 2003; Bodrova & Leong, 2001; Diamond, Barnett, Thomas, & Munro, 2007).

2.11 EMERGE Intervention

Classes were randomly assigned to three conditions. A control group used standard district practices, while the other two groups used either the *Building Blocks* curriculum (BB), or a synthesis of BB and scaffolding designed to promote executive function (SEF), especially through make-believe play. Similar to TRIAD, the two-main PD supports implemented in EMERGE were the sequence of training sessions and in-class coaching.

Teachers in the both groups were provided training in each of the 2 years of their involvement. Both groups received the same training on math education with BB, 2 days of during the first month of school in year 1, 2 days during the school day in the Fall and 2 days during the Spring (with the project paying for substitutes). Although this research-based plan is comprehensive, it was designed to be commensurate with support the program would receive under normal but adequate conditions of adoption. The course content included the following topics: learning trajectories for each math topic; using learning trajectories for observation and other authentic assessment strategies; supporting mathematical development in the classroom; recognizing and supporting math throughout the day; setting up math learning centers; teaching with computers (including use of the management system and research-based teaching strategies); small-group activities; and supporting mathematical development in the home. The main technological tool was the Building Blocks Learning Trajectory (BBLT) web application. BBLT provides scalable access to the learning trajectories via descriptions, videos, and commentaries. Each aspect of the learning trajectories—developmental progressions of children's thinking and connected instruction—are linked to the other (Clements, 2007; Clements & Sarama, 2008; Sarama & Clements, 2013).

The BBSEF group also included additional PD on the SEF component (i.e., they received twice as much PD). The SEF training, delivered by authors of TotM (Bodrova & Leong, 2007) and their colleagues, included an additional 6 days of training each of the 2 years of the teachers' involvement. The SEF training followed the general organizational structure as the BB training and included the topics of development of EF in early childhood, how dramatic play supports EF, and how teachers can scaffold mature and intentional dramatic play. Like BB PD, SEF PD combined building teachers' knowledge of young children's learning and development with helping teachers master effective instructional strategies designed to support this learning and development. The idea of mature dramatic play as a critical component in promoting EF is a new one to most teachers, and so is the idea that mature play may not emerge spontaneously but requires teacher scaffolding; therefore, SEF staff paid special attention to teachers' understanding of these new concepts. Videotapes illustrating various stages in play development as well as best practices of scaffolding play were discussed during training. In addition to SEF-specific training, the BBSEF teachers received training on modified BB instructional strategies redesigned to maximally promote EF. Further, the EF strategies added to the math curriculum combined focal activities implemented outside of the "math block" with instructional strategies embedded in existing math activities.

Coaching was similarly conducted throughout the 2 years. Coaching occurred within each teacher's classroom. Again, coaching was designed to be similar to that used in TRIAD (Clements, 2007, 2011; Clements & Sarama, 2008) except that coaches in EMERGE were from the district, not provided by the project. Coaches worked with teachers throughout the project, visiting teachers in their classrooms, with the initial expectation that they would visit each teacher no less than once per month. Coaches reported to a lead coach as well as mentors (in the second year) after each visit, who report to site coordinators to monitor the facilitation process. This type of real time, classroom-based observation, demonstration, coaching, and mentoring was again viewed as indispensable, and included shared decision making, one-to-one consultation, monitoring, and reinforcement.

2.12 EMERGE Coaches

Like that in TRIAD, this coaching model was intended to be dynamic in which additional attention by both mentors and coaches is given immediately until adequate fidelity of implementation is achieved. Coaching was provided to teachers in all three conditions. The coaches were hired to observe and give feedback on the implementation of BB and the self-regulation curriculum, as well as assist with materials acquisition and technical troubleshooting. Coaches agreed to participate in on-site coaching support for hour long, biweekly sessions, and provide study teachers with feedback using the structure of the classroom needs assessment (CNA). The CNAs were intended to be filled out during the observation period, and sent to the lead coach at the conclusion of each visit. Coaches were also tasked with providing off-site coaching support, or being available to teachers and research coordinators via email, phone, or fax. All three districts employed instructional coaches for mathematics, and the same coaches within each district worked with teachers in up to three of the study conditions.

There was one lead coach who was hired to supervise the coaches and accompany them on classroom visits to check for fidelity throughout the observation and feedback process. The lead coach completed CNAs documenting these visits with the coaches. There were two mentors who were hired to provide study teachers specific feedback on implementation of the BB curriculum. The mentors were active in working with teachers primarily during Spring 2011.

Coaching roles and responsibilities distributed during the first PD session listed the following:

- Sharing teaching strategies or information about early childhood teaching and learning mathematics in early childhood.
- Sharing information with the teacher about program procedures, guidelines, and expectations.
- Sharing information about children's thinking, including learning trajectories for mathematical topics.
- Sharing information about the curriculum components.
- Sharing information about effective use of computers.
- Linking the teacher to appropriate resources.
- Offering support by listening and by sharing your own experiences.
- Giving guidance and ideas about management, scheduling, planning organizing the day, and other topics.
- Assisting the teacher in arranging, organizing, and/or analyzing the physical setting.
- Counseling the teacher when difficulties arise.
- Allowing the teacher to observe you or other colleagues.
- Promoting self-observation and analysis.

While on-site with the teachers, coaches were expected to observe, offer suggestions, receive feedback from the teacher, and help the teachers with any issues they may be having or resources they may need.

As previously stated the coaching model for EMERGE was based on TRIAD (Clements, 2007, 2011; Clements & Sarama, 2008); however, coaches in EMERGE were from the district, not provided by the project. Coaches in EMERGE were primarily former administrators or teachers, and they continued to serve in these rolls part-time while also acting as coaches. Of the 21 coaches, 9 were employed as center administrators; 9 were resource teachers/lead teachers and/or pre-K teachers; 3 of the coaches hired were working as coaches for other entities (Preschool for All or a local School District).

2.13 Coach Training

Coaches in the EMERGE project were required to attend PD sessions throughout the study period as well as a ½ day Coaching Institute before the study. Coaches from each district participated in the same math and SEF PD as the teachers.

2.14 Implementation of Coaching in EMERGE

CNAs were coded over 2 years of EMERGE implementation to evaluate the extent to which coaching occurred as intended. CNAs provide information on the (1) duration and frequency of visits, (2) implementation fidelity, and (3) qualitative notes about the visit (Fig. 4).

2.15 Frequency and Duration

We first examine the frequency and duration of visits by coaches, the lead coach, and mentors as the quality of the teacher–coach relationship has been suggested to be dependent on consistent meetings and timely consultations. EMERGE teachers received varying amounts of coaching. Visits were recorded each time a CNAs was provided. Some teachers had as many as 15 visits from their coach, others as few as 1 visit, with the average being 4. However, it is uncertain if additional visits occurred in the absence of a completed CNA as they were not recorded elsewhere. The majority of the recorded visits was between 31 and 60 min. The lead coach and mentor CNA visits were generally longer than 60 min. Table 1 outlines the duration of visits by rater type.

Further, less than half (47%) of teachers had the same coach during the entire study (Fig. 5). The research on best practices would suggest that the coaching frequency and consistency in the current study may have posed challenges to building effective teacher–coach relationships and the effectiveness of coaches in facilitating changes in the classroom.

2.16 Fidelity Ratings

CNAs recorded ratings for the BB and SEF components based on the fidelity of teacher implementation of the intervention components. Table 2 explains the rating scale used by raters to complete the CNA. Figs. 6–8 display coach, lead coach, and mentor implementation ratings across intervention components. CNAs where raters listed a "0" or left blank are not included in these comparisons as interpretations to their meanings cannot be made; therefore, these figures will not add to 100%. First examining identification of

EMERGE Project Coaches

Classroom Needs Assessment Report

Coach: First Name ______________ Last Name ________________ District (choose one): ☐ SDUSD ☐ PUSD ☐ SBUSD

Teacher: First Name _____________ Last Name ________________ School/Site: ________________________

Date of Visit: ______/______/______ (Month / Day / Year) **Time of visit:** Start ____________ End____________

Teacher is on week _____ of *Building Blocks* curriculum **Classroom has _______ computers working for children's use**

Area		Rating	Comments
Building Blocks Implementation	Large Group		
	Small Group		
	Computer Center		
	Hands On Math Center		
	Reflection Time		
	Assessment / Small Group Record Sheet		
Self-Regulation Implementation			
Questions / Concerns			

*****Rating Key: 3-Good implementation/understanding 2-Work needed 1-Lead Coach/coach intervention**

Instructions: Please fill out the form completely for ***each*** teacher for ***each*** visit. Visits should occur for 1 hour every two weeks. All forms for each month should be mailed (1366 Chisholm Trail, San Marcos, CA 92069), faxed (760-294-6627), or emailed (cindyleso@cox.net) to Cindy Leso by the last day of each month in which the visits occur.

Fig. 4 Classroom needs assessment report.

Table 1 Duration of rater visit

Duration of Visit (min)	Coach CNA (#)	Lead Coach CNA (#)	Mentor CNA (#)
0–30	9	1	7
31–60	71	24	21
61–90	47	14	8
>90	40	14	20

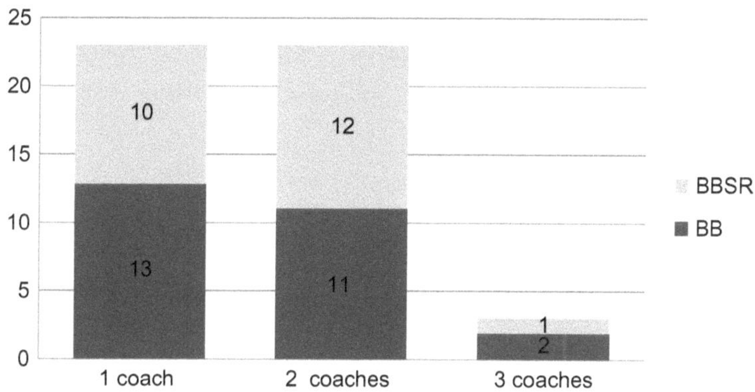

Fig. 5 Number of coaches a teacher had by research group.

Table 2 Classroom Needs Assessment Rating Scale

Rating	Explanation
1	"Lead coach/coach follow-up" was needed to assist the teacher struggling with implementation
2	"Work needed" to assist with implementation
3	"Good implementation/understanding" of the curriculum/a

struggling teachers, Fig. 6 shows that compared to coaches (12%) and the lead coach (7%), mentors most frequently (58%) rated the SEF implementation as needing "Lead Coach/Coach Follow-Up" to assist the teacher struggling with implementation. Recall that the CNAs are a primary source to provide real-time feedback to the lead coach and project directors on implementation issues. Additionally, recall that mentors were not conducting visits until the second year of implementation.

In the second year as mentors were included in observations, these raters were more likely to rate implementation across all intervention elements as

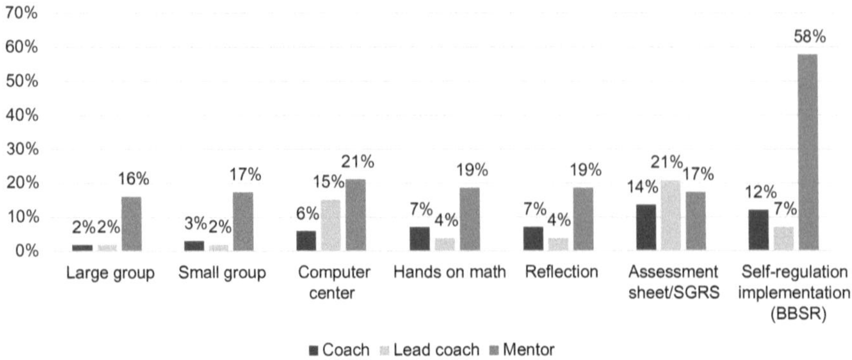

Fig. 6 "Lead coach/coach follow-up" to assist the teacher struggling with implementation.

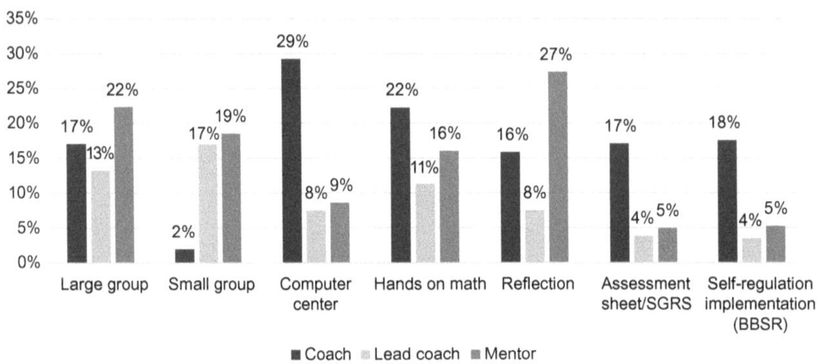

Fig. 7 "Work needed" to assist with implementation.

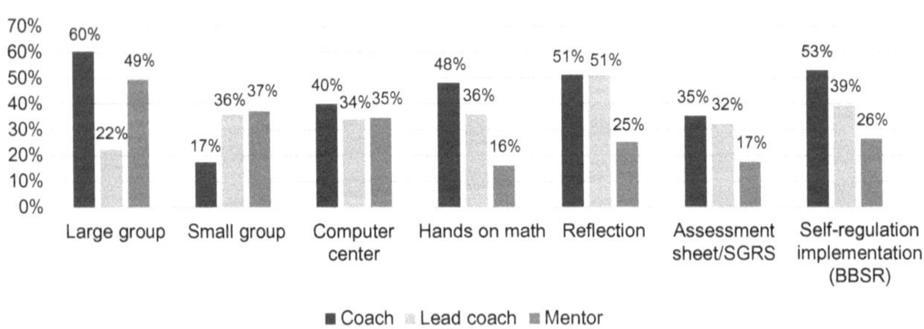

Fig. 8 "Good implementation/understanding" of the curriculum/a.

needing support. Finally, any teacher receiving a rating of 1 should have had a follow-up visit by the lead coach. However, it was found based on the CNA reports that the lead coach followed up with teachers only 20% of the time that a rating of 1 was provided on any component.

There was similar variability for rating 2 which indicated that a teacher needed assistance with implementation (Fig. 7). Coaches were most likely to rate computer centers (29%) as needing "work to assist with implementation." While the lead coach most frequently (17%) rated small group and mentors rated reflection (27%) as needing some assistance.

It is interesting when examining ratings for "good implementation" that coaches are nearly reverse in the SEF component (Fig. 8). On 53% of the CNAs coaches rated the SEF component as good, while the 58% of the mentor CNAs rated those teachers as struggling with implementation.

2.17 Qualitative Observation Notes

The coaches' qualitative notes about the visits were coded based on best practices in coaching.

1. Coaches should have both content and pedagogical knowledge for the domain being coached as well as the ability to cultivate a positive and trusting relationship with teachers (e.g., Campbell & Malkus, 2011; Garet et al., 2008). Coaching is multifaceted in that coaches need to be able to gauge both the complexity of what teachers are being instructed to implement, and the newness of the content in terms of previous knowledge and practice of the teachers (Joyce & Showers, 2002).
2. Coaches should assist teachers with lesson planning, providing resources, modeling, and coteaching. Technique modeling, coteaching, and acting as a resource provider are also essential (Fox et al., 2011; Neuman & Cunningham, 2009). These important, though often overlooked, aspects of the coach–teacher relationship are essential to effective coaching. For one, they help break the misconception that coaches are not (or should not be) an active part of the classroom. In fact, teachers and students see much more substantial gains when coaches move away from simple observation and feedback and move toward collaboration and active involvement (Neuman & Cunningham, 2009).
3. Coaches should also be continually working with teachers to set goals and adjust those goals to meet the teacher's current needs. For instance, as coaches and teachers enter into a professional relationship, curriculum and lesson planning, as well as goal setting, should be made priorities (Neuman & Cunningham, 2009; Skiffington et al., 2011).

4. Coaches should facilitate reflection. Effective coaches provide descriptive, not evaluative feedback using concrete observations from the classroom. Engaging the teacher in collaborative problem solving to improve practice (Guiney, 2001; Harwell-Kee, 1999).

The method was revised three times before the research assistants began coding. The assistants used the first 15 CNAs to test initial reliability. The remaining 290 CNAs were divided into four batches and assigned an assistant to code them and were then double coded. GRA's discussed codes of each batch to obtain 100% reliability. GRA's had to maintain 100% reliability before moving on to the subsequent batches of CNAs. This process was repeated for four batches until all 305 CNAs were coded and double coded with 100% reliability. The four variables on the CNA that were coded are defined below.

- *Number of direct coach interventions*—Rater engaged in a direct intervention during the observation period, including, but not limited to: coteaching, modeling, providing resources, suggesting related to implementation, using reflective questioning with the teacher, assisting the teacher in creating plans, debriefing on classroom activities, or other interactions related to curriculum implementation.
- *Presence of a need identified*—Rater describes that the teacher or classroom needs assistance with something in order to implement the curriculum properly.
- *Type of need identified*—Needs identified by rater were categorized by type of need: computer issues, materials needed, software/website issues information, schedule for computers, help curriculum implementation, classroom management help, assistance with students, planning, and training.
- *Identifying questions/concerns*—Indicates whether the rater used the questions/concerns section on the CNA form.

2.17.1 Direct Coaching Interventions

Effective coaches should be assisting teachers with lesson planning, providing resources, modeling, and coteaching. Coaches should also be continually working with teachers to set goals and adjust those goals to meet the teacher's current needs. The CNAs were coded for these types of direct coach interventions (DCI). DCIs were coded anytime qualitative notes on the CNAs suggested that coaches engaged in a direct intervention during the observation period, including, but not limited to: coteaching, modeling, providing resources, making a suggestion related to implementation, using reflective questioning with the teacher, assisting the teacher in creating plans, debriefing on classroom activities, or other interactions related to curriculum implementation. The number of DCIs reported on an individual CNA ranged from 0 to 14. Overall, data reflect few DCIs were provided to teachers in the current study with 81% of coach CNAs, 83% of lead coach CNAs, and 92% of mentor CNAs indicating 0–2 DCI were provided.

2.17.2 Identified Needs

CNAs were also coded for *Identified needs*. Identified needs were coded anytime qualitative notes on the CNAs noted implementation support was needed, including, but not limited to: computer issues, help with specific curricular components, and classroom management. The majority of needs identified on CNAs were related to issues with technology and computer software, a critical component of the synthesized curriculum teachers were implementing. Coded data indicated 41% of coach CNAs, 28% of lead coach, and 25% of mentor CNAs included identified needs. Of coach CNAs that were coded as having identified needs, 45% had an identified need that was continually identified. Of the lead coach CNAs that were coded as having identified needs, 38% had an identified need that was continually identified. Mentor CNAs did not appear to have identified needs that were continually reported. Nearly 50% of the time the same need was repeatedly identified on a CNA, which may reflect that while coaches and lead coaches were able to identify what teachers needed for successful implementation, they were not as successful at following-up on those needs and ensuring that they were addressed.

Identification of teacher needs was also captured on CNAs in the "Questions/Concerns" section. Lead coaches and mentors used the questions and concerns section of the CNA more frequently than coaches. Specifically, the lead coach included a question or concern 81% of the time, while coaches only included it 41% of the time.

Lastly, we examine the extent to which coaches participated in PD. Recall the only training coaches received on the intervention content was at the same time teachers received it. Fig. 9 displays that coach attendance at PD sessions was poor at best, with only 9% of coaches attending all eight PD sessions. Finally, examining the relation of frequency of rater visits as reported on the CNAs to the date of PD activities, we find that visits peak at or just after these sessions (Fig. 10). This speaks to the critical importance of having coaches involved in the PD sessions to increase engagement and connection with the intervention content and classrooms.

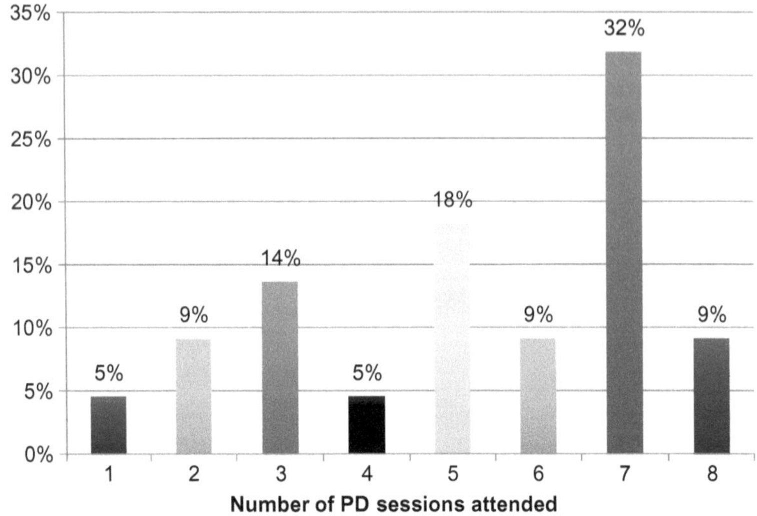

Fig. 9 Number of PD sessions attended by coaches.

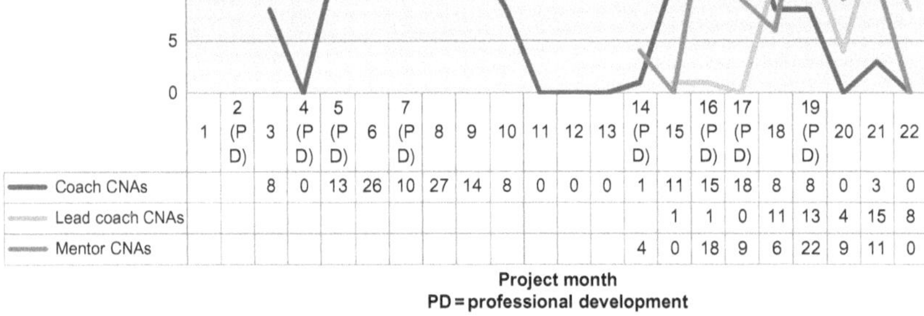

Fig. 10 Rater visits and PD.

2.18 Effectiveness of EMERGE as a PD Model

In summary, the CNA data demonstrate that implementation of coaching in EMERGE was less aligned with the overall TRIAD PD model than was the previously described TRIAD implementation. Moreover, the CNA data provide a glimpse at a coaching instantiation that did not meet all the effectiveness criteria from the greater literature on coaching. Certainly, the coaching component is but one piece of the overall PD model. However, if coaching is not delivered in a manner that meets research-based criteria then we may expect to see less positive results. In fact, Clements et al. (under review) found while the majority of the impact estimates for BB were positive, fewer were statistically significant. At the end of pre-K, only one of the four measures of executive function was significantly higher. At the end of kindergarten, with no additional treatment in kindergarten, two of the four executive function measures were significantly higher for the BB group and one of the two tasks of mathematics confidence was significantly higher for the BB group compared to either the BBSEF group or the control group. For BBSEF, the impact estimates were mixed in sign, small in magnitude and, all statistically insignificant. We conclude that while the effects on student-level measures do not make a strong case for students being positively affected by either intervention condition, they support two hypotheses: First, implementing multiple innovations simultaneously may limit desired outcomes. Second, early gains in both mathematics and EF competence can be mutually supportive and thus resist fade out.

Importantly for this paper, even though teachers achieved adequate fidelity and those measures differed in practically and significantly between groups in hypothesized directions (e.g., the BBSEF group outscored the other two groups on a measure of mature play), their fidelity scores were lower than in previous studies. We are conducting further analyses and suggest that future studies do so, to determine if the less satisfactory coaching can explain both the less adequate fidelity and the lower impact on child outcomes.

3. CONCLUDING THOUGHTS

This coaching implemented as part of the TRIAD and EMERGE projects highlights coaching that both adheres and strays from an intended model, and may serve as microcosm of the state of the practice—promising, but without sufficiently consistent and intense oversight to ensure teacher benefit. To simply say we have coaches or that there is coaching is not enough. Coaching still needs to be implemented with fidelity to a model that has been proven effective. We saw here the importance for coaches to be trained in both the content and process of coaching. Only 9% of EMERGE coaches attended all of the PD sessions, missing valuable content.

In comparison, TRIAD mentors received more training prior to the start of the study and participated in almost all the PD sessions. Neuman and Cunningham (2009) similarly found the importance for literacy coaches to be well versed in the PD content. The math coaches in the EMERGE study may have had less content expertise than the coaches in TRIAD. Perhaps this is somewhat related to the difference in how coaches were acquired for the studies. EMERGE coaches were part-time district employees who continued their role as administrator or teacher while trying to maintain coaching duties. In comparison, TRIAD mentors were not district employees and were identified by the TRIAD project staff. Perhaps this selection aspect as well as the TRIAD mentors not being doubly employed by districts in which they were coaching translated to better fidelity to the coaching model.

Overall, the data presented here suggest the importance of providing training and support for coaches on not just the content of coaching but also processes to achieve effective teacher–coach relationships. The data point to the need for more formal or standardized training for coaches to guarantee that enlisted coaches will follow best practices.

Despite these outlined "ideals" in coaching, the practice parallels teacher preparation in its high variance in training. Coaching, in itself, is often (though not always) implemented by experienced teachers. Regardless of any practical experience gained, little formal or standardized training for coaches exists. In fact, coaches often receive no more than a handful of multiple-day training sessions spread throughout the year *during* the time they are enlisted as coaches (Frost & Bean, 2006). The result has been coaches with a wide range of training, professional experiences, and qualifications. In spite of the strong support effective coaching has garnered, it is difficult to guarantee that enlisted coaches will follow a research-based model.

ACKNOWLEDGMENTS

This research was supported by the Institute of Education Sciences, US Department of Education through Grants R305A120813, R305K05157, and R305A110188. The opinions expressed are those of the authors and do not represent views of the US Department of Education. Although the research is concerned with theoretical issues, not particular curricula, a small component of the intervention used in this research have been published by the authors and their collaborators on the project, who thus could have a vested interest in the results. Researchers from an independent institution oversaw the research design, data collection, and analysis and confirmed findings and procedures. The authors wish to express appreciation to the school districts, teachers, and students who participated in this research.

REFERENCES

Ball, D. L., & Cohen, D. K. (1999). *Instruction, capacity, and improvement*. Philadelphia, PA: Consortium for Policy Research in Education, University of Pennsylvania.

Barlow, A. T., Burroughs, E. A., Harmon, S. E., Sutton, J. T., & Yopp, D. A. (2014). Assessing views of coaching via a video-based tool. *ZDM: Mathematics Education, 46*, 227–238.

Barnett, S. W., Yarosz, D. J., Thomas, J., & Hornbeck, A. (2006). *Educational effectiveness of a Vygotskian approach to preschool education: A randomized trial*. Rutgers, NJ: National Institute of Early Education Research.

Bay-Williams, J. M., & McGatha, M. B. (2014). *Mathematics coaching: Resources and tools for coaches and leaders*, K-12. Boston, MA: Pearson Education.

Berends, M., Kirby, S. N., Naftel, S., & McKelvey, C. (2001). *Implementation and performance in new American schools: Three years into scale-up*. Santa Monica, CA: Rand Education.

Biancarosa, G., & Bryk, A. S. (2011). Effect of literacy collaborative professional development: A summary of findings. *Journal of Reading Recovery, 10*, 25–32.

Bobis, J., Clarke, B. A., Clarke, D. M., Gill, T., Wright, R. J., Young-Loveridge, J. M., & Gould, P. (2005). Supporting teachers in the development of young children's mathematical thinking: Three large scale cases. *Mathematics Education Research Journal, 16*(3), 27–57.

Bodilly, S. J. (1998). *Lessons from new American schools' scale-up phase*. Santa Monica, CA: RAND Education.

Bodrova, E., & Leong, D. J. (2001). *The tools of the mind: A case study of implementing the Vygotskian approach in American early childhood and primary classrooms*. Geneva, Switzerland: International Bureau of Education.

Bodrova, E., & Leong, D. J. (2003). Self-regulation as a key to school readiness: How can early childhood teachers promote this critical competency? In M. Zaslow & I. Martinez-Beck (Eds.), *Critical issues in early childhood professional development* (pp. 203–224). Baltimore, MD: Brookes Publishing.

Bodrova, E., & Leong, D. J. (2005). Self-regulation as a key to school readiness: How can early childhood teachers promote this critical competency? In M. Zaslow & I. Martinez-Beck (Eds.), *Critical issues in early childhood professional development*. Baltimore, MD: Brookes Publishing.

Bodrova, E., & Leong, D. J. (2007). *Tools of the mind: The Vygotskian approach to early childhood education* (2nd ed.). New York, NY: Merrill/Prentice Hall.

Bredekamp, S. (2004). Standards for preschool and kindergarten mathematics education. In D. H. Clements, J. Sarama, & A.-M. DiBiase (Eds.), *Engaging young children in mathematics: Standards for early childhood mathematics education* (pp. 77–82). Mahwah, NJ: Erlbaum.

Bright, G. W., Bowman, A. H., & Vacc, N. N. (1997). Teachers' frameworks for understanding children's mathematical thinking. In E. Pehkonen (Ed.), *Proceedings of the 21st conference of the international group for the psychology of mathematics education: Vol. 2*. (pp. 105–112). Lahti, Finland: University of Helsinki.

Brooks-Gunn, J. (2003). Do you believe in magic? What we can expect from early childhood intervention programs. *Social Policy Report, 17*(1), 3–14. 1.

Buysse, V., Castro, D. C., & Peisner-Feinberg, E. (2010). Effects of a professional development program on classroom practices and outcomes for Latino dual language learners. *Early Childhood Research Quarterly, 25*(2), 194–206.

Campbell, P. F., & Malkus, N. N. (2011). The impact of elementary mathematics coaches on student achievement. *The Elementary School Journal, 111*(3), 430–454.

Carlisle, J. F., & Berebitsky, D. (2010). Literacy coaching as a component of professional development. *Reading and Writing, 24*(7), 773–800.

Carlisle, J. F., Kelcey, B., Rowan, B., & Phelps, G. (2011). Teachers' knowledge about early reading: Effects on students' gains in reading achievement. *Journal of Research on Educational Effectiveness, 4*(4), 289–321.

Carneiro, P., & Heckman, J. J. (2003). Human capital policy. In A. B. Krueger & J. J. Heckman (Eds.), *Inequality in America: What role for human capital policies?* (pp. 77–239). Cambridge, MA: MIT Press.

Carpenter, T. P., & Franke, M. L. (2004). Cognitively guided instruction: Challenging the core of educational practice. In T. K. Glennan, S. J. Bodilly, J. R. Galegher, & K. A. Kerr (Eds.), *Expanding the reach of education reforms: Perspectives from leaders in the scale-up of educational interventions* (pp. 41–80). Santa Monica, CA: RAND Corporation.

Case, R., Griffin, S., & Kelly, W. M. (1999). Socioeconomic gradients in mathematical ability and their responsiveness to intervention during early childhood. In D. P. Keating & C. Hertzman (Eds.), *Developmental health and the wealth of nations* (pp. 125–149). New York, NY: Guilford.

Claessens, A., Duncan, G. J., & Engel, M. (2007). *Kindergarten skills and fifth-grade achievement: Evidence from the ECLS-K.* Evanston, IL: Northwestern University.

Clements, D. H. (2002b). Linking research and curriculum development. In L. D. English (Ed.), *Handbook of international research in mathematics education* (pp. 599–636). Mahwah, NJ: Lawrence Erlbaum Associates.

Clements, D. H. (2007). Curriculum research: Toward a framework for 'research-based curricula'. *Journal for Research in Mathematics Education, 38*, 35–70.

Clements, D. H., & Sarama, J. (2004a). Learning trajectories in mathematics education. *Mathematical Thinking and Learning, 6*, 81–89.

Clements, D. H., & Sarama, J. (2004b). Building blocks for early childhood mathematics. *Early Childhood Research Quarterly, 19*, 181–189.

Clements, D. H., & Sarama, J. (Eds.). (2004c). *Mathematical thinking and learning: 6. Hypothetical learning trajectories* [Special issue]. 2, 81–260.

Clements, D. H., & Sarama, J. (2007). Effects of a preschool mathematics curriculum: Summative research on the *Building Blocks* project. *Journal for Research in Mathematics Education, 38*, 136–163.

Clements, D. H., & Sarama, J. (2008). Experimental evaluation of the effects of a research-based preschool mathematics curriculum. *American Educational Research Journal, 45*(2), 443–494.

Clements, D. H., & Sarama, J. (2009). *Learning and teaching early math: The learning trajectories approach.* New York, NY: Routledge.

Clements, D. H., & Sarama, J. (2011). Early childhood mathematics intervention. *Science, 333*(6045), 968–970. http://dx.doi.org/10.1126/science.1204537.

Clements, D. H., & Sarama, J. (2012). Learning and teaching early and elementary mathematics. In J. S. Carlson & J. R. Levin (Eds.), *Instructional strategies for improving students' learning: Focus on early reading and mathematics* (pp. 107–162). Charlotte, NC: Information Age Publishing.

Clements, D. H., Sarama, J., & DiBiase, A.-M. (2003). Preschool and kindergarten mathematics: A national conference. *Teaching Children Mathematics, 8*, 510–514.

Clements, D. H., Sarama, J., Layzer, C., Unlu, F., Germeroth, C., & Fesler, L. Effects on mathematics and executive function of a mathematics and scaffolded play intervention versus mathematics alone (under review).

Clements, D. H., Sarama, J., Spitler, M. E., Lange, A. A., & Wolfe, C. B. (2011). Mathematics learned by young children in an intervention based on learning trajectories: A large-scale cluster randomized trial. *Journal for Research in Mathematics Education, 42*(2), 127–166.

Clements, D. H., Sarama, J., Wolfe, C. B., & Spitler, M. E. (2013). Longitudinal evaluation of a scale-up model for teaching mathematics with trajectories and technologies: Persistence of effects in the third year. *American Educational Research Journal, 50*(4), 812–850. http://dx.doi.org/10.3102/0002831212469270.

Cohen, D. K. (1996). Rewarding teachers for student performance. In S. H. Fuhrman & J. A. O'Day (Eds.), *Rewards and reforms: Creating educational incentives that work* (pp. 61–112). San Francisco, CA: Jossey Bass.

Copley, J. V. (2004). The early childhood collaborative: A professional development model to communicate and implement the standards. In D. H. Clements, J. Sarama, & A.-M. DiBiase (Eds.), *Engaging young children in mathematics: Standards for early childhood mathematics education* (pp. 401–414). Mahwah, NJ: Erlbaum.

Costa, A., & Garmston, R. (1994). *The art of cognitive coaching: Supervision for intelligent teaching. Training syllabus, equity and choice.* Sacramento, CA: Institute for Intelligent Behavior.

Costa, A. L., & Garmston, R. J. (2002). *Cognitive coaching: A foundation for renaissance schools* (2nd ed.). Norwood, MA: Christopher-Gordon.

Darling-Hammond, L. (1996). Restructuring schools for high performance. In S. H. Fuhrman & J. A. O'Day (Eds.), *Rewards and reform: Creating educational incentives that work* (pp. 144–192). San Francisco: Jossey-Bass.

Darling-Hammond, L. (1997). *The right to learn: A blueprint for creating schools that work.* San Francisco, CA: Jossey-Bass.

Darling-Hammond, L., & Richardson, N. (2009). Teacher learning: What matters. *Educational Leadership, 66*(5), 46–53.

Denton, K., & West, J. (2002). *Children's reading and mathematics achievement in kindergarten and first grade.* Washington, DC: U.S. Department of Education, National Center for Education Statistics. Retrieved from, http://nces.ed.gov/pubsearch/pubsinfo.asp?pubid=2002125.

Diamond, A., Barnett, S. W., Thomas, J., & Munro, S. (2007). Executive function skills CAN be improved in preschoolers by regular classroom teachers. *Science, 318,* 1387–1388.

Doig, B., McCrae, B., & Rowe, K. (2003). *A good start to numeracy: Effective numeracy strategies from research and practice in early childhood.* Canberra, Australia: Australian Council for Educational Research.

Duncan, G. J., & Magnuson, K. (2011). The nature and impact of early achievement skills, attention skills, and behavior problems. In G. J. Duncan & K. Magnuson (Eds.), *Whither opportunity? Rising inequality and the uncertain life chances of low-income children* (pp. 47–70). New York, NY: Russell Sage Press.

Duncan, G. J., & Magnuson, K. A. (2013). Investing in preschool programs. *The Journal of Economic Perspectives, 27*(2), 109–131. http://dx.doi.org/10.1257/jep.27.2.109.

Elmore, R. F. (1996). Getting to scale with good educational practices. *Harvard Educational Review, 66,* 1–25.

Erchick, D. B., Brosnan, P., Forrest, D. B., Douglass, L., Grant, M., & Hughes, K. (2007). Findings from the first year of a K-6 Mathematics Coaching Project. *Paper presented at the National Council of Teachers of Mathematics,* Atlanta, GA.

Fixsen, D. L., Naoom, S. F., Blase, K. A., Friedman, R. M., & Wallace, F. (2005). *Implementation research: A synthesis of the literature (FMHI publication no. 231).* Tampa, FL: University of South Florida, Louis de la Parte Florida Mental Health Institute, the National Implementation Research Network.

Fox, L., Hemmeter, M. L., Snyder, P., Binder, D. P., & Clarke, S. (2011). Coaching early childhood special educators to implement a comprehensive model for promoting young children's social competence. *Topics in Early Childhood Special Education, 31*(3), 178–192. http://dx.doi.org/10.1177/0271121411404440.

Frost, S., & Bean, R. (2006). Qualifications for literacy coaches: Achieving the gold standard. *Literacy Coaching Clearinghouse*. Retrieved from www.literacycoachingonline.org.

Fullan, M. G. (1992). *Successful school improvement*. Philadelphia: Open University Press.

Fullan, M. G. (2000). The return of large-scale reform. *Journal of Educational Change*, *1*, 5–28.

Garet, M. S., Cronen, S., Eaton, M., Kurki, A., Ludwig, M., Jones, W., et al. (2008). *The impact of two professional development interventions on early reading instruction and achievement (NCEE 2008–4030)*. Washington, DC: National Center for Education Evaluation and Regional Assistance, Institute of Education Sciences, U.S. Department of Education.

Gersten, R., Dimino, J., Jayanthi, M., Kim, J. S., & Santoro, L. E. (2010). Teacher study group: Impact of the professional development model on reading instruction and student outcomes in first grade classrooms. *American Educational Research Journal*, *47*(3), 694–739.

Griffin, S., & Case, R. (1997). Re-thinking the primary school math curriculum: An approach based on cognitive science. *Issues in Education*, *3*, 1–49.

Griffin, S., Case, R., & Siegler, R. S. (1994). Rightstart: Providing the central conceptual prerequisites for first formal learning of arithmetic to students at risk for school failure. In K. McGilly (Ed.), *Classroom lessons: Integrating cognitive theory and classroom practice* (pp. 25–49). Cambridge, MA: MIT Press.

Guiney, E. (2001). Coaching isn't just for athletes: The role of teacher leaders. *Phi Delta Kappan*, *82*(10), 740–743.

Gupta, S. S., & Daniels, J. (2012). Coaching and professional development in early childhood classrooms: Current practices and recommendations for the future. *NHSA Dialog*, *15*(2), 206–220.

Guskey, T. R. (Ed.), (2000). *Evaluating professional development*. Thousand Oaks, CA: Corwin Press

Harwell-Kee, K. (1999). Coaching. *Journal of Staff Development*, *20*(3), 28–29.

Heck, D. J., Weiss, I. R., Boyd, S., & Howard, M. (2002). *Lessons learned about planning and implementing statewide systemic initiatives in mathematics and science education*. Retrieved from, http://www.horizon-research.com/presentations/2002/ssi_aera2002.pdf.

Hemmeter, M. L., Snyder, P., Kinder, K., & Artman, K. (2011). Impact of performance feedback delivered via electronic mail on preschool teachers' use of descriptive praise. *Early Childhood Research Quarterly*, *26*(1), 96–109.

Hiebert, J. C. (1999). Relationships between research and the NCTM standards. *Journal for Research in Mathematics Education*, *30*, 3–19.

Hsieh, W.-Y., Hemmeter, M. L., McCollum, J. A., & Ostrosky, M. M. (2009). Using coaching to increase preschool teachers' use of emergent literacy teaching strategies. *Early Childhood Research Quarterly*, *24*, 229–247.

Huberman, M. (1992). Critical introduction. In M. G. Fullan (Ed.), *Successful school improvement* (pp. 1–20). Philadelphia: Open University Press.

Hull, T. H., Balka, D. S., & Miles, R. H. (2009). *A guide to mathematics coaching: Processes for increasing student achievement*. Thousand Oaks, CA: Corwin, A Sage Publication.

Hyson, M., & Woods, T. (2014). Practices, knowledge, and beliefs about professional development. In H. P. Ginsburg, M. Hyson, & T. A. Woods (Eds.), *Preparing early childhood educators to teach math: Professional development that works*. Baltimore, MD: Brookes, pp. 29–52.

Institute of Medicine and National Research Council. (2015). *Transforming the workforce for children birth through age 8: A unifying foundation*. Washington, DC: The National Academies Press. *https://doi.org/10.17226/19401*.

Isner, T., Tout, K., Zaslow, M., Soli, M., Quinn, K., Rothenberg, L., et al. (2011). *Coaching in early care and education programs and quality rating and improvement systems (QRIS): Identifying promising features*. Washington, DC: Child Trends, Inc.

Jacobson, S. L., & Battaglia, C. F. (2001). Authentic forms of teacher assessment and staff development in the US. In D. Middlewood & C. Cardno (Eds.), *Managing teacher appraisal and performance a comparative approach* (pp. 75–89). London: Routledge Falmer.

Jacobson, S. L., Emihovich, C., Petrie, H., Helfrich, J., & Stevenson, R. (1998). *Transforming schools and schools of education: A new vision for preparing educators*. Newbury Park, CA: Corwin.

Joyce, B., & Showers, B. (2002). *Student achievement through staff development*. Alexandria, VA: Association for Supervision and Curriculum Development.

Kaser, J. S., Bourexis, P. S., Loucks-Horsley, S., & Raizen, S. A. (1999). *Enhancing program quality in science and mathematics*. Thousand Oaks, CA: Corwin.

Klein, A., Starkey, P., Clements, D. H., Sarama, J., & Iyer, R. (2008). Effects of a prekindergarten mathematics intervention: A randomized experiment. *Journal of Research on Educational Effectiveness*, *1*, 155–178.

Klingner, J. K., Ahwee, S., Pilonieta, P., & Menendez, R. (2003). Barriers and facilitators in scaling up research-based practices. *Exceptional Children*, *69*, 411–429.

Knight, J. (2007). *Instructional coaching: A partnership approach to improving instruction*. Thousand Oaks, CA: Corwin Press.

Kretlow, A. G., Wood, C. L., & Cooke, N. L. (2011). Using in-service and coaching to increase kindergarten teachers' accurate delivery of group instructional units. *The Journal of Special Education*, *44*(4), 234–246.

Lawless, K. A., & Pellegrino, J. W. (2007). Professional development in integrating technology into teaching and learning knowns, unknowns, and ways to pursue better questions and answers. *Review of Educational Research*, 77(4), 575–614.

Munby, H., Russell, T., & Martin, A. K. (2001). Teachers' knowledge and how it develops. In V. Richardson (Ed.), *Handbook of research on teaching*. (4th ed., pp. 877–905). Washington, DC: American Educational Research Association.

Mohrman, S. A., & Lawler, E. E., III. (1996). Motivation for school reform. In S. H. Fuhrman & J. A. O'Day (Eds.), *Rewards and reform: Creating educational incentives that work* (pp. 115–143). San Francisco: Jossey-Bass.

National Mathematics Advisory Panel. (2008). *Foundations for success: The final report of the National Mathematics Advisory Panel*. Washington, DC: U.S. Department of Education, Office of Planning, Evaluation, and Policy Development.

National Research Council. (2001). *Eager to learn: Educating our preschoolers*. Washington, DC: National Academy Press.

National Research Council. (2009). *Mathematics learning in early childhood: Paths toward excellence and equity*. Washington, DC: National Academies Press.

Natriello, G., McDill, E. L., & Pallas, A. M. (1990). *Schooling disadvantaged children: Racing against catastrophe*. New York, NY: Teachers College Press.

Nettles, S. M. (1993). Coaching in community settings. *Equity and Choice*, *9*(2), 35–37.

Neuman, S. B., & Cunningham, L. (2009). The impact of professional development and coaching on early language and literacy instructional practices. *American Educational Research Journal*, *46*(2), 532–566.

Obara, S., & Sloan, M. (2009). The evolving role of a mathematics coach during the implementation of performance standards. *The Professional Educator*, *33*(2), 11–23.

Pellegrino, J. W. (2007). From early reading to high school mathematics: Matching case studies of four educational innovations against principles for effective scale up. In B. Schneider & S.-K. McDonald (Eds.), *Scale up in practice* (pp. 131–139). Lanham, MD: Rowan & Littlefield.

Pianta, R., Mashburn, A., Downer, J., Hamre, B., & Justice, L. (2008). Effects of web-mediated professional development resources on teacher-child interactions in pre-kindergarten classrooms. *Early Childhood Research Quarterly*, *23*(4), 431–451.

Piasta, S. B., Logan, J. A. R., Pelatti, C. Y., Capps, J. L., & Petrill, S. A. (2015). Professional development for early childhood educators: Efforts to improve math and science learning opportunities in early childhood classrooms. *Journal of Educational Psychology, 107*(2), 407–422.

Poglinco, S., Bach, A., Hovde, K., Rosenblum, S., Saunders, M., & Supovitz, J. (2003). *The heart of the matter: The coaching model in America's choice schools*. Consortium for Policy Research in Education. Retrieved from http://repository.upenn.edu/cpre_researchreports/35.

Powell, D. R., & Diamond, K. E. (2011). Improving the outcomes of coaching-based professional development interventions. In S. B. Neuman & D. K. Dickinson (Eds.), *Vol. 3. Handbook of early literacy research* (pp. 295–307). New York: Guilford Press.

Powell, D. R., Diamond, K. E., Burchinal, M. R., & Koehler, M. J. (2010). Effects of an early literacy professional development intervention on Head start teachers and children. *Journal of Educational Psychology, 102*(2), 299–312.

Raudenbush, S. W. (2009). The Brown legacy and the O'Connor challenge: Transforming schools in the images of children's potential. *Educational Researcher, 38*(3), 169–180.

Richardson, V., & Placier, P. (2001). Teacher change. In V. Richardson (Ed.), *Handbook of research on teaching*. (4th ed., pp xiii, 1278 p). Washington, DC: American Educational Research Association.

Rudd, L. C., Lambert, M. C., Satterwhite, M., & Smith, C. H. (2009). Professional development + coaching = enhanced teaching: Increasing usage of math mediated language in preschool classrooms. *Early Childhood Education Journal, 37*(1), 63–69.

Sarama, J. (2002). Listening to teachers: Planning for professional development. *Teaching Children Mathematics, 9*, 36–39.

Sarama, J., & Clements, D. H. (2009). *Early childhood mathematics education research: Learning trajectories for young children*. New York, NY: Routledge.

Sarama, J., & Clements, D. H. (2013). Lessons learned in the implementation of the TRIAD scale-up model: Teaching early mathematics with trajectories and technologies. In T. G. Halle, A. J. Metz, & I. Martinez-Beck (Eds.), *Applying implementation science in early childhood programs and systems*. Baltimore, MD: Brookes, pp. 173–191.

Sarama, J., Clements, D. H., & Henry, J. J. (1998). Network of influences in an implementation of a mathematics curriculum innovation. *International Journal of Computers for Mathematical Learning, 3*, 113–148.

Sarama, J., Clements, D., Starkey, P., Klein, A., & Wakely, A. (2008). Scaling up the implementation of a pre-kindergarten mathematics curriculum: Teaching for understanding with trajectories and technologies. *Journal of Research on Educational Effectiveness, 1*(2), 89–119.

Sarama, J., Clements, D. H., Wolfe, C. B., & Spitler, M. E. (2012). Longitudinal evaluation of a scale-up model for teaching mathematics with trajectories and technologies. *Journal of Research on Educational Effectiveness, 5*(2), 105–135.

Sarama, J., Clements, D. H., Wolfe, C. B., & Spitler, M. E. (2016). Professional development in early mathematics: Effects of an intervention based on learning trajectories on teachers' practices. *Nordic Studies in Mathematics Education, 21*(4), 29–55.

Sarama, J., & DiBiase, A.-M. (2004). The professional development challenge in preschool mathematics. In D. H. Clements, J. Sarama, & A.-M. DiBiase (Eds.), *Engaging young children in mathematics: Standards for early childhood mathematics education* (pp. 415–446). Mahwah, NJ: Erlbaum.

Schoen, H. L., Cebulla, K. J., Finn, K. F., & Fi, C. (2003). Teacher variables that relate to student achievement when using a standards-based curriculum. *Journal for Research in Mathematics Education, 34*(3), 228–259.

Siegler, R. S. (1993). Adaptive and non-adaptive characteristics of low income children's strategy use. In L. A. Penner, G. M. Batsche, H. M. Knoff, & D. L. Nelson (Eds.),

Contributions of psychology to science and mathematics education (pp. 341–366). Washington, DC: American Psychological Association.

Sheridan, S. M., Edwards, C. P., Marvin, C. A., & Knoche, L. L. (2009). Professional development in early childhood programs: Process issues and research needs. *Early Education & Development, 20*(3), 377–401.

Shidler, L. (2009). The impact of time spent coaching for teacher efficacy on student achievement. *Early Childhood Education Journal, 36*(5), 453–460.

Showers, B., Joyce, B., & Bennett, B. (1987). Synthesis of research on staff development: A framework for future study and a state-of-the-art analysis. *Educational Leadership, 45*(3), 77–87.

Skiffington, S., Washburn, S., & Elliott, K. (2011). Instructional coaching: Helping preschool teachers reach their full potential. *YC Young Children, 66*(3), 12–19.

Sowder, J. T. (2007). The mathematical education and development of teachers. In F. K. Lester, Jr., (Ed.), *Vol. 1. Second handbook of research on mathematics teaching and learning* (pp. 157–223). New York, NY: Information Age Publishing.

Starkey, P., Klein, A., & Wakeley, A. (2004). Enhancing young children's mathematical knowledge through a pre-kindergarten mathematics intervention. *Early Childhood Research Quarterly, 19*, 99–120.

Stevenson, H. W., & Newman, R. S. (1986). Long-term prediction of achievement attitudes in mathematics and reading. *Child Development, 57*, 646–659.

Stigler, J. W., Lee, S.-Y., & Stevenson, H. W. (1990). *Mathematical knowledge of Japanese, Chinese, and American elementary school children*. Reston, VA: National Council of Teaching of Mathematics.

Thomson, S., Rowe, K., Underwood, C., & Peck, R. (2005). *Numeracy in the early years: Project good start*. Camberwell, Victoria, Australia: Australian Council for Educational Research.

Tyack, D., & Tobin, W. (1992). The "grammar" of schooling: Why has it been so hard to change? *American Educational Research Journal, 31*, 453–479.

Weiss, I. R. (2002). In *Systemic reform in mathematics education: What have we learned? Paper presented at the meeting of the research precession of the 80th annual meeting of the National Council of Teachers of Mathematics Las Vegas, NV.*

West, L., & Staub, F. (2003). *Content-focused coaching: Transforming mathematics lessons*. Portsmouth, NH: Heinemann.

Whitebook, M., & Bellm, D. (2013). *Supporting teachers as learners: A guide for mentors and coaches in early care and education*. Washington, DC: American Federation of Teachers.

Wright, R. J., Martland, J., Stafford, A. K., & Stanger, G. (2002). *Teaching number: Advancing children's skills and strategies*. London, UK: Paul Chapman/Sage.

Yuzawa, M., Bart, W. M., Kinne, L. J., Sukemune, S., & Kataoka, M. (1999). The effects of "origami" practice on size comparison strategy among young Japanese and American children. *Journal of Research in Childhood Education, 13*(2), 133–143.

Zaslow, M. (2014). General features of effective professional development. In H. P. Ginsburg, M. Hyson, & T. A. Woods (Eds.), *Preparing early childhood educators to teach math: Professional development that works*. Baltimore, MD: Brookes, pp. 97–115.

CHAPTER FIVE

Notice, Explore, and Talk About Mathematics: Making a Positive Difference for Preschool Children, Families, and Educators in Australian Communities That Experience Multiple Disadvantages

Ann Gervasoni[*,†,1], **Bob Perry**[*,†]
[*]Monash University, Melbourne, Vic, Australia
[†]Charles Sturt University, Albury, NSW, Australia
[1]Corresponding author: e-mail address: ann.gervasoni@monash.edu

Contents

Advances in Child Development and Behavior, Volume 53
ISSN 0065-2407
http://dx.doi.org/10.1016/bs.acdb.2017.04.001

Abstract

Let's Count is a preschool mathematics intervention implemented by The Smith Family from 2012 to the present in "disadvantaged" communities across Australia. It is based on current mathematics and early childhood education research and aligns with the Australian Early Years Learning Framework. *Let's Count* has been shown to be effective in enhancing mathematics learning and dispositions of young children, early childhood educators, and families through a longitudinal evaluation undertaken from 2012 to 2015. In this chapter, the authors explore the development, implementation, and evaluation of *Let's Count* and highlight the importance of adults noticing, exploring, and talking about children's mathematics. The findings from the longitudinal evaluation of *Let's Count* suggest that when adults notice children's mathematics, then children's learning thrives, and the positive dispositions and confidence of parents and educators increase. *Let's Count* has made a positive difference for many children and adults across Australia.

1. INTRODUCTION

Numerous international interventions have aimed at improving the educational fortunes of young children who live in communities experiencing multiple disadvantages. However, there is some debate about whether early interventions are able to overcome any educational disadvantage associated with young children living in financial and social disadvantage. Sarama and Clements (2015) question the longer term efficacy of such interventions and "hypothesise that most present educational contexts are unintentionally and perversely aligned against early interventions" (p. 153). Although it appears that early interventions in mathematics can be highly successful in promoting the mathematics learning of young children, at least in the short term, it seems that current systems of schooling can mitigate against the positive effects.

Let's Count is an Australian early intervention initiative that works toward a whole-community approach to enabling young children's mathematics learning to flourish (The Smith Family, 2015a). In this chapter, the authors present the theoretical underpinnings and development of *Let's Count*, the key findings from the longitudinal evaluation of the program, and propose a strategy to maintain the positive impact of *Let's Count* throughout children's transition to school.

2. IMPORTANCE OF YOUNG CHILDREN'S EARLY MATHEMATICS LEARNING

Young children encounter and learn a great deal about mathematics through their play and activities at home, in child care centers, and within their communities (Ginsburg, Lee, & Boyd, 2008; Hunting et al., 2012; Moss, Bruce, & Bobis, 2016; Perry, MacDonald, & Gervasoni, 2015). Indeed, before beginning school, many children, regardless of background, know much of the mathematics that they will be "taught" in the first year of school (Gervasoni, Perry, & Parish, 2015; Gould, 2012; Peter-Koop & Kollhoff, 2015; Sarama & Clements, 2015). For example, Gervasoni et al. (2015) compared 5-year-old children's number, geometry, and measurement knowledge at the end of preschool with the outcomes for the first year of school in the *Australian Curriculum: Mathematics* (Australian Curriculum, Assessment and Reporting Authority, ACARA, 2013) and found that many children met these curriculum expectations before they began school. This highlights that in order for many children's mathematics learning to flourish when they begin school, the initial school curriculum may need to be adjusted. Children's early mathematics knowledge is associated with their later achievement in primary school mathematics and literacy (Carmichael, MacDonald, & McFarland-Piazza, 2013; Duncan et al., 2007; Geary, Hoard, Nugent, & Bailey, 2013), so there is considerable community interest in ensuring that children's early mathematics learning flourishes. A positive disposition to learning mathematics (Carr, 2001) is a key factor influencing the ongoing success of children's mathematics learning. Dispositions such as curiosity, persistence, imagination, and inquisitiveness are important for mathematics learning (Carr, 2001; Ginsburg et al., 2008; Perry, Dockett, & Harley, 2007).

Children living in communities that are described by governments as "experiencing multiple disadvantages" are not expected, on average, to perform as well academically as children from more "advantaged" communities (Caro, 2009). This expectation extends to preschool children (Carmichael et al., 2013; Rimm-Kaufman, Pianta, Cox, & Bradley, 2003). Indeed, Carmichael et al. (2013, p. 16) found that "the socioeconomic status of the community in which the family resides was the strongest home microsystem predictor of numeracy performance, explaining 10.5% of the variance in the home-community microsystem model." This finding highlights a significant social justice issue related to children's future educational success and opportunities. Enabling all children to thrive mathematically, with a focus on including children living in communities characterized by multiple disadvantages, is the reason for The Smith Family commissioning the *Let's Count* initiative (The Smith Family, 2016).

2.1 Early Childhood and Learning

Children's experiences in their early childhood years are intrinsic to their health, development, education, and social wellbeing throughout life

(McCain & Mustard, 1999; Ochiltree & Moore, 2001). McCain and Mustard (1999) argue that the years from conception to age 6 have the most influence of any time in the life cycle on brain development and subsequent learning, behavior, and health. Such findings underpin the current interest in governments investing in the early childhood years to improve the health, educational achievement, and the social–emotional development of children. Further, early intervention aimed at the needs of the child and the family produces improved outcomes for those at greatest risk (Gervasoni et al., 2010; Peter-Koop & Kollhoff, 2015; Sarama & Clements, 2015; Shonkoff & Phillips, 2000).

There are many perspectives for describing the phenomenon of learning in early childhood. Currently, early childhood education is influenced strongly by sociocultural theories that position children as active and interactive participants in a variety of social and cultural learning contexts (Rogoff, 2003; Vygotsky, 1978). Developmental psychologists argue those children's responses to the family, the neighborhood, and their culture hinge significantly on genetically based ways of feeling, interpreting, and responding to environmental events. They argue that children respond to the environment in their own individual way, and parents, teachers, and other professionals interacting with children must take into account the individuality of each child to enhance their development appropriately. Recent research suggests that early intervention tailored to the characteristics of the child can shift the odds toward better developmental pathways in mathematics learning (Bailey, 2014) and that this is likely to occur in quality early childhood education and care programs (Australian Institute of Health and Welfare, 2015).

Neuroscience research has affirmed the importance of the early years for children's development in all domains (Doherty, 1997; McCain & Mustard, 1999; Ochiltree & Moore, 2001). For example, Ochiltree and Moore (2001) explain that children develop in several interacting domains, including the psychological, cognitive, and social-emotional. The rate of development is influenced by the social ecology into which children are born and is not simply dependent on the child maturing over time. Indeed, learning involves a complex interaction of many factors including those existing within the child, within the family, within the local community, and within the culture into which children are born and reared (Brofenbrenner, 1979; Ochiltree & Moore, 2001). Early interventions need to respond to the critical influences of a child's family, local community, and environment to be most effective.

2.2 The Role of Families in Young Children's Learning

Families play a critical role in creating the learning opportunities experienced by their children.

> *Most children grow up in families and, in their early years especially, families have a major impact on children's lives, influencing their developing patterns of thinking, feeling, and behavior. Families also have a critical role in instilling fundamental values in children, which in the vast majority of cases enables them to become ethical and law-abiding citizens. Although children may spend a lot of their time in contexts other than their family, the emotional intensity of families, and the durability of kinship ties mean that learning in this context is likely to be extensive and long lasting.*
>
> ***Bowes, Watson, and Pearson (2009), p. 91***

Put another way,

> *... children's learning is situated in the social and cultural contexts of their families and communities. As children observe family and community members, participate in daily events and engage in collaborative experiences, they learn about the processes, concepts, and practices that are valued within their community*
>
> ***Arthur, Beecher, Death, Dockett, and Farmer (2015), p. 37.***

In their extensive review, Henderson and Mapp (2002) note "a positive and convincing relationship between family involvement and benefits for students, including improved academic achievement. This relationship holds across families of all economic, racial/ethnic, and educational backgrounds and for students of all ages" (p. 64).

Children's experiences within their families influence their learning and their dispositions to learning. Resources provided to the child by the family (Kiernan & Mensah, 2011), home routines, and environment strongly predict educational and behavioral outcomes for children well into the primary years (Melhuish et al., 2008). For example, living in poverty can restrict families' abilities to provide materials and associated learning opportunities for their children (Magnuson & Shager, 2010). However, there are many examples of children living in poverty who have achieved positive educational outcomes. Often, this is due to particularly strong, positive approaches from family members (Kiernan & Mensah, 2011; Melhuish et al., 2008). It is not only the availability of physical resources in the family that makes a difference. Rather, it is the nature of the interactions among family members and wider social contacts that can make a major contribution to children's achievements.

> *Parenting in families with children 'succeeding against the odds' of disadvantage show that they encourage and facilitate academic success more directly through proximal learning processes they choose to offer their children, but also more indirectly through the opportunities they create for their children to engage in learning processes with others, and through the example they set their children through their own behavior and modeling.*
>
> ***Siraj-Blatchford et al. (2013), p. 17***

Despite the clear importance of family involvement in their children's education, there is evidence that many families—notably those described as having "disadvantaged" backgrounds—do not actively participate (Christenson, 2004; Miedel & Reynolds, 1999). Many reasons have been proposed to explain this lack of involvement, including both structural and psychological barriers (Brooker, 2003; Christenson, 2004). However, "the reality is that disadvantaged families are least likely to be involved and listened to in school environments" (Bernard Van Leer Foundation, 2007, p. 4). This is a significant issue and an impediment to strengthening the role of parents as the first mathematics educators of children (Gervasoni, Phillipson, & Sullivan, 2017).

Although the development of positive, collaborative, and productive partnerships with families can be quite time-consuming for early childhood educators, these partnerships are of great importance for children's learning and need to be nurtured.

2.3 Helping Families to Help Their Young Children Learn Mathematics

Family interactions have always provided the basis for young children's learning, but not all families are confident in their role as their children's first mathematics educators. Dockett et al. (2011) identified that initiatives that have been successful in engaging families with complex support needs "have built upon existing family strengths, avoided judgmental approaches, recognised families as experts in their own experiences and needs, and promoted positive relationships as the core to effective parenting" (p. 7). This finding highlights that new family-based interventions are best to apply a strengths-based approach, "a way of working with people, based on social justice values, that recognises people's and communities' strengths and facilitates their application to achieve self-determined goals" (Beilharz, 2002, p. 4).

Positive relationships among family members, children, and educators can have a substantial influence on learning. Many studies have shown a positive relationship between parents' involvement in their children's learning environments and their children's learning (Civil, Bratton, & Quintos, 2005; Young-Loveridge, Peters, & Carr, 1998). Some family members may be reluctant to engage with mathematics because of their perceived negative history with the subject. In such cases, interventions can encourage early childhood educators to provoke such engagements and assist families to notice the mathematical potential of their everyday activities with their children. Such provocations first require a strong positive relationship between the family member and the early childhood educator.

Clarke and Robbins (2004) conducted a study in which Australian families were given disposable cameras and asked to record literacy and numeracy activities that took place within the family. The families were given the opportunity to share their photographs with other families and to discuss their literacy and numeracy practices. There were many rich representations of what the families saw as numeracy practices in their family activities and routines. The researchers concluded:

> *this study demonstrated the extensive numeracy enactments occurring within many families in lower socioeconomic circumstances. The study highlighted not only the rich constructions of mathematics that individual children were achieving (personal focus of analysis), but also the important interpersonal relationships that were actively supporting and extending these constructions (interpersonal focus of analysis). Further, it evidenced the value placed by families within the study on supporting their children's learning in numeracy (community focus of analysis).*
>
> ***Clarke and Robbins (2004), pp. 180–181***

In contrast, other researchers have reported parent–income-related differences in the mathematics that parents notice and explore with their children. For example, a frequent family activity is the reading of picture storybooks, in which mathematics can always be found. One USA study (Vandermaas-Peeler, Nelson, Bumpass, & Sassine, 2009) explored differences between the ways in which high-income and low-income families drew mathematics out of their storybook reading. They found two types of mathematics interactions of interest: sociocultural mathematical exchanges that involved explaining the use and value of money or numbers in everyday activities such as shopping or cooking; and mathematical exchanges that included counting, quantity, or size comparisons. The results showed that there were no differences in the initiation of sociocultural numeracy exchanges between the two groups, but that parents with high incomes engaged in more *mathematical exchanges* during both storybook reading and play than did parents with low incomes (Vandermaas-Peeler et al., 2009). This reported difference in the focus of mathematical interactions within families is important to consider when designing new family-based interventions, but the contrast with the findings of Clarke and Robbins (2004) requires some careful deliberation. Clarke and Robbins did not classify mathematical exchanges in the same way as Vandermaas-Peeler et al., but provided strong evidence for the extensive numeracy enactments occurring within many (but not all) families living in lower socioeconomic circumstances. We hypothesize that the invitation for parents to take photographs of family numeracy activities and then discuss the mathematics shown in the photos with other families is important and provides parents with important opportunities for reflection across a broad set of experiences in ways that build their awareness of the mathematics involved in everyday experiences and the importance of discussing this mathematics with their young children. Two further questions are whether the mathematical interactions reported by Clarke and Robbins were more often sociocultural numeracy exchanges rather than mathematical exchanges, and whether or not this matters for children's mathematics learning. Regardless, it is important for family-based interventions to acknowledge parents' strengths as children's first mathematics educators, and also deepen all parents' awareness of the range of mathematics that can be noticed, explored, and talked about when families read stories together, and engage in everyday activities.

3. NOTICING, EXPLORING, AND TALKING ABOUT MATHEMATICS

Family engagement is a key pillar of the *Let's Count* initiative, based on the premise that a child's experiences within the family are critical for development. This includes a focus on deepening all parents' awareness of their influence on their young children's mathematics learning, even though some may feel reluctant to participate in explicitly mathematical activities due to previous negative experiences. Thus, *Let's Count* enlists the help of educators and the partnerships that develop between parents and early years educators to *nudge* parents toward noticing the mathematics in everyday contexts, exploring this mathematics with their young children, and talking about it with them. The key principle on which *Let's Count* is based is that when parents and educators engage in noticing, exploring, and talking about mathematics with young children in everyday contexts, then children's mathematics learning and dispositions for learning will flourish

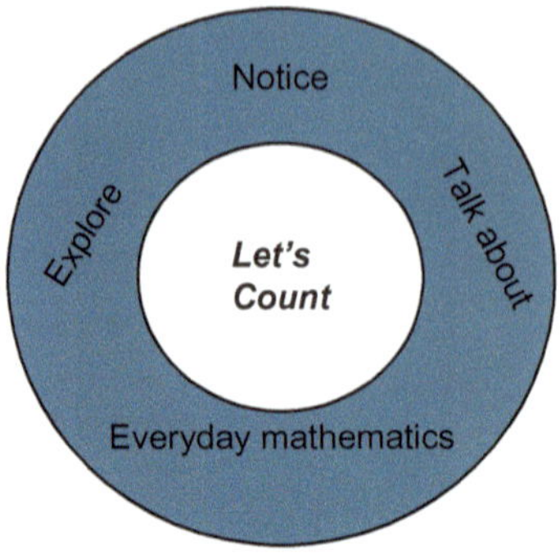

Fig. 1 Representation of the *Let's Count* mantra—notice, explore, and talk about mathematics in everyday contexts.

(Gervasoni & Perry, 2015b). This principle underpins the *Let's Count* mantra depicted in Fig. 1.

This mantra is not a revolutionary idea in mathematics education. Indeed, the study of how educators *recognize* what is happening in children's mathematical learning has been extensive (Breen, McCluskey, Meehan, O'Donovan, & O'Shea, 2014; Jacobs, Lamb, & Philipp, 2010; Mason, 2002; Thomas et al., 2014/2015; van Es & Sherin, 2008; Wager, 2014). Such recognition is sometimes described as "noticing" children's mathematics.

Most of the emphasis in this area has been on teachers noticing children's mathematical learning and various analytical frameworks have been suggested. For example, Mason (2011) suggests that

> *The discipline of noticing is a collection of techniques for (a)* preparing *to notice in the moment, that is, to have come to mind appropriately, and (b)* post-paring *by reflecting on the recent past to select what you want to notice or be sensitized to particularly, in order to* pare, *that is, to notice in the moment and so be enabled to act freshly rather than habitually (pp. 37–38).*

A *Learning to Notice Framework* has been developed by van Es and Sherin (2008, p. 245) consisting of three main components: "(a) identifying what is important in a teaching situation; (b) using what one knows about the context to reason about a situation; and (c) making connections between specific events and broader principles of teaching and learning." In turn, Jacobs et al. (2010, p. 172) have extended this framework to develop the notion of "professional noticing of children's mathematical thinking" with "three related skills: attending to children's strategies, interpreting children's understandings, and deciding how to respond on the basis of children's understandings." More recently, Wager (2014) applied the Jacobs et al. framework to examine ways that noticing mathematical thinking is related to equitable pedagogy.

Our development of the *Let's Count* mantra: notice, explore, and talk about mathematics in everyday contexts is based on the earlier frameworks but uses language that is both rigorous and understandable for early childhood educators, families of young children, and the young children themselves (Gervasoni & Perry, 2015a). "Noticing" is both reflective and metacognitive for all involved with known impacts on the quality of young children's learning (Ginsburg et al., 2008; Whitebread & Coltman, 2010). Notice, explore, and talk about mathematics in everyday contexts provided the framework for developing the *Let's Count* initiative.

3.1 *Let's Count*

Let's Count is an early mathematics intervention that promotes positive mathematical experiences for young children (3–5 years) living in socially and financially disadvantaged communities through building partnerships between parents and early years educators and child care professionals. The program was commissioned by The Smith Family, an Australian children's charity, "helping disadvantaged Australian children to get the most out of their education, so they can create better futures for themselves" (The Smith Family, 2016). The *Let's Count* initiative strengthens partnership between families and early childhood educators to foster children's engagement with the mathematics they encounter every day, and explore and talk about this mathematics in ways that are interesting and relevant. This enables children to learn powerful mathematical ideas in ways that develop positive dispositions to learning and mathematical knowledge and skills.

Let's Count includes two professional learning modules for early childhood educators that strengthen their capacity to engage with families around mathematics. Each module comprises 6 h of engagement. Educators receive a 50-page *Let's Count Educator's Handbook* (Perry & Gervasoni, 2012) that provides an overview of *Let's Count* and explores the research, issues, ideas, and activities considered during the two professional learning modules. The handbook provides educators with the opportunity for further professional reading and reflection.

Module 1 focuses on the theme of noticing and exploring everyday opportunities for mathematics. The professional learning begins with educators spending 10 min taking photos of the mathematics found in the environment around them. The educators then discuss these photos with their colleagues and consider the opportunities that might arise for families to notice, explore, and talk about mathematics in this context. Our experience is that following this activity, educators are always amazed at the richness of the mathematics surrounding them that typically goes unnoticed. This leads to the opportunity to explore a framework of powerful mathematical ideas (Perry & Dockett, 2008) that are relevant for young children, and to use this framework to analyze the mathematics identified in the educators' photos. Educators also consider the six mathematical activities in which, Bishop (1997) claims, people from all cultures engage. These activities are counting, locating, measuring, designing, playing, and explaining. Noticing, exploring, and talking about these six activities from different cultural perspectives enables educators to recognize the richness that culture contributes to children's mathematics learning. During Module 1 educators also explore a range of games, children's literature, and songs as contexts for families to notice, explore, and talk about mathematics. These activities include discussing the range of number, geometry, measurement, and statistical concepts that underpin these experiences. Further topics discussed in Module 1 are the role of the affective domain in mathematics learning and the influence this may have on parents' engagement with mathematics. The final topic enables educators to share and discuss successful approaches for building partnerships between educators and families, and provides educators with the opportunity to begin planning approaches that may be relevant for introducing *Let's Count* in their community setting.

Following Module 1, educators introduce *Let's Count* to their communities and initiate strategies to strengthen partnerships with parents. Before describing this intervention phase, Module 2 is presented.

Module 2 focuses on the theme of celebrating and building on the mathematics that educators and parents notice children exploring. Each educator presents a brief report or narrative about the strategies they used to strengthen their partnerships with families through initiating *Let's Count*, and the outcomes of these strategies. The educators provide feedback and suggestions to extend the initiatives. Based on the feedback, the educators spend time designing an action plan for future activities related to *Let's Count*. Other topics include: examining educators' findings and documentation of the mathematics used and explored by children in their settings; discussing the findings and implications of the *Let's Count* Longitudinal Evaluation, particularly in relation to children's mathematics knowledge;

and further exploring the contexts presented in literature and songs for exploring mathematics.

Between the two modules, the educators connect with families to consider ways that they can all encourage children to *notice, explore, and talk about the mathematics* that children encounter in everyday situations, including through games, stories, and songs. The educators use a range of strategies for connecting with parents, depending on what was most suitable for their community. These ranged from one-off events and sending home resources to more day-by-day strategies such as incorporating mathematical learning into their everyday dialogue with families. The approach used in one community is explained in the evaluation interview excerpt that follows.

Every child gets a chance to take this bag home. So it's got so much stuff in it, and plays the games with their families and really having the conversations with the families and getting them more confident in the fact that they can help their children with their maths.

Interviewer: How have the families responded to the bags?

Oh they've loved it. Loved it. We've had some fantastic comments. In the back of … I've done up a little book to go home and in the front it has like an introductory letter and then it's got instructions to all the games and then at the back it's got parent comments. And we've had some fantastic comments through there about what the children have learned, what the children have been doing. One little girl went home and measured absolutely everything in the house, including the dog. [Educator]

Another approach is described in the following interview excerpt with an educator.

When they come to the information session, we were going to set up a few things that we might do with the children here and get them to participate in that. One of our educators is actually going to read a book that isn't about maths to the families but showing them how she draws the maths out of it even though it's not a counting or number book. And then we were just going to have some discussions and we were going to talk to them about how things can be done in a play-based, fun way, when you're already doing them. Letting them [parents] know they don't have to sit down and do maths activities as such to get them ready for school. [Educator]

Another educator explained that their center used a range of prompts to raise awareness among parents about how they might engage with their children around mathematics.

I think the main difference that it made was the way we engaged the parents in it and we didn't do a lot, it was just little things like putting notices out, putting little newsletters out about it and also we had a board out the front where we just put a little maths problem on there and the parents could sort of get involved. It was just something they could do on the way home or something they could do on the way in. Like counting buses or plan your trip somewhere, things like that. It was just little problems that we posed on the board. And that sort of got the parents really interested and talking about maths a lot more. And so I think that was probably the main thing but it made us more aware within the centre, even though we do integrate it quite well I think it was being conscious of using the language, the maths language with the children because we do play the games and we do all the mathematical concepts but we weren't using the language, so I guess that's what we seem to be a bit more aware of.

These examples illustrate the different ways in which educators engaged with parents and the impact for building parents' awareness of how they might support their children's learning. It was clear that these strategies focused on building communication and engagement between educators and parents.

4. EVALUATING THE IMPACT OF *LET'S COUNT*

The *Let's Count* intervention was piloted in 2011 in five sites across Australia whose communities were identified as experiencing social and economic disadvantage. In 2012 and 2013, The Smith Family delivered a revised *Let's Count* program in six additional sites that also participated in a longitudinal evaluation of the program (Gervasoni & Perry, 2015b). The longitudinal evaluation was conducted by the authors. Its aim was to ascertain the success of the *Let's Count* intervention in bringing early childhood educators, parents, and other family members together, and enhancing children's mathematical engagement, learning, and dispositions. The three key research questions framing the evaluation were:

1. How does participation in *Let's Count* impact children's numeracy knowledge and dispositions as they make the transition to school?
2. What is the impact of *Let's Count* on participating educators' knowledge, interest, and confidence in mathematics learning and teaching?
3. What is the impact of *Let's Count* on the participating families' confidence, and knowledge about noticing, investigating, and discussing mathematics with their children?

4.1 Longitudinal Evaluation Method

The *Let's Count* longitudinal evaluation concentrated first on two regions of Australia in 2013 (one in coastal New South Wales (NSW) and the other in western Victoria) and then, in 2014, continued in these two regions and expanded to include Perth, Western Australia (WA) and a regional city in southern NSW.

The longitudinal evaluation used a multimethods approach, particularly for the adult participants who were involved in both surveys and telephone interviews. The children who participated in the evaluation completed individualized task-based assessment interviews at the beginning and end of each year that assessed their whole number, measurement, and geometry knowledge. This instrument, called the Mathematics Assessment Interview (MAI), is derived from previous Australian work (Clarke et al., 2002). The MAI was originally developed as part of the *Early Numeracy Research Project* (ENRP) (Clarke et al., 2002) and refined in 2010–2011 (Gervasoni et al., 2011, 2010).

The MAI was designed for young children, is task-based and interactive, derived from extensive research, and enables mathematical learning to be measured in nine domains. All tasks are presented orally. One section of the assessment focuses on early mathematics concepts for children who are beginning school (the Foundation Detour).

4.2 Data Collection: Adult Participants

Educators in the study undertook a brief survey on two occasions in the year in which they participated in *Let's Count*—immediately prior to the commencement of the first professional learning workshop and immediately after the completion of the second workshop. Telephone interviews with educators were conducted on three occasions—within 3 weeks after the first workshop (postworkshop 1), mid-year, and near the end of their *Let's Count* year (postprogram). There was a small reduction in the numbers involved in both surveys and interviews as the year went on as well as a few educators who did not complete the earlier survey or interview but joined in later. A small number of parents was interviewed twice during 2013. In 2014, a larger number of parents was interviewed up to three times using a similar schedule to that of the educators. In total, there were 133 surveys and 111 telephone interviews completed by educators and 125 telephone interviews completed by parents.

The evaluation project had ethics approval from both Charles Sturt University (CSU) and Australian Catholic University (ACU) and from the participating centers, educators, parents, and family members. No participant was required to be involved in the evaluation of *Let's Count* and participation in the evaluation was not a condition of participation in the *Let's Count* program. There were separate consent procedures for the surveys and the interviews for adults and for parents/caregivers to consent to children's participation in the MAI assessment. In addition, the children were asked for their assent before each MAI was administered (Dockett, Einarsdóttir, & Perry, 2012).

4.3 Data Collection: Child Participants

In November 2012 (near the end of the Australian school year), the mathematical knowledge of 125 children eligible to start school in 2013 was assessed using the MAI (Gervasoni et al., 2011). These children formed the quasi-experimental comparison group for the study, also known as the 2012 Comparison Group. All children attended the same set of 10 early childhood centers that participated in the *Let's Count* intervention the following year and all were eligible to attend school in 2013. This ensured that the children in the comparison and *Let's Count* groups lived in the same communities and that children were of equivalent stages in their education. This maximized the opportunity for the comparison and *Let's Count* groups to have similar characteristics.

In early 2013, 142 children eligible to start school in 2014 and who were going to participate in *Let's Count* during 2013 were assessed using the MAI. At the end of 2013, 117 of these children were assessed again using the same instrument. In early 2014, another 195 children eligible to start school in 2015 and who were going to participate in *Let's Count* during 2014 were assessed using the MAI. At the end of 2014, 172 of these children were reassessed using the same instrument. The small reduction in numbers of children reassessed in December was either due to children leaving the center during the year or not being present during the December assessment period.

The MAI includes the number domains of Counting, Place Value, Addition, and Subtraction Strategies, and Multiplication and Division Strategies; the measurement domains of Time, Length, and Mass; and the geometry domains of Properties of Shape and Visualization. There is also a Foundation Detour that focuses on early mathematics concepts. The assessment tasks in the interview took between 30 and 45 min for each child and were administered in this study by independent, trained assessors who followed a detailed script and recorded responses on a standardized record sheet. Each child completed about 30 tasks in total, and given success with one task, the assessor continued with the next tasks in a domain for as long as a child was successful, according to the script.

4.4 Methods of Analysis

4.4.1 Children

Children's responses to the MAI tasks were recorded on a detailed record sheet by the assessors and then independently coded to determine whether a response was correct and to identify the strategy each child used to complete a task. This information was entered into an SPSS database for analysis. Simple statistical analysis was carried out, including frequency counts, crosstabs, and χ^2 calculations to measure for statistical significance of key differences.

4.4.2 Adult Participants

Qualitative data from the surveys were used in conjunction with interview data to provide a clear picture of the respondents' attitudes to mathematics and mathematics pedagogies. All the interviews were audio-recorded and transcribed for analysis. An initial coding frame was established for the interview data using a coding procedure based on the principles of grounded theory (Charmaz, 2004; Strauss & Corbin, 1998). Each interview was then double coded using this coding frame as a starting point. Any discrepancies in coding were resolved through discussion and agreement between the coders. The coded data were entered into NVivo. The qualitative analysis was undertaken in this software environment.

5. The IMPACT OF *LET'S COUNT* ON CHILDREN'S MATHEMATICS KNOWLEDGE

An important aspect of the longitudinal evaluation was investigating whether there was any discernible impact of families' participation in *Let's Count* and the extent of their children's mathematics knowledge, assessed with the MAI. It must be emphasized that the children did not experience any formal mathematics program during *Let's Count*. Instead, families were encouraged to notice, explore, and talk about mathematics with their children in everyday contexts. The next section presents a summary of children's responses to the tasks presented during the MAI.

5.1 Tasks With Small Sets

Table 1 shows children's success with tasks involving small sets of objects, usually small plastic teddies. The performance of the 2013 and 2014 *Let's Count* (LC) cohorts on each task was compared with those from the comparison group (Comp) to determine whether children's participation in *Let's*

Table 1 Percentage Success on Tasks With Small Sets

Tasks	Significance: Comparison (Dec, 2012) to (Dec, 2013) (χ^2, P)	Significance: Comparison (Dec, 2012) to (Dec, 2014) (χ^2, P)	Percentage of Children Successful With Task		
			Comp Dec, 2012 ($n = 125$)	LC Dec, 2013 ($n = 117$)	LC Dec, 2014 ($n = 172$)
Tasks with small sets					
Count a collection of 4 teddies	NS	NS	95	96	97
Identify one of two groups as "more"	NS	NS	90	92	87
Make a set of 5 teddies when asked	7.043, $P < 0.01$	10.735, $P < 0.01$	77	90	91
Conserve 5 when rearranged by child	NS	6.748, $P < 0.01$	79	88	90
Conserve 5 when hidden				77	86
Combine 5 + 3 blue teddies and total	NS	NS	75	71	79
Make collection of 7 (when shown number 7)	11.016, $P < 0.01$	23.852, $P < 0.01$	63	84 ($n = 92$)	89 ($n = 135$)
Knows one less than 7 when 1 teddy removed	12.018, $P < 0.01$	24.804, $P < 0.01$	61	82 ($n = 85$)	88 ($n = 131$)
Knows one less than 7 without recounting	12.018, $P < 0.01$	24.804, $P < 0.01$	25	40 ($n = 85$)	33 ($n = 131$)

Part–part whole tasks					
Show 6 fingers (usually 5 and 1)	NS	NS	79	78	82
Show 6 fingers 2nd way	4.566, $P < 0.05$	3.870, $P < 0.05$	27	40	18
Show 6 fingers 3rd way	NS	NS	10	15	8
One–one correspondence task					
Know 5 straws needed to put 1 straw in each of 5 cups	NS	NS	88	87	95

NS, not significant.

Count was associated with improved performance. The results show that the *Let's Count* cohorts at the end of 2013 and 2014 were significantly different to the comparison group with respect to four tasks involving making sets of objects and reducing them by one. The December 2014 *Let's Count* group was also significantly different from the comparison group in being able to conserve a group of five teddies. Importantly, differences between the comparison and *Let's Count* intervention groups were more apparent for the more complex tasks.

5.2 Recognizing and Matching Quantities and Numerals

The results in Table 2 provide insights about children's abilities to recognize the number of dots on a card without counting them (subitizing), and their ability to match a numeral to the number of dots on a card. These results highlight that the majority of preschool children can recognize quantities up to about four items without counting and about one-tenth of children at the end of preschool can recognize nine dots without counting. This ability to recognize quantities without counting is an important ability for educators to build upon when planning learning experiences and is an important basis for exploring pattern and structure in the whole number domain.

Table 2 shows that there was little difference between the *Let's Count* groups and comparison groups for the subitizing tasks, except when children recognize 0, 2, or 3 dots. The 2013 *Let's Count* group was more likely than the comparison group to recognize 0 and 2 dots, and the 2014 group was more successful at subitizing 3 dots. There were also some statistically significant differences between the 2014 *Let's Count* and comparison groups in matching numerals to more than three dots. Overall, the vast majority of children in all groups at the end of preschool could match numerals to the number of dots on a card, although nine was much harder to match than the other numbers.

5.3 Pattern and Structure

The importance of pattern and structure in young children's mathematical learning is gaining increased attention (Mulligan, Mitchelmore, English, & Crevensten, 2013). The Australian Curriculum Foundation proficiencies of fluency and reasoning focus on continuing and creating patterns. The results in Table 3 show that almost all children can identify colors in patterns. However, children are less likely to be able to match a pattern or continue a pattern. This appears to be a potential area for teaching and learning.

Table 2 Percentage Success in Subitizing Tasks and Matching Numerals to Dots

Tasks	Significance: Comparison (Dec, 2012) to (Dec, 2013) (χ^2, P)	Significance: Comparison (Dec, 2012) to (Dec, 2014) (χ^2, P)	Percentage of Children Successful With Task		
			Comp Dec, 2012 ($n = 125$)	LC Dec, 2013 ($n = 117$)	LC Dec, 2014 ($n = 172$)
Subitizing tasks					
Recognize 0 without counting	4.690, $P < 0.05$	NS	81	91	87
Recognize 2 without counting	5.133, $P < 0.05$	NS	94	99	97
Recognize 3 without counting	NS	6.976, $P < 0.01$	83	91	93
Recognize random 3 without counting	NS	NS	86	87	87
Recognize 4 without counting	NS	NS	70	79	74
Recognize random 4 without counting	NS	NS	50	53	54
Recognize 5 without counting	NS	NS	44	41	48
Recognize 9 without counting	NS	NS	16	9	12
Matching numerals to dots tasks					
Match numeral to 0 dots	NS	8.925, $P < 0.01$	73	81	87
Match numeral to 2 dots	NS	NS	90	89	92
Match numeral to 3 dots	NS	14.902, $P < 0.01$	73	83	90
Match numeral to 3 random dots	NS	NS	82	83	85
Match numeral to 4 dots	NS	8.629, $P < 0.01$	73	75	87
Match numeral to 4 random dots	NS	7.813, $P < 0.01$	69	76	83
Match numeral to 5 dots	NS	3.891, $P < 0.05$	65	68	76
Match numeral to 9 dots	NS	4.250, $P < 0.05$	38	47	51

NS, not significant.

Table 3 Percentage Success in Pattern Tasks

Tasks	Significance: Comparison (Dec, 2012) to (Dec, 2013) (χ^2, P)	Significance: Comparison (Dec, 2012) to (Dec, 2014) (χ^2, P)	Percentage of Children Successful With Task		
			Comp Dec, 2012 (n = 125)	LC Dec, 2013 (n = 117)	LC Dec, 2014 (n = 172)
Pattern tasks					
Name colors in pattern	NS	NS	98	99	96
Match pattern	5.623, $P < 0.05$	8.824, $P < 0.01$	72	85	86
Continue pattern	5.102, $P < 0.05$	14.765, $P < 0.01$	34	48	56
Explain pattern	NS	8.464, $P < 0.01$	34	42	51

NS, not significant.

The data suggest that about three-quarters of children can match patterns when they begin school, and about one-third of children can continue and explain a pattern (Table 3). However, the December success rate on these tasks is statistically greater for children in both the 2013 and 2014 *Let's Count* groups. This is an important outcome for the *Let's Count* children and highlights the importance of noticing, exploring, and talking about patterns (Clements & Sarama, 2014; Mulligan et al., 2013; Papic, Mulligan, & Mitchelmore, 2011).

5.4 Counting and Ordering Numerals

Although continuing, creating, and describing patterns is likely to be a profitable aspect of instruction for most children in their preschool and beginning year at school, many children need greater mathematical challenge right from the start of school. The standard for the first year of school in the *Australian Curriculum: Mathematics* (Australian Curriculum, Assessment and Reporting Authority, 2013) focuses on children counting to and from 20 and ordering small collections within this number range. Several tasks in the MAI focused on sequence counting, counting a collection of at least 20 items, and ordering numerals. The percentage of students able to complete these tasks is presented in Table 4.

Table 4 Percentage Success With Counting and Ordering Numerals

Tasks	Significance: Comparison (Dec, 2012) to (Dec, 2013) (χ^2, P)	Significance: Comparison (Dec, 2012) to (Dec, 2014) (χ^2, P)	Percentage of Children Successful With Task		
			Comp Dec, 2012 ($n = 125$)	LC Dec, 2013 ($n = 117$)	LC Dec, 2014 ($n = 172$)
Counting tasks					
Rote count to 10	NS	NS	93	93	93
Rote count to 20	NS	6.117, $P < 0.05$	45	55	59
Count a collection of at least 20	8.079, $P < 0.05$	13.165, $P < 0.01$	37	55	58
Count a collection of at least 20 and when one item is removed knows total without recounting	8.079, $P < 0.05$	13.165, $P < 0.01$	8	16	11
Ordering numbers tasks					
Order numeral cards 1–5	—	—	NA	68	74
Order numeral cards 1–9	NS	NS	48	60	54
Order numeral cards 0–9	10.354, $P < 0.01$	5.924, $P < 0.05$	31	52	45
Orders 3 one-digit numbers	NS	9.777, $P < 0.01$	47	52	68
Orders 3 two-digit numbers	NS	NS	14	15	23
Orders 3 three-digit number	NS	4.002, $P < 0.05$		1	3

NS, not significant; NA, not asked.

The results in Table 4 suggest that the majority of children at the end of preschool can count to 10 and that many can forward count to 20, but not back from 20. However, few students could count 20 teddies successfully nor identify how many teddies remained when one teddy was removed from the set. The results suggest that exploring the cardinal value of 20 is a profitable area for instruction in the first year at school, but it is not highlighted in the Foundation Standard in the Australian Curriculum. The tests of significance show that the *Let's Count* 2013 and 2014 cohorts were more likely to count a collection of at least 20 teddies and order one-digit numbers than was the 2012 comparison group.

5.5 Measuring Time and Length

Several tasks in the interview focused on children measuring length and time. Table 5 highlights that many children by the beginning of school were able to compare and order lengths and are also aware of the purpose of a clock. The 2014 *Let's Count* group was significantly more successful in ordering four candles, comparing two lengths, and using informal units to measure a drinking straw. However, this was not apparent for the 2013 *Let's Count* cohort. Seventeen percent of children in the 2012 comparison group could read 2 o'clock and knew some days of the weeks and months. Only 5% of *Let's Count* children were successful with these tasks at the end of the year in 2013, and 10% at the end of the year in 2014. These were the only items in the assessment in which the comparison group outperformed the *Let's Count* cohorts. We hypothesize that *Let's Count* had the effect of expanding the mathematics that families and educators explored with children at the expense of teaching young children to tell the time.

5.6 Properties of Shapes and Spatial Visualization

Spatial reasoning is a key aspect of learning mathematics (The Spatial Reasoning Study Group, 2015; Verdine, Golinkoff, Hirsh-Pasek, & Newcombe, 2017). The data presented in Table 6 show children's success with tasks involving describing and interpreting locations, recognizing the properties of shapes, and using mental imagery to manipulate shapes. The results suggest that almost all children were successful with these tasks. Participation in *Let's Count* was associated with more children being able to demonstrate the meaning of the location term "behind." The most difficult task in this group of tasks involved tracing a partially hidden shape, and only

Table 5 Percentage Success With Length and Time Measurement Tasks

Tasks	Significance: Comparison (Dec, 2012) to (Dec, 2013) (χ^2, P)	Significance: Comparison (Dec, 2012) to (Dec, 2014) (χ^2, P)	Percentage of Children Successful With Task: Comp Dec, 2012 (n = 125)	LC Dec, 2013 (n = 117)	LC Dec, 2014 (n = 172)
Ordering length tasks					
Ordering 3 candles small to largest	NS	NS	73	78	82
Ordering 4 candles small to largest	NS	4.148, $P < 0.05$	54	63	66
Length measurement tasks			($n = 125$)	($n = 60$)	($n = 95$)
Accurately compares two lengths string and stick	NS	15.690, $P < 0.01$	65	73	87
Measures length—informal units	NS	15.690, $P < 0.01$	8	7	17
Time measurement tasks			($n = 125$)	($n = 60$)	($n = 81$)
Aware of the purpose of a clock	NS	NS	83	83	81
Knows some days/months and 2 o'clock	NS	NS	17	5	10

NS, not significant.

16%–21% of the children assessed were successful. This is likely a profitable area for learning and teaching during children's first year at school.

5.7 Calculating

The MAI includes a range of tasks involving increasingly complex calculations, although few children progressed far in these domains. Results for four

Table 6 Percentage Success on Spatial Tasks

Tasks	Significance: Comparison (Dec, 2012) to (Dec, 2013) (χ^2, P)	Significance: Comparison (Dec, 2012) to (Dec, 2014) (χ^2, P)	Percentage of Children Successful With Task: Comp Dec, 2012 (n = 125)	LC Dec, 2013 (n = 117)	LC Dec, 2014 (n = 172)
Language of location tasks					
Beside	NS	NS	94	96	94
Behind	4.304, $P < 0.05$	4.423, $P < 0.05$	87	96	94
In front of	NS	NS	91	88	87
Properties of shapes tasks			($n = 125$)	($n = 61$)	($n = 81$)
Knows square	NS	NS	85	92	92
Knows circle	NS	NS	92	95	97
Knows rectangle	NS	NS	74	76	80
Knows some triangles	9.548, $P < 0.01$	9.404, $P < 0.01$	83	95	94
Knows all triangles			63	58	58
Visualization tasks			($n = 125$)	($n = 58$)	($n = 95$)
Recognizes static images in embedded situations	7.894, $P < 0.05$	NS	90	95	94
Identifies a reoriented rectangle in room; and traces possible shapes when a shape is partially hidden			16	17	21

NS, not significant.

calculation tasks are presented in Table 7. All tasks were presented orally and involved the use of materials.

The two addition tasks required children to combine two small sets of teddies (5 + 3 and 9 + 4) and work out the total when the larger set was

Table 7 Percentage Success on Calculation Tasks Involving Materials (Teddies)

Tasks	Significance: Comparison (Dec, 2012) to (Dec, 2013) (χ^2, P)	Significance: Comparison (Dec, 2012) to (Dec, 2014) (χ^2, P)	Percentage of Children Successful With Task: Comp Dec, 2012 ($n = 125$)	LC Dec, 2013 ($n = 117$)	LC Dec, 2014 ($n = 172$)
m					
Adds 5 + 3 when screen over 5 removed	NS	17.081, $P < 0.01$	49	63	72
Adds 9 + 4 when screen over 9 removed	9.664, $P < 0.01$	7.627, $P < 0.05$	25	42	40
Calculates total for 2 teddies in 4 cars	NS	12.005, $P < 0.01$	58	64	76
Divides 12 teddies between 4 mats	15.497, $P < 0.01$	NS	31	61	44

NS, not significant.

screened. The screen was removed if the child needed to see all the teddies to work out the total. Most children required the screens to be lifted and counted the teddies one by one. A small number of students used the count-on strategy when the teddies were screened. For example, to work out the total for 5 teddies (screened) and 3 teddies, children imagined the hidden set of 5 teddies and then increased the set one teddy at a time, saying "6, 7, 8."

Most children who solved the division task that involved distributing 12 teddies between 4 mats used grouping strategies rather than sharing by ones. Most children solved the multiplication task by counting all the objects one by one. This task involving placing two teddies in each of four cars and then working out the total number of teddies in the four cars.

The results presented in Table 7 indicate that a large group of children were successful with these calculations. It was statistically significant that the 2013 *Let's Count* group were more likely than the 2012 comparison group to successfully add 9 teddies and 4 teddies and successfully divide 12 teddies between 4 mats. The 2014 *Let's Count* group were more likely than the 2012 comparison group to successfully add 9 + 4 teddies and successfully calculate the total for two teddies in each of four cars.

5.8 Summary

Overall, these results show many young children know a lot of mathematics before they begin school and can demonstrate their learning in responses to tasks in a formal MAI. Further, participation in *Let's Count* was often associated with mathematics learning that was significantly greater than for a comparison group. These findings are a strong endorsement of the power of families helping their children to notice, explore, and talk about mathematics in everyday contexts.

6. EDUCATORS NOTICING, EXPLORING, AND TALKING ABOUT CHILDREN'S MATHEMATICS

As part of the *Let's Count* evaluation process, educators at the four evaluation sites (two in 2013 and four in 2014) were invited to participate in a series of three interviews about the impact of the intervention. Across the sites and years, there were 41 who agreed to be interviewed within 3 weeks of the first workshop. Of these, 35 were interviewed for a second time within 3 weeks of the second workshop. As well, 35 educators were interviewed close to the end of the calendar year in which they were involved with *Let's Count*. The interviews were undertaken via telephone, transcribed, and the transcripts analyzed to establish the key themes that emerged.

Seven themes emerged from the grounded theory analysis (Charmaz, 2004) of the educator interviews. Each of these themes indicates that *Let's Count* has had an impact on educators' pedagogical practice, while also highlighting the challenges educators face as they navigate how the program works in their individual settings. The seven themes are:

1. Engaging families with mathematical learning and *Let's Count*;
2. Continuity of mathematical learning between early childhood setting and home;
3. Impact of *Let's Count* on educator confidence, professional identity, and pedagogical practice;
4. Awareness of the potential of everyday tasks for prompting mathematics discussion;
5. Sustainability of *Let's Count* over time;
6. Children's engagement with mathematical learning and mathematical concepts; and
7. Importance of mathematical language.

Illustrative interview excerpts and interpretive statements for each of these themes follow.

6.1 Engaging Families With Mathematical Learning and *Let's Count*

Educators used a range of strategies to engage families in *Let's Count*. There were hands on activities, one-off events, and maths resources sent home. As well, day-to-day strategies designed to incorporate mathematical learning were introduced. Special "one-off" events were initially really successful for launching *Let's Count*, but sustained dialogue with families about their children's mathematical learning was a challenge for many of the educators.

> *It's been really nice to hear parents saying that it's been beautiful to spend that sort of quality family time together and they'd forgotten how much fun it was to play games, board games and things with the children. How they'd gotten out in the environment and looked for things. Like, there was a lot more around to do with maths than what they'd realised. And so it's been really nice. And one of the parents actually nabbed me the other day and she said how fantastic it was, they had a fantastic time with the bag over the weekend and it was a great idea.*
>
> *At one of the first* Let's Count *sessions, there was a stick with beads and it was a little challenge between two children. They roll the dice and they took whatever number came up on the dice across the stick. It's just like a little game between siblings even at home. I had a parent telling me they were taking them to bed and hiding them under the pillow. They were enjoying them that much.*
>
> *I will talk about* Let's Count *and show them that this is what we're doing to encourage mathematics. But a lot of them as I said before, one mum said 'Oh I did that with the slice of bread' but she didn't actually realise she was doing maths. She was 'just saying look, we've got two parts now and they're equal, rather than one big one' and I said 'well that's what we're doing but now we're bringing it to the front, to say Oh you've just learned a little bit of maths, that's fantastic'.*
>
> *On one of our boards we're going to have a big almost like a mind map of noticing maths during our normal day, not our actual intentional teaching moments but just things that happen during the day. Say we're outside playing and we notice what's floating and what's sinking. And taking those photos and adding the children's words and showing examples, visually, of what is maths. And hoping to sort of take some of that stigma away of that fear of maths if other people perhaps are like me and only have had that sort of negative feeling about it.*
>
> *I think [*Let's Count*] was very beneficial to the non-English speaking families. When I demonstrated how to play just your very basic games like Snap and Dominoes, they had no idea, they had never played those games before so in actually demonstrating how to play them they were very enthusiastic about going home with them. They've said how much they've enjoyed playing [the games] with their children.*

The above excerpts from educator interviews suggest that educators have tried many different strategies through which to engage families initially but that the *Let's Count* program activities have sustained and enhanced this engagement. Although there were some challenges met by educators as they endeavored to engage all parents of children in their settings, in general, *Let's Count* was an effective vehicle for achieving the engagement of the majority of the parents.

6.2 Continuity of Mathematical Learning Between Early Childhood Setting and Home

Many of the interviews discussed the mathematics that was being met in the early childhood center and how that was related to the mathematics the children were able to experience in their family situations. These discussions recognized the potential continuity of mathematical learning between home and preschool and the importance of establishing communication strategies among educators and parents.

> *Oh just one little boy came in today and said 'I really want to measure my bed'. So we made a measuring tape for him. I said 'You could use your hands' and he said 'No, I want a measuring tape'. So we made a measuring tape. ... the information came from his mother first and then we discussed it with the child. The mother came in and said 'Oh he really wants to measure his bed' and I went Ok, we can do that, we can work out a way to do that for you. So sometimes ... It depends on developing a rapport between the educator and the parent.*
>
> *Well I'm trying to look at new ways of presenting things for the children's interest. We're looking at the children's cultures at the moment. ... we have one child from Nigeria, a child from Spain and another one who is from South Africa and we've been looking at numerals and things. And using items from their culture, like their animals and how they pronounce their letters and things. So we're trying to make it more, maths more cultural with regards to their family like, so that they can go home and try and share it with their family.*
>
> *Well we've sent out emails on a regular basis with our parents. And so they've been emailing things that have been happening in their home. We also have a feedback journal-type thing that parents can write things up in, in the mornings or and we pose questions to the parents relating to maths. You know, we might just pop on a question, you know, what did you do over the weekend that involved mathematics or that kind of thing. ... So they're able to see what the children are interested in doing here and then maybe continue that on. So I think that's been a good way of doing it. Rather than having portfolios that go home at the end of the year and they go 'Oh I didn't realise you liked that'.*
>
> *I can definitely see and notice a real difference in their maths comprehension. Even the parents are referring back to us saying how they know this and this and this at home, which they haven't taught them. So like with the fractions, we did an activity where the children made their own sort of ... It was like their honey bread and they had to eat a quarter at a time. And then when they've taken that home and demonstrated it to their parents, 'Oh I've eaten a half of my sandwich' and sort of that way.*

> *[Mum] was just gobsmacked that her child at 4 even knew that there were sixty seconds in a minute and sixty minutes in an hour just from us talking about it in the everyday experiences. ... It's sinking in and it's making a difference, which is really, really nice. We've had lots of parents emailing and lots of parents bringing in photos and showing us photos on their phones and telling us all these wonderful stories about mathematical things that their children are doing. It's just amazing that they're taking that home and they're kind of teaching their parents and their families as well. So it's been really, really good.*

These comments provide strong evidence of the educators noticing changes in the ways in which the families are engaging with mathematics in their homes and communities. Potentially, parents and educators have much to talk about, but during *Let's Count* it was mathematics that enhanced their conversations about the children's learning.

6.3 Impact of *Let's Count* on Educator Confidence, Professional Identity, and Pedagogical Practice

Let's Count did not set out to change educators' practice per se. Rather, the aim was to have the educators work with the parents in order that the parents would notice their children's everyday mathematics. However, there were many educators who reported changes in their pedagogical practice, identity, and dispositions toward mathematics. Selected quotes from interviews reinforce these claims. It appears that the *Let's Count* professional learning and the process of implementing *Let's Count* assisted educators to be more confident with mathematics, more positive about mathematics learning, more aware of the mathematics encountered during everyday life, and more likely to notice children using mathematics during their play, and to explore and talk about this mathematics.

> *I said you want to be a mathematician for life not just while you're in school. Like we're readers for life, I think it's very important for children to be mathematicians for life now as well.*
>
> *As I said, I think the* Let's Count *program has made me more aware of maths in my language and in my encouraging propositions for children. So I'm asking them to solve it. But because I know about it I can direct it. So I think the* Let's Count *program has made me aware, as I said, of maths in our society. Or how children can be involved in developing the process of thinking.*
>
> *It's just been overwhelming how staff once they get that concept in their head how they're looking for it everywhere. It's been really beneficial to enlighten staff who might not have actually thought of activities in that way before. So it's been really fantastic in that respect. And Rosa and I are always looking for new ways to invent, both outside and inside, to create that maths in our kinder environment.*

I'm probably more confident with doing things now with the kids where before I would have been like Oh that's just too hard. But it's not and I suppose the measurement side of things we've been really working on with the kids and water volume, and things. We've been looking at fractions and different things and the kids really love it. And I suppose if you're confident in being able to present that, the kids just take more out of it than if you're not so confident they won't be as interested. Does that make sense? I guess if you're not captivating your audience they're going to lose interest quickly.

I can remember learning all this when I was at university and learning to be a teacher so I think for me it was really good to revisit that because we don't really talk as much in the ... Because of the new framework and that the language has changed with what we're doing. So it is nice to still talk sort of about mathematics rather than ... Because it's always literacy and numeracy which sounds like it's just all about numbers all the time and that children just need to learn to count. But when it's mathematics it covers everything and that's the lovely thing, because I'm not sure that it's being taught in university the way it used to be when I was doing it a long time ago. I really don't think it's there and I think this program is filling that gap. I think it needs to get out to all the teachers and staff that are out there, that mathematics is everywhere. And it is, it's a very valuable thing to be doing within your program. Because I think we hide it so well that we don't even realise we're doing it. And it needs to be on more of a conscious level and I think that's where it's ... I think it's very worthwhile. I think everyone should be doing it.

There were many educators who talked about *Let's Count* being a stimulus for a "turnaround" in the ways they thought and felt about mathematics. Many recalled negative feelings about mathematics in school, particularly secondary school, and some about negativity in their teacher education programs. For most of the participants in the longitudinal evaluation, *Let's Count* was actively changing these dispositions and, as a consequence, changing practice.

6.4 Awareness of the Potential of Everyday Tasks for Prompting Mathematics Discussion

The notion that there is mathematics in everything has clearly been accepted by many of the educators. Over the life of *Let's Count*, the depth of the mathematics being noticed has increased and mathematics has broadened beyond the study of counting and numbers. In some cases, "noticing mathematics" has become an integral part of educators' normal daily procedures. This is a very positive aspect that many of the educators have drawn from their experiences with *Let's Count*.

I had a conversation a little while ago about setting up activities for the children and we were talking about students and I said it's really important that we actually ask the questions of students what's the learning that can happen within this activity and I just reeled off a few things that could be learnt out of just a simple activity of having farm animals on the table. We started talking about how many legs the farm animals might have and those sorts of things. So I think it's really made me think about the learning that can take place within every activity that happens that may have some sort of maths attached.

Whatever we're involved in, even with the gardening outside, we look at how many plants we have planted, we look at the pictures of them when their fruits develop, what shape is it. Colours and everything, whatever we do we're attempting to incorporate maths throughout.

I'm getting better at incorporating it, yes. I like maths anyway but some of it is … It's back to … I was telling the girls, it's back into things like we're always going 'Wash your hands, they've got germs on them,' 'Don't touch that, it's got germs on it,' 'Don't forget to swish and swallow and clean your teeth' but we're getting at incorporating maths in a 'Oh look how tall that is, it's not as tall as something else and sort of taking those moments of … It might not be a huge big lesson planned but it is pointing something out or … If they find a long stick 'Do you think that's as long as the garbage bin,' so we're predicting and like doing size comparisons and doing those observations and things like that. So it's sort of becoming more second nature now, as opposed to sort of 'Oh my God, I've got to do this maths stuff'.

So everything we do, literally everything we do, even cutting up the streamers today to put out to blow in the wind was about how long we will we make them. I've got a big piece of material, how will I make many of them. Well, we can cut it in half and then we can cut it in half again. So like, everything you do become a maths experience. Today one of the children brought in oranges from her tree. So we talked about healthy eating but we also talked about how many oranges there were and what was happening if we were going to share one. Because that was our learning from today was approaching the issue of sharing. So we shared it between the children. What will we do next, we cut it in half. Was that enough? No, it's not, we need to do more. What will we do? Cut it in half again. That gives us what? … They're quarters and these four quarters will make a whole. Oh a new word, Ok. So you know, that's how we're going along. It just happens. It's just heightens awareness.

I probably feel more comfortable. Let's just say, it's something that we've always used and I've always felt Ok using but just the different ways I think it was sort of more of an eye opening, how much we bring it into our gardening experiences and just generally day to day. It's so easy to add terms and different things to just about everything we do. Even just down to children talking about the shapes and sizes of their sandwiches or the number of packages in their lunchbox. I've noticed even with the kiddies there, when they're making a circle at group time they'll describe whether it's not a circle and the shape they feel it is. So yeah, it's definitely something that I feel quite comfortable doing now. I've never been an overly confident mathematical person but I'm finding it quite easy to do it with the children, which is great. And just how frequently we were actually doing it anyway. So that's good.

6.5 Sustainability of *Let's Count* Over Time

The sustainability of programs like *Let's Count* depends not only on the quality of the initial professional learning but also on the continued enthusiasm and drive of the participants. Some of the educators interviewed were very keen to see a sustainable future for *Let's Count* in their settings. Many of those interviewed felt that *Let's Count* would provide its own momentum to continue. For the educators, seeing the results of the program for the children, parents, and themselves was a strong motivation to keep going.

> *I think with the current staff we have got it will continue very well but again, if that staff changes it would depend on the abilities and the interests of those staff I suppose. It would be certainly something that I would like for anyone to continue but it would just really depend on how that kindergarten teacher wanted to work and their own knowledge of those things. And usually you find when a program is introduced like this it works really well but when the momentum stops so does the program.*
>
> *Every year we set a program of things that we want to ensure that we have focused on or thought of while we're developing our program. So we will definitely put it in there as one of those things that we want to keep in our mind while we're developing our program and running projects and things, because mathematics can go into any project that you're doing whatsoever. So it would go onto there and it can be something that we can continue to talk about at our staff meetings and things like that, so that those who have done the program have passed that on. So even if they were to leave, the other educators that are still at the centre would have the skills to continue.*
>
> *I think that we'll probably just keep going with what we're doing because it's making a difference and it's probably not to the extent that all the parent nights and all that sort of stuff, but just incorporate the maths program, like our literacy program and everything and just keep on going with it. ... and if we give it another 12 months it will be ingrained like the literacy thing and our sustainability stuff, that we won't even know we're doing it and it will make a big change.*
>
> *I think because we're all so positive about it, I think that is going to be a big part. I think if we were to go 'Ok we've done our bit now. We've done our two sessions, our job is done'. I think the interest would go. ... It's very, very much a part of our day, so I think that because we're doing it almost on auto ... well, not on auto pilot but almost without thinking, we're doing it now as part of all day every day. The children are going home and talking about it and they're wanting to share their new knowledge with their parents. So they're really excited. That excitement, I think, will help keep the families motivated as well, because they're going to be going home and talking about it and wanting to do it and wanting to show them and wanting to come in and tell us what else they know. So I think that will be a big part of it. So we're hoping it will anyway.*
>
> *No, we're actually looking forward to doing it again next year, just so that we can move on and further ... Because although it's mainly for our ones going off to*

school our younger ones have been involved in it too so just to see where we can take them after 2 years of doing it. So we're just sort of looking forward to what they're going to come up with, having been involved in it twice, to see basically what they're going to get out of it and what other concepts we can do and what challenges lie ahead really.

Of course, another vitally important aspect of sustainability for a new intervention is to have it accepted and backed as a government initiative, funded for scale, and accepted as part of the fabric of early childhood education. In Australia, both the Australian Federal Government and the Victorian State Government have provided substantial funding to implement *Let's Count* more widely in their jurisdictions and to incorporate it into the everyday practice of their educators.

6.6 Children's Engagement With Mathematical Learning and Mathematical Concepts

So much of educators' knowledge of what children are capable of doing, how they engage with activities and how they are disposed to this learning comes from their observations of children during play and other, perhaps more structured, contexts. The *Let's Count* educators interviewed often described their awareness of advances in children's mathematics learning that they have noticed. In some circumstances, these advances have been quite dramatic. Many of the educators put changes of the children's engagement down to changes in confidence in both the children and the educators, and the educators' increased ability to provide meaningful, mathematical feedback to the children. There was little mention by educators of any impact increased engagement with mathematical ideas at home might be having on the children's engagement with mathematics in the centers.

So far I think it's fantastic. I'm really seeing the children ... Just their knowledge is just blown me away, of what concepts they're understanding. Their understanding of like symmetry and patterns. And now it's starting to be more about adding. Last week we worked out that 10 × 3 is 30 and that was from a story book: '10 Red Apples' from Dr. Seuss. They had noticed that there were 10 on each of the animal's heads and then I chose three children to show me 10 fingers. And then they were able to count along and find out that that actually meant 30 apples in total. And that all came from the children ... So they just have been absorbing the maths and really extending them further with just like the slightest bit of encouragement, which is fantastic.

The child that's really taken to the graphs is one that I would not have expected.... It's a little boy, but he's also getting other friends involved too, both boys and girls. And so it started off as being just him and a female friend, writing their names and arguing about who had the most letters. … But then we ended up measuring feet, so it was two of the girls talking about who had the prettier and bigger shoes. Yeah, so then we chased around and made a graph of everyone's feet. And he then came along and wanted to do that as well because it would go with his graph on his name.

So it's sort of been, like the graph part, has really taken on them on board but it's been tailored to obviously what they're interested in, which has then brought them into the maths concept better too, because it's not just 'these are numbers, this is … '. It's whatever topic it is.

I suppose they've all grown through confidence. I guess there's a couple that from the start of the year would never, ever have had a go. Because we usually do a little counting game at the start of the kinder session and they're now the ones that have always got their hand up first to get in there and start the counting game. So that's a big improvement, just children's self-confidence, which sometimes they don't even know that they're doing or interacting in an activity that you're getting a lot of positives out of.

I think because we are sort of recognizing and celebrating those moments and those times, talking about them being mathematicians and things like that. And seeing them sort of then taking that on board and then sort of wanting to show you … So you might be discussing something at the table with them, something that they're doing and then turn away for a few minutes and then they'll be wanting to show you something more that they've done. They're extending their own experiences I think because of that; because of that feedback that they're getting back from us.

My children have built so much over this year that their concepts all around are starting to grow. I mean, they're transitioning to school so really we've studied a lot of maths this year so probably there's one that I've noticed is how they ask for the maths to be incorporated. If we're doing something they ask for the maths. Like have we divided the groups for the day or can we play one of the games that are maths components. … I have got one little boy who is very negative at the beginning of the year about actually engaging in games, full stop. And now I've actually got him requesting to stay in the group that has the games each day and actually sharing in the maths with other children and yeah, look, that's just been brilliant. He's so excited about playing these games of maths, with our support of course, because he is from probably a less academic environment, if I can say that, and so for him to embrace the games and really take on board the maths is just brilliant. So it's really good.

I think they surprise us all the time, in a good way and I think they surprise each other too because they're kind of constantly sort of sitting there and wanting to share their ideas with their friends as well. So I think then their friends don't want to be … They want to be sharing their ideas too so you kind of get this … Not competition, but you get this kind of springboard that keeps getting moved up and up and up because they're just thirsty for everything. Yeah, we can't keep up sometimes with what they want to know and how to find out and so yeah, it's been really, really good. I think that confidence has been one of the best things that's come out of it, is their willingness to ask and ask and ask and ask some more. They want to know more, they want to know this, they want to how to do this and how this will work and what about if we try this. So quite often it's them kind of taking on the role of the teacher sometimes and they're the ones leading it, which is really, really good.

Not only are the educators noticing children using a greater range of mathematics but they are also noticing a lot more mathematical development in these same children, as they explore and talk about their mathematics in everyday life.

6.7 Importance of Mathematical Language

The importance of mathematical language was a stronger theme emerging from the analysis of 2014 data than 2013 data. Mathematical language had been mentioned in conjunction with learning in 2013, but in 2014, it was regularly discussed as an end in itself and an important addition to the pedagogies of educators. This difference may be because of an increased emphasis on mathematical language in 2014 but there may be other unknown factors at play. Although children's use of mathematical language does not necessarily show that a particular concept has been learned, it does show awareness of particular aspects of communication of mathematical ideas and has been recognized for a long time as an important entry point to the mathematical thinking of young children (Hughes, 1986; Purpura, Napoli, Wehrspann, & Gold, 2016).

Well they do a lot of maths but they don't realize it. We talk about how long … We give them the words, the language for what they're doing. One little boy this morning said to me 'Look what I've made, come and see, we've made a really long thing' and I said 'How long is it?' and he said 'Well it's longer than this'. We kept going on about it, I said 'What have you used?', he said 'I made some long and some short blocks' and I said 'Well what else can we do with it?', sort of thing. So it's just conversation.

Oh yes, definitely over the year. I just feel that yeah, I'm just using more mathematical language and that's getting passed on to the children. So yeah, and just like velocity and distance. It's just using it how we would speak in everyday to an adult, so it's great.

The Let's Count *training we had gave us a different approach as well, where we can talk to parents and talk with children. … We use mathematical language, a lot more mathematical language, so the children know this is a part of our vocabulary but it's also part of real life. Instead of just counting twenty children, or twenty-five children, who have turned up for the day … [we can say] how many boys and how many girls. So if there were two away, how many … So we can use that as a mathematical game as well.*

It is, just how simple and important it is to mention that terminology. I suppose that's my biggest thing, is you've always done mathematics stuff with the children but not necessarily saying well actually, you know what, this is mathematics and this is sorting and classifying and that sort of stuff. Just actually really using the words. I suppose that's been my biggest change. It used to be just like rote counting or just this or that sort of thing.

Yes, there's two little boys who are just amazing. They just pick up your language and they're coming out with the language of maths. The quarter, it's how long, longer than or higher than. They're then doing estimation. Because we're trying to teach them that as well, to say you don't have to count first, you can make a judgment, you can make an estimation. If there's a lot, if there's less as to how many there are. And they're doing that. They're saying 'I think there's 6' and then I say now you need to check, so they're checking sort of thing. So there's two little boys who are just … Their language of maths has just come out in them. It just increased their vocabulary.

We've all just become more aware of taking advantage of when we can use that [mathematical] language in other areas of learning. In those little times and just using the language that we didn't do maybe as much before. So that's probably been a really big thing. There's always something every day. Like we have a section in our program dedicated to Let's Count *and numeracy and mathematics. So every day in the program that fills up really easy. We have no problem with putting different mathematical elements into that. I think it's just that awareness of how you can incorporate mathematical language into everyday things. And we're all getting quite good at it, so we're all just like 'Oh I've got another thing'. So the actual program itself hasn't changed, just our awareness of those moments has become stronger.*

7. PARENTS NOTICING, EXPLORING, AND TALKING ABOUT CHILDREN'S MATHEMATICS

In 2013, a small number of parents were interviewed on two occasions, once at the beginning of their involvement in the program and again near the end of the year. On the first occasion, four parents were interviewed in each of the two evaluation sites while, at the end of 2013, these parents plus another one in each site were interviewed. All interviews were conducted by phone. In 2014, a much larger number of parents were interviewed on three different occasions: the first shortly after they had begun in the program (38 parents); the second around July 2014 (36) and the third at the end of 2014 (33). These parents were spread over only three of the evaluation sites.

7.1 Themes Emerging From Parent Interviews

Six themes have been identified through the grounded theory (Charmaz, 2004) analysis of the parent interviews: some of these relate to the impact of *Let's Count*, while others relate to the processes around its implementation. Some of these themes are similar to themes that arose from the educators' data.

The identified themes are:

1. Noticing children's mathematical learning and facilitating that learning in the everyday;
2. Parent–educator communication about mathematics and *Let's Count*, with an emphasis on strengths of all involved;
3. Children's growing confidence, knowledge, and enjoyment of/engagement with mathematics;
4. Importance of mathematical language;
5. Positive impacts within families, extending to older and younger siblings' inclusion in mathematical activities at home; and
6. Sustainability of *Let's Count* over time.

As for the educators, selected excerpts are presented next to illustrate each of the themes.

7.2 Noticing Children's Mathematical Learning and Facilitating That Learning in the Everyday

Every parent interviewed talked about their increased abilities to "notice" mathematical concepts as part of their everyday interactions with their children, and to extend those concepts when children showed interest. Many recognized that there was "mathematics in everything" and that part of their role in their children's mathematics learning was to notice, explore and talk about this mathematics in their everyday lives. Although this noticing of mathematics was not always attributed to their family's involvement in *Let's Count*, in many cases, parents indicated that this was an influential aspect of the program. Pseudonyms are used throughout.

> *The major difference I think has been I'm much more aware of how she can learn from everyday things. An example of that was yesterday my husband brought home a little thermometer, he works in refrigeration, and she wanted to know how it worked. And I was just trying to explain and I couldn't be bothered, and then I thought 'Oh put it in the fridge'. And then she put it in the fridge and we looked at the degrees and she wanted to put it in the freezer and look at the differences in temperature. Yeah, from that it kind of snowballed into looking at why were there different numbers, what's Fahrenheit, what's Celsius, all that kind of stuff. So I think it was good. At the start of the year I wouldn't have bothered, I wouldn't have even thought about it but it just occurred to me like, this is a good moment for her to explore it. Whereas before I wouldn't have done that and I would have just said 'I don't know, I can't be bothered teaching you that.*

I guess looking at it, he will say 'What do these two numbers make mummy'. So he's looking at double-digit numbers. So say the numbers on our letterbox. He'll say there's a 7 and a 3, although it's 3 and a 7, he just wants to know what do those two numbers make mummy and so for him to actually ask me that, I think that's pretty good. And then I'll say it doesn't really make 37 but that's what he's asking, is what number is it joined together so I'll say 37 and he's like Oh Ok. So to be interested and eager to know, that's what surprised me at the moment.

So the other morning Mia came in with her mega sketcher to me and she said 'Let's make a pattern'. She directed the game, obviously she was the teacher and she was saying 'First I want a flower'. So I did a flower, then circle, a square, a triangle, and then she said 'Now a flower again, a circle, a triangle' and then she realised, she's like 'No, that's not right' and she said 'Let's start again' because she self-corrected that she'd forgotten the square. So we went back again and we started again and we repeated the pattern three times. And she's like 'Great work mum, great work'. And off we went for breakfast or whatever happened after that. So that was fun.

We were outside the other day because it was a nice day and he was jumping on the trampoline and he was counting. He counted past 100 and I'm like 'Wow, that's pretty good'. Every time he jumped he counted how many times he jumped. I think he got to 140, I'm like 'Oh my goodness, I didn't know he could count that high'. I am probably more interested when he starts talking about it or he's playing a game that involves numbers. I do, I get involved with him and listen to him and I'll tell him stuff if I need to.

Probably one category would be more intentional so whether it'd be sitting down and playing a game of Uno or a game of dominos where we're focusing on that maths. And he has also developed an interest in dot-to-dots and stuff like that. So that was what I'd say more intentional whilst other opportunities just sort of a spontaneous. So whether it's like he's helping me set the table, well how many forks are we going to need for our family or just things that coincidentally pop up in our everyday lives. Like swimming groups this week he noticed that there were numbers on the side of the pool so he wanted to know what it meant, so we talked about depth and then he went on his own tangent of measuring the depth of the pool in different areas, on his own body. So how high the water would reach. So yeah, I guess it's a whole range of experiences, some are planned for and others have just cropped up coincidentally throughout each day, everyday living really.

His skills have certainly picked right up. As I said, he's always been a bit interested in maths type things anyway but certainly he's very good at … Like he's at a point now where he counts forward and backwards, he can count well over 100. He has started to do things like negative subtractions and stuff like that, so he'll be 'Mum what's 2–8' and all of those sorts of things, so there's a lot of those concepts coming in. And he's very good with his concepts around big, medium size and small, front end and middle and all of those sorts of sequencing type things. So yeah, he definitely has come on board in leaps and bounds this year with his understanding about a lot of those concepts.

Parents expressed confidence in their own abilities to "notice" their children's mathematics and satisfaction when they did this. As a result, they built on their successes and broadened their mathematical scope of their "noticing." Similarly, the educators participating in *Let's Count* found new enthusiasm for mathematics and were much more comfortable in "noticing" it in the lives of the children and themselves. With both parents and educators "noticing" mathematics, there was fertile content on which to build enhanced communication.

7.3 Parent–Educator Communication About Mathematics and *Let's Count*, With an Emphasis on Strengths of All Involved

Having educators and parents/family members talk about the mathematics the children experience both at home and in the center is central to *Let's Count*. The effectiveness of parent–educator communication, both about children's mathematical knowledge and experiences and about *Let's Count* in general, varied markedly across the sites and across individuals in the sites. For most parents, the level and intensity of this communication increased across the year in which *Let's Count* was implemented, as did the parents' satisfaction with it. In some cases, poorer levels of communication around the mathematics children were doing and *Let's Count* itself, were attributed to parents' own acknowledgement of being time poor or their child not attending the center regularly.

> *If something pertinent to the* Let's Count *thing crops up then they will mention it. Probably like on the Facebook page they're seeing what we're doing at home so I guess they're learning more about us as well, through a different way than just chatting. Because I mean, pick-ups and drop-offs are always so busy, you don't always have that opportunity, so I think that it is giving them a little bit more insight into each child. It's probably giving them a greater awareness of each child's strengths and needs as well because maybe they're getting surprised by some of the stuff that the kids do know. Or seeing areas where they could focus on more.*
>
> *Earlier in the year our kinder teacher ran a night with ideas about reminding families about how you can utilise your home environment and what's around you to introduce these basic maths concepts to children. And more than just your simple counting. There's just so many other sorts of things with spatial awareness and patterns and so forth. And so there were various activities that were run on the night following a little power point presentation from the kindergarten teacher. And activities that the children engaged in with parents supporting. And then there were some resources that were taken home. So it was probably the catalyst for everything that took place in kinder this year. And the kinder teacher has touched base on that occasionally throughout the year but it was probably mainly for parents just to see what we can do at home.*

Oh I think it's fabulous, that's exactly how I think kids should learn most things, particularly when it can start at home from such a young age and not just at school in a formal setting. And I think sometimes you don't realise as a parent that you're actually doing it, quite often, much more than you probably think. And I think it's great that education and programs are going in this direction and trying to educate parents too, on how to teach maths and use it every day in a much more holistic approach rather than just 'Let's Count to 10'.

Well I definitely think that my relationship with Lisa, who is the one who is heading the Let's Count *with Clare, like I just talk to her so much more. Like we're engaging so much more. Even with other parents, you know. Because we have this Facebook page as well we're all communicating, we're all uplifting each other. Every day I come in and Clare actually has been amazing. Like, she has done so much in the room. They've got this little mathematics table where they're constantly changing things. They've got scales, they've got estimation, they've got all these types of things and she's so into it that it kind of is … What's the word I'm looking for? Like you take it on board. It's awesome. It's so much fun. And we talk so much more. Like on an every other day basis she's like 'Oh I did this with the kids' and I'm like 'Oh my gosh, it's awesome'. And she'll mention something that I've done on Facebook and she'll be like 'It was so cute'. The language is open. The communication is open. It's great.*

Evidence of two-way communication between educators and parents as well as enhanced communication among parents about the mathematics the children were experiencing seems to be associated with participation in the *Let's Count* program. Although there was an increased emphasis at the stage of introducing *Let's Count*, both educators and parents have worked to maintain satisfactory levels of communication throughout the year. This is important as both work together to sustain the program in the early childhood centers.

7.4 Children's Growing Confidence, Knowledge, and Enjoyment of/Engagement With Mathematics

Parents reported their genuine surprise at their children's increased mathematical capabilities and, particularly, their children's confidence in trying out new mathematical ideas. In some cases, this mathematical development exceeded the parents' expectations of children's capabilities at their age. For the most part, the parents attributed these increases to the emphasis on mathematics in both the center and the home, as a result of the introduction of *Let's Count*.

He is more mathematically literate than he was, which is really good. In particular, when he did his primary school screener to see if he was school-ready, they commented on his mathematics understanding as a really positive thing that he was

> *quite excelling in. … that was really good feedback for us too. And we knew as soon as we heard that we went 'Oh we know why that's happened, because that's the* Let's Count *program'. So that's really good, because the screener that he did was really comprehensive. It was a two and a half hours [assessment that] covered a whole heap of things.*
>
> *And he'll come home and tell me about it and talk about it. Like, they had a rain gauge and it was measuring the water and he was telling me how many mLs were in the rain gauge. I'm like, 'Oh wow'. So they are interested in all the things that they've been doing at kinder.*
>
> *Yes, we were playing … We've got a kids' Monopoly game and I was … Oh well actually dad and I were quite surprised at how quickly she could break down numbers, so with money changes. She likes to have every denomination in front of her and if she has to pay like $5 or something she likes to use a 3 and a 2 or a 4 and a 1 or 4 you know, 5, instead of taking that last five dollar note. And she can do it very quickly.*
>
> *Just knowing that she can probably count 1 to 100 is like a big thing to me. But the other day she picked up counting by 10s to 100s. So that kind of threw me off and I went 'wow, now let's just go with this' because this could be the easier way to count to 100. That's one of those things that she said to me that's kind of spun me out, when she learnt how to count by 10s, that was cool.*
>
> *She comes out with things every day, basically. Something that really surprised me … Oh, Maddie and I were talking about my birthday and that I'm turning 22 and she said 'Oh mummy, you're turning 22, isn't that two two', as in like 2–2' and I was like 'Yes, that's a number' and then she's just like 'So how do we add … ' like 'What do we do to get to that number', like … You know. She was just trying to work out how to get to twenty-two, like all different scenarios on how to get to the number 22.*
>
> *For me personally, with Emma, things that I've noticed are things like she always says … For setting the dinner table, she's started to do that now and she's like 'How many forks do I need? I need two kids' spoons, two big spoons that equals four spoons'. She says things like that. She says things like, her sister is there with her and she's like 'I've only got five clips and I need six to make it match together'. Things like the other day we were driving and she's like 'What's the distance from Newcastle to home?' Just thinking of other examples, we also had this boomerang at home and she's asked me for a measuring tape so she could ….And she lined the tape up from one end to the other and she began to read the numbers. So she was like '5, 6, 7, 8, 9' and then she's like 'It's 300.' Which was a bit whacked but … Not obviously correct but it was about her using the lingo and the skills and all of those things that she's obviously picking up from the program.*

Just as the educators were sometimes amazed with the knowledge and confidence children displayed in their mathematics, so, too, were the parents. Not only did this amazement lead to many exciting conversations among educators and parents, it also impacted on educators' practices in the early childhood settings. The surprise of one mother was very clear, "I am totally gobsmacked by what she comes out with."

7.5 Importance of Mathematical Language

Children's mathematical language is a critical part of their mathematical learning. *Let's Count* emphasizes the importance of the use of appropriate mathematical language in discussions about mathematical activities as well as ways in which such language can be introduced incidentally, in everyday and fun approaches. Educators participating in *Let's Count* were often "relieved" that *Let's Count* had given them permission to use mathematical language with the children. The parents interviewed seem also to have developed an understanding of the importance of mathematical language and how they can assist their children to develop it.

> *Before the* Let's Count *program started, I used to take Sophie shopping and work with a budget and everything and she used to help me remember the numbers and stuff like that. So we've always sort of had that sort of learning mathematics basis but it's definitely just made me a little more aware of using the more technical terms with her. Because she's going to learn what they mean eventually so it's really important to start using those straight away.*
>
> *Maybe I'm using that word mathematics with him more. So like perhaps before we might say he's helping me, we're making muffins or something and we're measuring half the cup, before I might have just said to him 'Ok, this is half a cup, let's measure it'. Whilst now I would highlight to him that this is actually mathematics, so that he puts the connection together. So yeah, just perhaps using that terminology more often … so that he's comfortable with it. That's probably the main thing that I've changed.*
>
> *He does things like we went into a surf shop, I was looking for new boardies for him and he saw the surfboards and he was talking about how they were almost … He said 'They weigh almost as much as dad', but what he meant was they're almost as tall as dad. So he's using the technical terms that he wouldn't have normally talked about. And we, my husband and I, have been more explicit with our maths with him as well. And he's starting to get more the concept of addition and subtraction. So he and I have been talking more about his numbers. His individual number recognition is getting better and his understanding of numbers representing groups and adding and subtracting from those groups. He's exploring that a lot more than he has done before. And we've always done a lot of cooking and things with him. So we've been talking about wholes and halves and thirds and all that sort of stuff with him for that. But yeah, he has had definitely an emergence in his mathematical and scientific literacy, which is really good.*
>
> *It's been a lot of noticing things in her surroundings that I don't think she would have noticed before, based on the fact that she has had exposure to the words and the language and the concepts. So you know, like an example was the other day she just noticed a clock they had at Bunnings [hardware store] and then she was trying to tell me the time and talking about the hands and things like that. Even playing games she uses language like halves and that's a quarter. Before she would never have, never, you know, been talking like that.*

Although *Let's Count* does recommend the use of "correct" mathematical language as appropriate, the emphasis is on the experiences which might lead to the use of such language rather than the language itself. We cannot be certain that all parents have been able to make such pedagogical decisions.

Both educators and parents noted the importance of using correct mathematical language with children and each other during their conversations with the evaluation team. This use of language was not just routine adherence to mathematical vocabulary but meaningful acceptance of mathematical language being used powerfully to describe and label developing concepts. For example, the use of terms such as "half" and "quarter" were associated with demonstrations of the meanings attached to these terms.

7.6 Positive Impacts Within Families, Extending to Older, and Younger Siblings' Inclusion in Mathematical Activities at Home

While *Let's Count* has clearly impacted the children and the parents who have been immediately involved with it, there is some evidence to suggest that some impact has also been felt by other siblings, both older and younger than the preschool child. The approaches drawn from the program seem to have impacted the overall position of mathematics and mathematical learning in the family. This augurs well for the future sustainability of *Let's Count.*

> *Well he just ... I suppose we had to change doing things with him because he's further along than my older son was as well. So he wants you to help, he wants you to do stuff with him rather than just not caring. He'd rather sit there and do stuff. We've got two school-aged boys. And you know there's a lot of focus on literacy and not necessarily numeracy. So you know, they bring home their home readers but there's never really anything numeracy in that. So it's been good for us all to kind of do that, measurements, counting the time. Yeah they've definitely participated in it.*
>
> *He tends to use it every day, whether it's counting the steps up in a building or yeah, just every time when he's buying things. And he's also encouraging his brother to do the same, which is good. So yeah, he's telling him and also commenting and when he does something right he goes 'Yes, that's right!'. Yeah, it's really good. ... And he was trying to show him how to play dominoes too. Yeah, it's really good. I've noticed the little one actually taking things in more so than Xavier because ... Yeah, he seems to be noticing numbers because Xavier is showing him which numbers are what and how to count. It's really good.*
>
> *Even if it's just being in the back of my mind. It's really, really taking advantage of moments where she can learn and enjoy maths without having to be sitting in a classroom, learning. That's been really good, because then I've thought about it,*

the rest of the family have thought about it. Her brother has really benefited because he gets excited because she is, and he's only 3. So I think it's been really good.

Let's Count has the aim of engaging parents to notice, explore and talk about their preschool children's mathematics in everyday life. There is evidence that it is much more than this, allowing all members of families to experience mathematics in ways that are beneficial.

7.7 Sustainability of *Let's Count* Over Time

Although many of the educators involved with *Let's Count* were quite positive about the steps that they were taking in their centers to maintain the program beyond the year of initial implementation, parents were not always as forthcoming. Communication around the mathematics children were engaging with in both the centers and homes seemed to be maintained or, in some cases, intensified, but some parents felt that "organisational information" was lacking. Often, parents expressed that there was a feeling of an initial flurry of information when the program commenced followed by a relative lack of input into what parents might do next. This situation is likely to impinge on the sustainability of the *Let's Count* within families.

Well, in regards to the Let's Count *program that really was promoted by the preschool that Thomas attended many months ago when it all came out and when the bags were given to parents and so forth. From then it was a primary focus and everyone was doing it and it was reported on by the preschool educators on the day sheets that we were receiving. But obviously now it being six months post that and it really went out the window. It's not really continuing. Everyone is back to their same routine.*

At the beginning they would talk about the Let's Count *program to me, in the first initial stages, and they would tell me about how he was progressing but nothing since the beginning.*

So it's not always me who picks up, sometimes it's my husband, sometimes it's me, sometimes it's my sister-in-law, I think maybe the thing that's lacking is the take home part of it, for me. I don't really know the connection between home and what they're doing at the centre. So that's the thing that is confusing me. I do know they're doing a lot at the centre but as far as what I'm meant to be doing at home, if I'm meant to be doing something, I have no idea what that is.

There were also more positive data to report about the sustained impact of *Let's Count*. Two parents, who had children involved in both the 2013 and 2014 *Let's Count* cohorts, spoke in glowing terms about the "start" that *Let's Count* had given their children as they began school and how this start had endured during the first year of school. Other positive comments about the ongoing impact of *Let's Count* included the following.

> *And it was exciting because he had his own little bag with something in it that was going to help him and he had to learn it and he really enjoyed that. Actually, he still looks ... They've got the bags hanging up in the centre there ready to go and he still looks at it each time he goes in there, to see if there's something new. So it doesn't stop.*
>
> *They [educators] really follow up the kids' interest I and let the parents know what's going on ... That's one thing I've noticed different over the year. Much more feedback, specific feedback so that we can then follow that up at home and continue it.*

8. DISCUSSION

Three research questions framed the longitudinal evaluation of *Let's Count*:

1. How does participation in *Let's Count* impact on children's numeracy knowledge and dispositions as they make the transition to school?
2. What is the impact of *Let's Count* on the educator participants' knowledge, interest, and confidence in mathematics learning and teaching? and
3. What is the impact of *Let's Count* on the participating families' confidence, and knowledge about noticing, investigating, and discussing mathematics with their children?

In this discussion, we take each of these questions in turn and use the themes derived from the data from children, educators, and parents along with the literature reviewed earlier in the chapter to provide evidence-based answers.

8.1 How Does Participation in *Let's Count* Impact on Children's Numeracy Knowledge and Dispositions as They Make the Transition to School?

The assessment of children's formal mathematics knowledge for the 2013 *Let's Count* cohort of children using the MAI yielded highly positive results in terms of growth in most aspects of mathematical knowledge and skill when compared with those of the "comparison" group. On almost every measure, the *Let's Count* cohort bettered the scores of the comparison group, with many comparisons showing statistically significant differences (The Smith Family, 2015b).

These highly encouraging results for the 2013 *Let's Count* cohort were replicated in the results for the 2014 *Let's Count* cohort. A larger sample of children, across four, rather than two, evaluation sites was assessed using the

MAI and reported substantially similar results to those of the 2013 cohort. This consistency is a compelling sign of the impact of *Let's Count* on young children's mathematical development in the year before they begin formal schooling.

Although the "comparison" group is not matched exactly with the *Let's Count* cohorts, both groups are of a similar age, come from the same sites as the 2013 cohort, and are at the same point of their education (end of the preschool year before they start school). On the basis of these results over 2 years, we claim that *Let's Count* does assist young children in learning some mathematical concepts (as measured by the MAI) to a significantly greater level than for young children who have not had the benefit of *Let's Count* and for almost all other concepts to at least the same level.

Another observation is that, for a number of tasks in the assessment interview administered in December, just prior to children beginning school, a very large proportion of children in all cohorts were successful. These tasks included: matching numerals (1–9) with dot collections, matching repeating patterns, counting to 10, ordering three objects by length, comparing two items by lengths, naming common shapes, following directions to position, and calculating totals for 5 and 3 teddies and for 4 groups of 2 teddies. Many of these tasks represent mathematical ideas that are scheduled to be met in the first year of school, according to the *Australian Curriculum: Mathematics* (ACARA, 2013). The extent of children's knowledge has been surprising for many educators and provides further evidence beyond that from Gervasoni and Perry (2015a), Gould (2012), and Hunting et al. (2012). This evidence suggests that a great deal of care needs to be taken when children start school to determine their mathematical skills and knowledge, and design appropriate learning experiences for them.

In terms of children's dispositions toward mathematics, there is evidence that *Let's Count* makes a real difference. One of the key themes identified in the analysis of both the educator and the parent interview data was that the *Let's Count* children showed growing confidence in, knowledge of, enjoyment of, and engagement with mathematics. These data include many examples of educators and parents noticing that *Let's Count* children seek out and explore the mathematics in their everyday experiences. The mantra of *notice, explore, and talk about* mathematics really does seem to make a difference. A number of parents specifically commented on changes that have occurred in their children's positive dispositions toward mathematics and, almost always, claim *Let's Count* as the cause of these changes.

In a similar vein, educators, in their interviews, have commented on the increasingly positive dispositions of the children toward mathematics. The educators also noted some of the "amazing mathematics" that their children do. There are many examples of educators being surprised by the level of mathematics that the *Let's Count* children undertake either at their own volition or in activities initiated by the educators. Such positive dispositions are important as children start school, as they are likely to impact on future learning (Carr, 2001; Ginsburg et al., 2008). The *Let's Count* parents interviewed were feeling quite confident about their children's transition to learning mathematics at school.

Given what is already known about the impact of enhanced mathematical knowledge and dispositions in young children as they start school on their prospects as mathematics learners in later life (Clements & Sarama, 2007; Ginsburg, Klein, & Starkey, 1998; Sarama & Clements, 2015), it would seem that *Let's Count* can claim an important role in future considerations about approaches to enhance the mathematical learning of young children. There seems to be little doubt that *Let's Count* has had an important, positive impact on children's mathematical knowledge and dispositions as they make the transition to school.

8.2 What Is the Impact of *Let's Count* on the Educator Participants' Knowledge, Interest, and Confidence in Mathematics Learning and Teaching?

One of the key themes identified in the analysis of the *Let's Count* educator interviews highlights the impact of *Let's Count* on educator confidence, professional identity, and pedagogical practice. Many of the educators commented on how they feel much more knowledgeable and confident in both their own mathematics and also in understanding how their mathematical education practice can be improved. It seems clear that participation in *Let's Count* has made a substantial positive impact on the educator participants' knowledge, interest, and confidence in mathematics learning and teaching. Given what is already known about the impact of educators' knowledge, skills, and confidence in mathematics on their children's mathematical learning (Anthony & Walshaw, 2007; Dunphy, 2009; Hunting et al., 2012), it would seem that *Let's Count* has done some very important work in this area.

8.3 What Is the Impact of *Let's Count* on the Participating Families' Confidence, and Knowledge About Noticing, Investigating, and Discussing Mathematics With Their Children?

Many parents and many educators commented in the evaluation interviews that families are much more able to notice mathematics in their and their children's environments and more willing to talk about mathematics with both the children and educators since they engaged with *Let's Count*. Both educators and parents commented on the much greater levels of communication between parents and educators about their children's mathematics that has arisen from participation in *Let's Count*. One of the key themes to emerge from analysis of the educator interviews was the ways in which educators and families could engage in conversations much more positively and regularly than they did prior to the introduction of *Let's Count*. Data drawn from the interviews suggest that *Let's Count* provided educators with many opportunities to enhance the mathematical outcomes of children and their families.

Another theme concerned the ways in which mathematics learning in homes and early childhood settings were able to be more related. *Let's Count* impacted on the abilities of family members and educators to follow up on the activities that had been undertaken in the other setting. This provided a reason for communication that was embraced by many educators and parents, thereby building relationships and increasing the confidence of both groups.

Many parents reported on the impact of *Let's Count* on their ability and confidence to identify the mathematics that their children were noticing and to extend their mathematics learning through questioning and the use of mathematical language. On a number of occasions, self-confessed "mathematics phobics" found that not only could they assist their children notice, explore, and talk about mathematics but that they were enjoying doing so and would continue. *Let's Count* clearly has had major positive impacts on the participating families' confidence, and knowledge about noticing, exploring, and talking about mathematics with their children.

9. CONCLUSION

Let's Count has been evaluated over 2 years and was found to be effective for enhancing children's mathematics learning in the years prior to beginning school. The positive results led the Australian Federal Government to provide funding to The Smith Family to extend the program's implementation in disadvantaged communities across Australia. Although we can anticipate the positive impact that is possible from this extension to many more families, the challenge remains to identify whether the positive gains for children and family members are maintained once children transition to school.

We argue that the positive gains would be maximized through *Let's Count* being extended into the first years of schooling. In a new school setting, once parents hear the familiar mantra of *notice, explore, and talk about mathematics with children during every day activities*, they may feel more confident about their role and ability to discuss their children's mathematics with a new teacher in a new environment. Further, they will already have stories to tell the new teacher about how their child engages with mathematics, and how they promote their mathematical understanding at home. This has great potential to build relationships between parents and teachers in the new educational setting. The school could promote the *Let's Count* mantra through their website, newsletters, and posters, and provide examples of how parents can continue to engage with their children as their understanding of mathematics grows at school.

There are examples of parents who participated in the longitudinal evaluation claiming that *Let's Count* assisted their child's transition to school. We believe that the implementation of *Let's Count* in the first year of school should be trialed and evaluated against a comparison group. This will further test whether the positive effects of *Let's Count* continue, extend, or fade after children's transition to school.

The findings reported in this chapter also demonstrate that, prior to beginning school, many Australian children have constructed powerful mathematical ideas involving number, measurement, and geometry. It is vital that both preschool and primary school teachers recognize the extent of the mathematics that children know and use so that they can build upon and extend this knowledge during children's play, explorations, and mathematics lessons. This may also require some adjustment to the mathematics program children typically encounter during the first year of school.

Perhaps the most important feature of the *Let's Count* intervention is that it acknowledges and build's on parents' strengths and capabilities as their children's first mathematics educators (Gervasoni et al., 2017). The findings demonstrate that when early years educators nudge parents and build partnerships centered on families becoming confident to notice, explore, and talk about mathematics in everyday activities, then their young children's learning flourishes (Gervasoni, 2017). Noticing, exploring and talking about mathematics with children during everyday activities require no special materials or resources. Rather, what is needed is awareness and confidence for both parents and educators about how their engagements and discussions with their young children and each other really matter and amplify children's educational opportunities.

The evidence presented in this chapter suggests that the *Let's Count* intervention has been successful. We conclude with the "voices" of a parent and an educator, who alongside other evidence presented in this chapter, illustrate that *Let's Count* is making a positive difference for preschool children, educators, and families in Australian communities that experience multiple disadvantages.

> *I loved it after the first couple of months of it. They don't do any structured real teaching at kinder but stuff like this, just to get the kids interested and thinking about numbers is a really good way for them to get comfortable with it without being scared of it. Because sometimes numbers can really intimidate kids if they don't have any background of it when they get to the school level, so I think it's really … Because even just playing around with it is such a good way to get them comfortable with using numbers and the concept of maths. And even just hearing the language and stuff has to be positive for them getting a good head start at school.*
>
> ***Parent (2014)***
>
> *I wish that everybody had the opportunity to do the* Let's Count, *like as in the whole of New South Wales, because I've just seen such big improvement in the children's concepts of maths and preparing them for school next year … if they've got one more ability … to comprehend maths and how it works then they're going to be so much more comfortable in a new setting and really participate to their fullest.*
>
> ***Educator (2014)***

ACKNOWLEDGMENTS

Let's Count was commissioned by The Smith Family, is supported by the Origin Foundation and was developed in partnership with Blackrock Investment Management. The authors of this chapter gratefully acknowledge the important contribution of all family members, children, educators, and colleagues who contributed to *Let's Count* and its Longitudinal Evaluation.

REFERENCES

Anthony, G., & Walshaw, M. (2007). *Effective pedagogy in mathematics/pāngarau*. Wellington, NZ: Ministry of Education.

Arthur, L., Beecher, B., Death, E., Dockett, S., & Farmer, S. (2015). *Programming and planning in early childhood settings* (6th ed.). Melbourne: Cengage Learning.

Australian Curriculum, Assessment and Reporting Authority (ACARA). (2013). *The Australian curriculum: Mathematics v2.4.* Retrieved from http://www.australiancurriculum.edu.au/Mathematics/Curriculum/F-10.

Australian Institute of Health and Welfare. (2015). *Literature review of the impact of early childhood education and care on learning and development.* Working paper Canberra: Australian Institute of Health and Welfare.

Bailey, D. H. (2014). *What's the point of teaching math in preschool?* Washington, DC: Brookings. Retrieved from http://www.brookings.edu/research/papers/2014/11/13-chalkboard-preschool-math-bailey.

Beilharz, L. (Ed.), (2002). *Building community: The shared action experience.* Brisbane: Solutions Press.

Bernard Van Leer Foundation. (2007). Successful transitions: The continuum from home to school. *Early Childhood Matters, 107,* 23–28.

Bishop, A. (1997). What can culture tell us about numeracy? *Learning Matters, 2*(3), 12–14.

Bowes, J., Watson, J., & Pearson, E. (2009). Families as a context for children. In J. Bowes & R. Grace (Eds.), *Children, families and communities: Contexts and consequences* (3rd ed., pp. 91–110). Melbourne: Oxford University Press.

Breen, S., McCluskey, A., Meehan, M., O'Donovan, J., & O'Shea, A. (2014). A year of engaging with the discipline of noticing: Five mathematics lecturers' reflections. *Teaching in Higher Education, 19*(3), 289–300.

Brofenbrenner, U. (1979). *The ecology of human beings.* Cambridge: Harvard University Press.

Brooker, E. (2003). Learning how to learn: Parental ethnotheories and young children's preparation for school. *International Journal of Early Years Education, 11*(2), 117–128.

Carmichael, C., MacDonald, A., & McFarland-Piazza, L. (2013). Predictors of numeracy performance in national testing programs: Insights from the longitudinal study of Australian children. *British Educational Research Journal.* http://dx.doi.org/10.1002/berj.3104.

Caro, D. H. (2009). Socio-economic status and academic achievement trajectories from childhood to adolescence. *Canadian Journal of Education, 32*(3), 558–590.

Carr, M. (2001). *Assessment in early childhood settings: Learning stories.* London: Paul Chapman.

Charmaz, K. (2004). Grounded theory. In S. N. Hesse-Biber & P. Leavy (Eds.), *Approaches to qualitative research* (pp. 496–521). New York: Oxford University Press.

Christenson, S. L. (2004). The family-school partnership: An opportunity to promote the learning competence of all students. *School Psychology Review, 33*(1), 83–104.

Civil, M., Bratton, J., & Quintos, B. (2005). Parents and mathematics education in a Latino community: Redefining parental participation. *Multicultural Education, 13*(2), 60–64.

Clarke, D., Cheeseman, J., Gervasoni, A., Gronn, D., Horne, M., McDonough, A., … Rowley, G. (2002). *ENRP final report.* Melbourne: ACU.

Clarke, B., & Robbins, J. (2004). Numeracy enacted: Preschool families' conceptions of their children's engagements with numeracy. In I. Putt, R. Faragher, & M. McLean (Eds.), *Mathematics education for the third millennium: Towards 2010.* Proceedings of the 27th annual conference of the mathematics education research group of Australasia (pp. 175–182). Townsville: MERGA.

Clements, D. H., & Sarama, J. (2007). Effects of a preschool mathematics curriculum: Summative research on the building blocks project. *Journal for Research in Mathematics Education, 38,* 136–163.

Clements, D. H., & Sarama, J. (2014). *Learning and teaching early math: The learning trajectories approach* (2nd ed.). New York: Taylor and Francis.

Dockett, S., Einarsdóttir, J., & Perry, B. (2012). Young children's decisions about research participation: Opting out. *International Journal of Early Years Education, 20,* 244–256. http://dx.doi.org/10.1080/09669760.2012.715405.

Dockett, S., Perry, B., Kearney, E., Hampshire, A., Mason, J., & Schmied, V. (2011). *Facilitating children's transition to school from families with complex support needs.* Albury-Wodonga: Research Institute for Professional Practice, Learning and Education, Charles Sturt University. Retrieved from http://www.csu.edu.au/__data/assets/pdf_file/0004/517036/Facilitating-Childrens-Trans-School.pdf.

Doherty, G. (1997). *Zero to six: The basis for school readiness.* Ottawa: Human Resources Development Canada, Applied research branch. Retrieved from http://www.hrsdc.gc.ca/eng/cs/sp/sdc/pkrf/publications/1997-002557/SP-241-02-01E.pdf.

Duncan, G. J., Dowsett, C. J., Claessens, A., Magnuson, K., Huston, A. C., Klebanov, P., Pagani, L., Feinstein, L., Engel, M., Brooks-Gunn, J., Sexton, H., Duckworth, K., & Japel, C. (2007). School readiness and later achievement. *Developmental Psychology, 43,* 1428–1446.

Dunphy, E. (2009). Early childhood mathematics teaching: Challenges, difficulties and priorities of teachers of young children in primary schools in Ireland. *International Journal of Early Years Education, 17*(1), 3–16.

Geary, D. C., Hoard, M. K., Nugent, I., & Bailey, D. H. (2013). Adolescents' functional numeracy is predicted by their school entry number system knowledge. *PloS One, 8.* http://dx.doi.org/10.1371/journal.pone.0054651.

Gervasoni, A. (2017). Bringing families and preschool educators together to support young children's learning through noticing, exploring and talking about mathematics. In S. Phillipson, A. Gervasoni, & P. Sullivan (Eds.), *Engaging families as children's first mathematics educators* (pp. 199–216). Singapore: Springer.

Gervasoni, A., Parish, L., Hadden, T., Turkenburg, K., Bevan, K., Livesey, C., & Croswell, M. (2011). Insights about children's understanding of 2-digit and 3-digit numbers. In J. Clark, B. Kissane, J. Mousley, T. Spencer, & S. Thornton (Eds.), *Vol. 1. Mathematics: Traditions and [new] practices. Proceedings of the 23rd biennial conference of the AAMT and the 34th annual conference of the mathematics education research group of Australasia* (pp. 315–323). Alice Springs: MERGA/AAMT.

Gervasoni, A., Parish, L., Upton, C., Hadden, T., Turkenburg, K., Bevan, K., … Southwell, J. (2010). Bridging the numeracy gap for students in low SES communities: The power of a whole school approach. In L. Sparrow, B. Kissane, & C. Hurst (Eds.), *Shaping the future of mathematics education. Proceedings of the 33rd annual conference of the mathematics education research group of Australasia* (pp. 202–209). MERGA: Fremantle.

Gervasoni, A., & Perry, B. (2015a). Children's mathematical knowledge prior to starting school and implications for transition. In B. Perry, A. MacDonald, & A. Gervasoni (Eds.), *Mathematics and transition to school—International perspectives* (pp. 47–64). Dordrecht: Springer.

Gervasoni, A., & Perry, B. (2015b). *Let's Count longitudinal evaluation: Final report.* Unpublished report. Sydney: The Smith Family.

Gervasoni, A., Perry, B., & Parish, L. (2015). The impact of *Let's Count* on children's mathematics learning. In M. Marshman, V. Geiger, & A. Bennison (Eds.), *Mathematics education in the margins. Proceedings of the 38th annual conference of the mathematics education research Group of Australasia* (pp. 253–260). Sunshine Coast: MERGA.

Gervasoni, A., Phillipson, S., & Sullivan, P. (2017). Insights for engaging families as the first mathematics educators of children. In S. Phillipson, A. Gervasoni, & P. Sullivan (Eds.), *Engaging families as children's first mathematics educators* (pp. 249–261). Singapore: Springer.

Ginsburg, H., Klein, A., & Starkey, P. (1998). The development of children's mathematical thinking: Connecting research with practice. In I. E. Siegel & K. A. Renninger (Eds.), *Child psychology in practice: Vol. 4. Handbook of child psychology* (pp. 23–26). New York: John Wiley & Sons.

Ginsburg, H. P., Lee, J. S., & Boyd, J. S. (2008). Mathematics education for young children: What it is and how to promote it. *Society for Research in Child Development Social Policy Report, 22*(1).
Gould, P. (2012). What number knowledge do children have when starting kindergarten in NSW? *Australasian Journal of Early Childhood, 37*(3), 105–110.
Henderson, A. T., & Mapp, K. L. (2002). *A new wave of evidence: The impact of school, family and community connections on student achievement*. Austin, Texas: National Center for Family and Community Connections with Schools. Retrieved from, http://www.sedl.org/connections/resources/evidence.pdf.
Hughes, M. (1986). *Children and number*. Oxford: Wiley-Blackwell.
Hunting, R., Bobis, J., Doig, B., English, L., Mousley, J., Mulligan, J., ... Young-Loveridge, J. (2012). *Mathematical thinking of preschool children in rural and regional Australia: Research and practice*. Melbourne: Australian Council for Educational Research.
Jacobs, V. R., Lamb, L. L., & Philipp, R. A. (2010). Professional noticing of children's mathematical thinking. *Journal for Research in Mathematics Education, 41*(2), 169–202.
Kiernan, K. E., & Mensah, F. K. (2011). Poverty, family resources and children's educational attainment: The mediating role of parenting. *British Journal of Educational Research, 37*(2), 317–336.
Magnuson, K., & Shager, H. (2010). Early education: Progress and promise for children from low-income families. *Child and Youth Services Review, 32*, 1186–1198. http://dx.doi.org/10.1016/j.childyouth.2010.03.006.
Mason, J. (2002). *Researching your own practice: The discipline of noticing*. London: Routledge Falmer.
Mason, J. (2011). Noticing: Roots and branches. In M. G. Sherin, V. R. Jacobs, & R. A. Philipp (Eds.), *Mathematics teacher noticing: Seeing through teachers' eyes* (pp. 35–50). New York: Routledge.
McCain, M., & Mustard, F. (1999). *Reversing the real brain drain: Early years study final report*. Toronto: Ontario Children's Secretariat.
Melhuish, E. C., Phan, M. B., Sylva, K., Sammons, P., Siraj-Blatchford, I., & Taggart, B. (2008). Effects of the home learning environment and preschool center experience upon literacy and numeracy development in early primary school. *Journal of Social Issues, 64*(1), 95–114.
Miedel, W. T., & Reynolds, A. J. (1999). Parent involvement in early intervention for disadvantaged children: Does it matter? *Journal of Social Psychology, 37*(4), 379–402.
Moss, J., Bruce, C. D., & Bobis, J. (2016). Young children's access to powerful mathematical ideas: A review of current challenges and new developments in the early years. In L. English & D. Kirshner (Eds.), *Handbook of international research in mathematics education* (3rd ed., pp. 153–190). Abingdon, Oxon: Routledge.
Mulligan, J. T., Mitchelmore, M. C., English, L. D., & Crevensten, N. (2013). Reconceptualizing early mathematics learning: The fundamental role of pattern and structure. In L. D. English & J. T. Mulligan (Eds.), *Reconceptualizing early mathematics learning* (pp. 47–66). Dordrecht: Springer.
Ochiltree, G., & Moore, T. (2001). *Best start for children: The evidence base underlying investment in the early years (children 0–8 years)*. Retrieved from http://on-line/. http://www.eduweb.vic.gov.au/edulibrary/public/beststart/evidence_base_project_1002v1.2.pdf.
Papic, M., Mulligan, J. T., & Mitchelmore, M. C. (2011). Assessing the development of preschoolers' mathematical patterning. *Journal for Research in Mathematics Education, 42*, 237–268.
Perry, B., & Dockett, S. (2008). Young children's access to powerful mathematical ideas. In L. D. English (Ed.), *Handbook of international research in mathematics education* (2nd ed., pp. 75–108). New York: Routledge.

Perry, B., Dockett, S., & Harley, E. (2007). Learning stories and children's powerful mathematics. *Early Childhood Research and Practice, 9*(2). Retrieved from http://ecrp.uiuc.edu/v9n2/perry.html.

Perry, B., & Gervasoni, A. (2012). *Let's Count educators handbook*. Sydney: The Smith Family.

Perry, B., MacDonald, A., & Gervasoni, A. (Eds.), (2015). *Mathematics and transition to school—International perspectives*. Dordrecht: Springer.

Peter-Koop, A., & Kollhoff, S. (2015). Transition to school: Prior to school mathematical skills and knowledge of low-achieving children at the end of grade 1. In B. Perry, A. MacDonald, & A. Gervasoni (Eds.), *Mathematics and transition to school—International perspectives* (pp. 65–83). Dordrecht: Springer.

Purpura, D. J., Napoli, A. R., Wehrspann, E. A., & Gold, Z. S. (2016). Causal connections between mathematical language and mathematical knowledge: A dialogic reading intervention. *Journal of Research on Educational Effectiveness, 10*, 116–137.

Rimm-Kaufman, S. E., Pianta, R. C., Cox, M. J., & Bradley, R. H. (2003). Teacher-rated family involvement and children's social and academic outcomes in kindergarten. *Early Education & Development, 14*(2), 179–198.

Rogoff, B. (2003). *The cultural nature of human development*. Oxford: Oxford University Press.

Sarama, J., & Clements, D. H. (2015). Scaling up early childhood mathematics interventions: Transitioning with trajectories and technologies. In B. Perry, A. MacDonald, & A. Gervasoni (Eds.), *Mathematics and transition to school* (pp. 153–169). Dordrecht: Springer.

Shonkoff, J. P., & Phillips, D. (Eds.), (2000). *From neurons to neighborhoods: The science of early childhood development*. Washington, DC: National Academy Press.

Siraj-Blatchford, I., Mayo, A., Melhuish, E., Taggart, B., Sammons, P., & Sylva, K. (2013). The learning life course of at 'risk' children aged 3–16: Perceptions of students and parents about 'succeeding against the odds'. *Scottish Educational Review, 45*(2), 5–17.

Strauss, A. L., & Corbin, J. (1998). *Basics of qualitative research: Grounded theory procedures and techniques*. Thousand Oaks, CA: Sage.

The Smith Family. (2015a). *Let's count*. Sydney: The Smith Family. Retrieved from. https://www.thesmithfamily.com.au/what-we-do/our-work/at-school/early-years-and-primary/lets-count.

The Smith Family. (2015b). *Strengthening early numeracy learning: The Let's Count program*. Sydney: The Smith Family. Retrieved from, https://www.thesmithfamily.com.au/~/media/Files/Research%20and%20Advocacy%20PDFs/Research%20and%20Evaluation%20page%20PDFs/Lets-Count-Research-Report.ashx.

The Smith Family. (2016). *Who we are*. Retrieved from, https://www.thesmithfamily.com.au/.

The Spatial Reasoning Study Group. (2015). *Spatial reasoning in the early years: Principles, assertions, and speculations*. New York, NY: Routledge.

Thomas, J. N., Eisenhardt, S., Fisher, M. H., Schack, E. O., Tassell, J., & Yoder, M. (2014/2015). Professional noticing: Developing responsive mathematics teaching. *Teaching Children Mathematics, 21*(5), 294–303.

van Es, E. A., & Sherin, M. G. (2008). Mathematics teachers "learning to notice" in the context of a video club. *Teaching and Teacher Education, 24*, 244–276.

Vandermaas-Peeler, M., Nelson, J., Bumpass, C., & Sassine, B. (2009). Numeracy-related exchanges in joint storybook reading and play. *International Journal of Early Years Education, 17*(1), 67–84.

Verdine, B. N., Golinkoff, R. M., Hirsh-Pasek, K., & Newcombe, N. S. (2017). Links between spatial and mathematical skills across the preschool years. *Monographs of the Society for Research in Child Development, 82*(1), 1–150. Serial No. 324. Retrieved from, http://onlinelibrary.wiley.com/doi/10.1111/mono.v82.1/issuetoc.

Vygotsky, L. S. (1978). *Mind in society*. Cambridge: Harvard University Press.

Wager, A. (2014). Noticing children's participation: Insights into teacher positionality towards equitable mathematics pedagogy. *Journal for Research in Mathematics Education, 45*(3), 312–350.

Whitebread, D., & Coltman, P. (2010). Aspects of pedagogy supporting metacognition and self-regulation in mathematical learning of young children: Evidence from an observational study. *ZDM, 42*(2), 163–178.

Young-Loveridge, J., Peters, S., & Carr, M. (1998). Enhancing the mathematics of four-year olds: An overview of the EMI-4s study. *Journal for Australian Research in Early Childhood Education, 2*, 82–93.

CHAPTER SIX

Designing Studies to Test Causal Questions About Early Math: The Development of Making Pre-K Count

Shira K. Mattera*[,1], Pamela A. Morris*[,†], Robin Jacob[‡], Michelle Maier*, Natalia Rojas[†]

*MDRC, New York, NY, United States
[†]New York University, New York, NY, United States
[‡]University of Michigan, Ann Arbor, MI, United States
[1]Corresponding author: e-mail address: shira.mattera@mdrc.org

Contents

Abstract

A growing literature has demonstrated that early math skills are associated with later outcomes for children. This research has generated interest in improving children's early math competencies as a pathway to improved outcomes for children in elementary school. The Making Pre-K Count study was designed to test the effects of an early math intervention for preschoolers. Its design was unique in that, in addition to causally testing the effects of early math skills, it also allowed for the examination of a number of additional questions about scale-up, the influence of contextual factors and the counterfactual environment, the mechanism of long-term fade-out, and the role of measurement in early childhood intervention findings. This chapter outlines some of the design

Advances in Child Development and Behavior, Volume 53
ISSN 0065-2407
http://dx.doi.org/10.1016/bs.acdb.2017.04.002

considerations and decisions put in place to create a rigorous test of the causal effects of early math skills that is also able to answer these questions in early childhood mathematics and intervention. The study serves as a potential model for how to advance science in the fields of preschool intervention and early mathematics.

Carefully conducted longitudinal research has led many researchers in the field to argue that preschool children's math competencies are a foundational outcome that may be a pathway to improving a broader set of outcomes for children into elementary school. More specifically, such research has suggested that math may be a vehicle to improving a broad set of children's competencies in addition to math, including language, early reading, and executive function (made up of several factors, including inhibition, cognitive flexibility, and working memory). And theory has been brought to bear to support these claims from longitudinal research: math may build language skills because math learning expands and enriches children's vocabulary—for example, when children learn about comparisons such as "more" and "less"—and math requires children to use language to express and to justify mathematical thinking (Ginsburg, Lee, & Boyd, 2008). In addition, the computational demands of math may build children's working memory and problem-solving skills, both components of children's executive function (Diamond, 2013; Duncan et al., 2007).

Although it is clear that early math skills are *associated* with later positive outcomes, less evidence exists about the long-term causal effects of intervening directly to improve children's math abilities. In other words, if you actively change children's early math skills, does that causally lead to changes in those associated long-term outcomes? A recent paper by Greg Duncan and colleagues, who conducted much of the meta-analytic work demonstrating the predictive power of early math, raises this exact issue (Bailey, Duncan, Odgers, & Yu, 2017). Analysis in the paper demonstrates that associations and quasi-experimental estimates of the relationship between early math and later outcomes are consistently larger than experimental estimates from smaller studies. As a result, Duncan and his colleagues (Bailey et al., 2017) urge the field to consider whether predictive associations between early skills and later outcomes should be interpreted as causal in nature. In doing so, they identify a number of explanations that would account for correlational associations between math and later outcomes. Although one plausible explanation for the correlation is that teaching and improving measureable early math skills directly affects later outcomes, others they propose would suggest that the correlation between early math skills and later outcomes is driven by one or more underlying internal or external stable traits, contextual factors, or aptitudes that account for both early and later outcomes (Bailey et al., 2017).

This chapter describes an effort, called Making Pre-K Count (MPC), that was launched to rigorously address the question about the long-term effects of early math skills. The large-scale randomized controlled trial of early math programs was unique in that it accomplished two goals. First, it contributed to the evidence base around the causal effects of the long-term, cross-domain associations that had been identified in the literature (Duncan et al., 2007). Causal tests that explore whether improving early math abilities at scale in the real world can improve later outcomes can help home in on math's potential as a target of intervention and a public policy lever. Second, in addition to causally testing the effects of early math skills, it also allowed for the examination of a number of additional questions about scaling up promising programs, the influence of contextual factors and the counterfactual environment, the mechanism of long-term fade-out, and the role of measurement in early childhood intervention findings. What differentiates this effort from others that preceded it is that it was intended *not* only as a test of the efficacy of a particular early childhood program. Rather, it was explicitly designed as a test of the predictive nature of early math skills across *domains* and across *time*, at scale in a complex real-world *counterfactual* condition. In doing so, it may provide a model about the ways in which some of the most rigorous designs—randomized trials—might be brought to bear to test long-standing questions about the ability to scale-up programs in the real world, the role of context in program impacts, theories about developmental trajectories of learning and fade-out, and issues of measurement in our field—and, in this way, advance research more broadly in the field of early childhood education.

1. MAKING PRE-K COUNT AND HIGH 5s

The MPC and High 5s studies were led by the Robin Hood Foundation and MDRC, in collaboration with Bank Street College of Education and RTI International. MPC is the first study of the Robin Hood Early Childhood Research Initiative, which was designed to identify and to rigorously test promising early childhood interventions. The initiative is a partnership between Robin Hood, one of New York City's leading antipoverty organizations, and MDRC, a nonprofit, nonpartisan education, and social policy research organization. MPC is also supported with lead funding from the Heising-Simons Foundation, the Overdeck Family Foundation, and the Richard W. Goldman Family Foundation.

MPC tests whether implementing an evidence-based math curriculum (Building Blocks; Clements & Sarama, 2007) along with training and coaching of teachers would improve 4-year-olds' short- and long-term outcomes, relative to the typical pre-k math experience in New York City. This complex counterfactual condition provides the opportunity for a deeper examination of the role of contextual factors, such as classroom math practices and program and student characteristics, for program effects. High 5s builds on the MPC study, aligning children's math experiences from pre-k through the end of kindergarten in an attempt to sustain gains from the pre-k experience and learn about the role of fade-out in early childhood programming. High 5s provides small-group math "clubs" three times a week to some of the kindergartners who had received the MPC math curriculum the year before. The two studies rigorously examine the short- and long-term impacts of math interventions in preschool and kindergarten and also provided much-needed information about the role of measurement in the detection of effects across time. Follow-up data collection planned on a wide array of domains of child learning and development, through early elementary school, allows for the study of any short-term improvements of gains in math learning on children's long-term developmental trajectories.

2. DESIGN CONSIDERATIONS FOR BUILDING A STRONG TEST OF THE EFFECTS OF MATH ON CHILDREN'S LONG-TERM OUTCOMES

This chapter details the most consequential design elements that were put in place as part of the MPC study to build the strongest possible test of the strength of early math skills in the real world that is also able to answer new and emerging questions in early childhood mathematics and intervention. These include several features of the intervention and the counterfactual: (a) *identifying the most promising intervention* to be tested; (b) ensuring adequate *support for strong implementation* at scale; and (c) taking into account the "*business as usual*" *condition*. They also include a number of features of the design: (d) *the need for a rigorous random assignment design* that was unlikely to lead to differences in the intervention or control sites at the start of the study; (e) a large sample of children assessed in *multiple domains and followed well into elementary school*—enabling us to capture the longer term, cross-domain effects of the intervention and thereby gain some perspective on the likely impact of the intervention on later outcomes. Finally, the design included an innovative additional randomization study on top of the original one, to address the potential "fade-out" of effects. We address each of these in turn, later.

2.1 Selecting the Strongest Intervention

First and foremost, a key component of this study of early math skills required the selection of a program with strong potential to increase children's math learning. And, evidence of efficacy was needed across diverse sites and populations, especially populations that would most closely resemble those likely to be found in the selected site for this test—New York City. Secondarily, to ensure that the program could be well-implemented by a research team that were not program developers, the program that was selected needed to be ready for scale.

A structured review process was created for the purpose of selecting the math curriculum that could best meet the goals of the study. First, the team reviewed the early childhood education literature, read evidence-based practice guides, and spoke with experts in the early childhood education field to identify promising instructional models for preschool mathematics. Nine math instructional models were identified in 2011. To identify the strongest possible intervention to test, these potential math curricula for the study were evaluated in three key areas: (1) efficacy evidence, (2) curriculum content and program delivery, and (3) the implementation requirements and readiness for scale-up.

2.1.1 Review of Efficacy Evidence

The first step involved an assessment of the empirical evidence of efficacy, indicated by findings of at least one randomized controlled trial with preschool children; particular attention given to research with samples of children from low-income communities in New York City or other large urban areas, and to replication of findings across studies. The research literature on each of the nine math models was reviewed to determine whether there was sufficient evidence of efficacy to warrant further consideration for the first initiative. Based on this minimum standard of efficacy evidence, six of these models were excluded from consideration for the first initiative. Three remaining curricula had sufficient efficacy evidence for further consideration: *Building Blocks* (Clements & Sarama, 2007), *Big Math for Little Kids* (*BMLKs*; Ginsburg, Greenes, & Balfanz, 2003), and *Pre-K Mathematics*

(Klein, Starkey, & Ramirez, 2002). However, in a head-to-head test of *Building Blocks* and *Pre-K Mathematics*, *Building Blocks* had larger impacts. For that reason, only two early childhood math curricula were included in the more intensive review: *Building Blocks* (Clements & Sarama, 2007) and *BMLKs* (Ginsburg et al., 2003). See Table 1 for the review findings based on studies available in 2011.

The two remaining curricula were reviewed for more detailed evidence of efficacy for classrooms/teachers and for children. The review examined whether each curriculum's efficacy evidence from at least one large, multisite random assignment study took place in a major metropolitan area with a sample with characteristics similar to MPC's targeted sample (a high percentage of racial and ethnic-minority families and low-income families). Within the relevant studies, the review looked at whether there were significant effects on child math outcomes after preschool of at least 0.2 SD on a nationally normed measure or at least 0.5 SD on a developer measure. As can be seen in the top panel of Table 1, both programs had evidence of effectiveness for a relevant sample and context. Building Blocks had a larger evidence base, but BMLKs evidence came from a New York City-based trial.

2.1.2 Review of Content and Program Delivery

The second step included a review of curriculum materials for alignment with early math standards, strategic sequencing choices to maximize child learning, integrated assessment, and program delivery modes that were appropriate for the preschool models that were included in the study. The team conducted an extensive review of the lessons and activities in each curriculum, in order to understand the breadth of mathematical content as well as depth of coverage of each concept within the curriculum. Each curriculum was analyzed to identify the percentage of total curriculum time in which each math content area and each of its subtopics was addressed as a primary focus.

In addition, the curricula were reviewed to see whether the approach to program delivery and sequencing was based upon a clearly articulated theory of children's math learning development, the program emphasized teacher-guided learning with interactive participation of children, the program had an instructional software component that is integrated with the curriculum, and that assessments were integrated into the curriculum and used for leveling of activities and individualization. The second panel of Table 1 shows that, at the time, both programs met these requirements. *BMLKs* covered

Table 1 Review of Building Blocks and Big Math for Little Kids Pre-K Math Curricula

Criteria	**BMLKs**	**BB**
Efficacy evidence		
1. Efficacy evidence from at least one large, multisite random assignment study in a major metropolitan area, including a high percentage of minority families and low-income families	✓	✓+
2. At least one efficacy study found significant effects on child math outcomes after preschool of at least 0.2 SD on a nationally normed measure or at least 0.5 SD on a developer measure	a	✓+
Content and delivery		
3. Coverage of a range of math content areas in alignment with NCTM standards, including thorough instruction in basic early math skills (e.g., number, cardinality, shape), as well as complex early math skills (e.g., patterns, operations, measurement)	✓+	✓
4. The approach to program delivery and sequencing is based upon a clearly articulated theory of child development	✓	✓+
5. Program emphasizes teacher-guided learning with interactive participation of children	✓	✓
6. Program has an instructional software component that is integrated with the curriculum	b	✓
7. Assessments are integrated into the curriculum, and used for leveling of activities and individualization	✓	✓+
Implementation requirements and readiness for scale-up		
8. Program has an established professional development model	✓	✓
9. Meets all or most math-related standards in the New York State Early Learning Standards and Common Core Learning Standards for Prekindergarten, and also the Head Start Child Outcomes Framework and Head Start Performance Standards	✓	✓
10. Spanish language version of the curriculum is available	✓	c

Note: In this criteria table:

- A check mark (✓) indicates that the curriculum met the criterion in a satisfactory way
- A check plus mark (✓+) indicates that the curriculum exceeded the criterion, such as having efficacy evidence from multiple studies rather than just one
- A letter indicates that the curriculum did not meet the criterion fully, but there are footnotes below with additional information that is relevant to the criterion
  a. While *BMLKs* did not meet the second efficacy criterion for the preschool year, this criterion was met for the kindergarten version of the curriculum.
  b. *BMLKs* did not have integrated software in 2011, but it was in development.
  c. Spanish versions of the software and some curriculum materials were available, but the full *BB* curriculum had not been translated into Spanish.

a broader range of content areas than *BB*, specifically touching on each of the areas included in the NCTM math standards. *BB* focused more on number and operations and a bit less on some math topics such as measurement and representing data. Still, *BB* did cover most of the topics in the NCTM standards, and the developer had a developmental rationale for the emphasis on certain topic areas that were included in the curriculum. In addition, the sequencing of *BB* was based upon child learning trajectories in math content areas that extend through sixth grade, and topics are reinforced throughout the year as the level of difficulty progresses, whereas BMLKs relied on discrete units. At the time of the review, BMLKs also did not have a finalized computer program.

2.1.3 Review of Implementation Requirements and Readiness for Scale-Up

Finally, curriculum materials were reviewed and program developers were interviewed to determine whether the curriculum model was ready for large-scale implementation in the study context. A central question for this study was to examine the effects of a math intervention in the complex, real-world context of New York City pre-ks. Identifying curricula that not only focused on the appropriate math domains but also had the infrastructure to support teachers' practices in those math domains was essential for addressing questions about program implementation and scale-up. The review included information about the level of training, coaching, technical assistance, and other preparations necessary, and the presence of a standardized curriculum manual to achieve significant effects on teacher's math practices in a large-scale implementation, as well as the developer's availability to support such a large-scale implementation of this nature. The team also examined whether the curricula were appropriate for the study context (New York City) in meeting New York's math-related standards and having appropriate supports for the City's diverse student population by having Spanish-language materials available. Both programs had manualized curricular materials, although those from *BB* were ready for dissemination to teachers on a large scale. At the time of the review, Building Blocks had some Spanish-language materials but was not fully translated into Spanish. Also at the time, BMLKs did not have a manualized professional development model and had not previously used coaches to support implementation.

Based on this in-depth review, the team selected the *Building Blocks* math curriculum (Clements & Sarama, 2007) to include in its study of children's long-term math learning. The curriculum had strong efficacy evidence for teachers and children across a number of studies and diverse setting. In addition, the curriculum was sequenced in a way that experts believe may be important for child learning, cycling to reinforce content over the year while building difficulty over time. Finally, the program had a high level of manualization, with formal, written lesson-plans for teachers and documents to fully support the training, coaching, and technical assistance appropriate for use in diverse contexts. These three characteristics suggested that Building Blocks was highly likely to lead to changes in teachers' math practices and children's math outcomes in the preschool year when scaled-up across New York City preschools, allowing for a strong test of the effects of gains in math learning in preschool on math and other domains of learning in the short- and long-term and further investigation of variation in real-world implementation within the context of New York City.

2.2 Designing tor Scale: Ensuring Strong Implementation

The work reviewed in the curriculum selection stage demonstrated that curricular enhancements can be implemented effectively to produce moderate to large improvements in children's math skills, at least in the short term. Yet, despite this promising research, there is still a need for strong evidence that these gains can be achieved when the models are tested at scale amid the complexity of New York City's service delivery system. To support the Building Blocks curriculum in this study and address the question of scale-up, an implementation infrastructure was created in New York City. The infrastructure included ongoing training provided by the developer-approved trainers, in-classroom coaching provided by a local high-quality provider (Bank Street College of Education), and technical assistance supported by a management information system.

Developer-created training content was provided across 2 years to over 170 lead and assistant teachers. In the first year, lead and assistant teachers were offered 6 days of Beginner Training, which focused on understanding the curricular content and activities. In the second year, lead teachers were offered 5 days of Advanced Training, focusing on using child data to differentiate instruction to children at different skill levels. To address turnover between years, the Beginner Training was also reoffered in the second year to any new lead teachers. Curriculum developers had a small cadre of trained staff who had previously led teacher training. As part of scaling up, curriculum developers also trained additional local trainers to ensure there were enough skilled staff to provide training.

A coaching model was developed in collaboration with Bank Street, based on best practices in the field and the coaching model previously used to support the curriculum (TRIAD; Clements, Sarama, Spitler, Lange, & Wolfe, 2011). Pairing teacher training sessions with ongoing, in-classroom coaching has been shown to have the best effect on supporting teachers' transfer of what they have learned in training to their work with children in the classroom (Joyce & Showers, 2002; Sheridan, Edwards, Marvin, & Knoche, 2009). Bank Street hired, trained, and cosupervised the MPC coaches. The coaching model required that coaching occur in the classroom, when the coach could observe teachers implementing the Building Blocks curriculum and support implementation in real time as needed. Coaches also met with the lead and assistant teachers to reflect on how implementation had been going in the classroom that week and offer implementation support. In the first year, teachers received an average of 149 out of the expected 180 min of coaching weekly. In the second year, teachers received 99 out of the expected 120 min of coaching every other week. Nearly, all coaching sessions that were expected to occur took place (96% in the first year, 98% in the second year).

Finally, technical assistance was offered by MDRC, the research organization leading the study. A management information system was designed to track coaching and implementation dosage and quality through online logs completed by the coaches. Coach logs collected information about how often the Building Blocks activities were implemented in the classroom, the quality of each individual curricular component, and the overall quality of implementation. For example, the logs asked about how often Building Blocks whole-group activities were conducted that week and the quality of whole-group implementation as seen by the coach. The logs also asked more holistically about implementation, using Likert-scale type questions such as "This class looks like a Building Blocks classroom." The technical assistance team monitored these logs in real time, comparing data against a set of prespecified benchmarks created in collaboration with the curriculum developers. Table 2, drawn from the MPC pre-k report, outlines the prespecified benchmarks for each curricular component. For example, Building Blocks includes a whole-group math activity for each day of its 30-week curriculum. The whole-group benchmark specified that a core

Table 2 Building Blocks Curricular Component Definitions and Technical Assistance Benchmarks (Morris, Mattera, & Maier, 2016)

Curricular Component	Definition	Technical Assistance Weekly Benchmark
Main component		
Whole group	Activity led by a teacher and conducted with the majority of children in a class.	A core whole-group activity is completed most days (at least 66%) that children are in attendance in a week.
Small group[a]	Activity led by a teacher and conducted with three to four children in a class.	At least 75% of all children participate in small group.
Hands on math centers	Activities or manipulatives[b] for children to work and play with independently, or with a small group of children, with or without a teacher.	Math center activities are available most days (at least 66%) that children are in attendance in a week.
Computer	Activities made available to children through the BB web-based computer software.	At least 75% of all children participate in computer activities.
Supplementary component		
Small group record sheet	Template for teachers to record children's participation in and response to small group activities.	Teachers fully complete at least one small group record sheet for the week.
Family letters	Ready-made letters in English or Spanish that are sent home with children to help parents reinforce BB content at home.	Family letters "sent home" or "not sent home but did not need to" during the week.
ConnectED	Teachers' version of BB software that allows them to assign computer activities to children and to review reports of children's activity completion and progress.	Teachers access ConnectED during the week.

[a]Each week, the Building Blocks (BB) curriculum included one or two small group activities. In Year 1 (2013–2014), the research team asked teachers to conduct at least one small group activity on a weekly basis. In Year 2 (2014–2015), teachers were asked to conduct both small group activities if more than one was listed.

[b]Manipulatives are hands on objects that allow children to explore abstract math concepts concretely. Clements and Sarama (2013).

whole-group activity should be completed most days (at least 66%) that children are in attendance in a week. The benchmarks were designed to take into account developer expectations for implementation and help identify programs in need of additional support and technical assistance. The team met weekly with coaching staff to troubleshoot implementation challenges and to identify and provide support to persistently low-implementing classrooms.

These three components were essential to addressing the questions about early math *at scale*. The training and coaching put in place in this study was comparable to the implementation supports put in place in other studies of Building Blocks as well as real-world implementation of Building Blocks in Boston Public schools (Clements et al., 2011; Weiland & Yoshikawa, 2013). In other words, the implementation levels achieved by this implementation infrastructure created conditions similar to those that would be expected with real-world implementation outside of a research design. But most importantly for the causal effects of math—the model that was put in place was designed to have a good chance of changing math practices among preschool teachers as a vehicle to improving the early math skills of young children.

2.3 Considering the Counterfactual Condition and Context

A key consideration for developing a strong test of children's early math skills is ensuring that the intervention makes a difference over and above what children would normally receive in preschool—that is, over and above the counterfactual condition. The impact of an intervention is produced by the program's treatment contrast, which is the difference between the actual services received by participants and their counterfactual service receipt (what they would have received if they had not participated). That condition is defined by a combination of the program or services already in place, the characteristics of the participants, and the context in which these participants receive these services. In this case, given our interest in improving math skills, our focus was on ensuring that any program that was implemented in the program group classrooms would be better in terms of supporting math skills, than that which children would receive in the absence of the treatment.

In the past, math instruction has seemed particularly ripe for this kind of a test because math instruction has historically been underemphasized in preschool. Earlier descriptive studies have found that preschool teachers rated math as the lowest priority for children, behind preliteracy and social–emotional development (Lobman, Ryan, & McLaughlin, 2005). When researchers conducted focus groups with these same preschool teachers, teachers had "no substantive ideas about how teachers could be prepared to teach [math]" (Lobman et al., 2005). And, this is highly consistent with findings on classroom instruction: research has shown that when teachers do provide any lessons in math, they have often focused on quite simple aspects, such as the names of shapes and numbers from 1 to 20, without the richness of mathematical reasoning, inferences, metacognition (explaining how one arrived at an answer), and complex vocabulary that characterize many of the successful, developmentally appropriate curricula that experts recommend (Ginsburg et al., 2008; National Research Council, 2009). Clifford et al. (2005) found that only 8% of classroom time was spent on math activities involving counting, time, shapes, sorting, while 21% was devoted to literacy activities. Thus, it was thought that providing teachers with training in delivering more math instruction might represent a substantial shift in children's preschool experiences compared with what the typical preschool has provided (and thus a resulting change in children's math learning).

Interestingly, "business as usual" pre-k programming in the MPC study turned out to be a shifting target. The years of the study (2012–2015), including a pilot and 2 years of implementation, took place during a time of major change for New York City pre-K. In 2011, the city began rolling out their preschool Common Core learning standards, requiring programs to implement a Core-aligned curriculum of their choice and implementing math and literacy tasks embedded in thematic units. In 2012, the city established consistent program quality requirements across all sites through EarlyLearn, its consolidation of funding streams. And in 2014, Mayor Bill deBlasio's signature campaign promise of full-day pre-k for all 4-year-old children led to the creation of an additional 34,000 pre-k slots.

Although none of these initiatives was a direct math intervention, they led to growing scrutiny of pre-k and a focus on instructional quality. This context was different than the pre-k landscape of 10 years ago. While in theory this may or may not have led to growing math content being delivered in preschool classrooms, in practice, this translated into a *much larger* amount of math—35 min of math instruction—taking place in New York City pre-ks compared to prior studies of Building Blocks (and, indeed, other math-focused preschool curricula, as well). This context, which included a much greater amount of time on math than has been seen in other studies, suggests that the nature of math instruction may be changing and math interventions may have a higher bar to clear to make a difference for young children.

A key aspect of understanding this context was data collected in the study on teachers' math practices, classroom climate, and the structure and content of instructional time during the day. These data were used to measure impacts on teacher practices and classroom processes, and therefore were collected not only in the program group, but in the control group as well. Collecting detailed, longitudinal data about the counterfactual condition was a critical component of the study—because it provided not just an analysis of program impacts but a deeper examination of the ways in which the historical and geographical context for the study may have played a role in producing such program impacts.

2.4 Building a Rigorous Design

To investigate the relationship between early math and later math, language, and executive function outcomes, it was critical to implement a rigorous causal design. As such, a cluster-randomized controlled trial was implemented, in which full pre-k centers were either offered the math program or assigned to a control group using a lottery-like process. This randomization process (a) ensured minimal, if any, spill-over from one group of teachers to another (which was especially important for this intervention, which calls for teachers to implement the program for 2 years in order to get comfortable with the program); (b) strengthened the impacts on teacher practice by espousing "whole school reform" efforts; and (c) most importantly, allowed for the calculation of unbiased estimates of treatment that can be leveraged for other causal estimates of the effects of math on other aspects of child learning and development (Bloom, Bos, & Lee, 1999; Gennetian, Magnuson, & Morris, 2008).

Thirty-five of 69 pre-k sites serving predominantly low-income children were assigned to receive the math curriculum, training, and coaching (the program group) over 2 years. Teachers in these 35 program sites spent the first "gentle introduction" year learning the curricular components and activities and receiving weekly coaching to support classroom implementation, with the second implementation year focused on high-quality implementation. The other 34 sites were assigned to continue their typical programming (the "pre-k-as-usual" control group) during the same 2 years. While teachers implemented the curriculum and received coaching and training for 2 years, as designed by the program developer, impacts on child outcomes were assessed for the children who were in the classrooms in the second year of implementation. This design allowed teachers to become familiar with the program, while still testing the effects of 1 year of the Building Blocks curriculum for children.

Notably, we blocked pre-k sites into groups of four to six sites before randomization. This blocking was critical to ensure a strong "match" between treatment and control sites in a constrained sample. We did not want the lottery-like process of randomization to result in all program sites in a single borough or serving a particular demographic group of children, and all program sites in another or serving a different demographic group, as we discuss later. Sites were selected to reflect the geographical, racial, and ethnic diversity of New York City's low-income population. Approximately two-thirds ($n = 47$) of sites were in public schools, while the remainder ($n = 22$) were in community-based organizations. Sites were selected in four of New York City's five boroughs. Sites were blocked based on three criteria: venue (CBO vs public school), borough, and racial-ethnic composition (sites that served a primarily Hispanic population vs other). Blocking of sites was conducted for two key reasons: first, it provided for some protection against a "bad" draw in which treatment and control sites are poorly matched. This is particularly likely to occur in studies such as this one in which there are few units at the level of randomization (i.e., only 60–70 sites). Second, blocking (as compared to "pairing" sites) provided some protection against the loss of sites between randomization and the beginning of the "impact" or second year of study. That is, given the lengthy period of implementation, sites could have dropped out of the study before the second year of implementation—although in practice, this did not happen. Given the challenge of holding onto a matched waitlist of sites for a lengthy period, blocks of four or larger sites made it such that the block could be part of the analysis even if a single center had dropped out of the study.

The blocked, cluster-randomized controlled trial spanning 2 years of implementation ensured the most rigorous test of the effects of an early math intervention. Blocking sites to account for a 2-year implementation maximized teachers' time and familiarity with the curriculum by allowing them time to grow comfortable with implementing the curriculum (and thus ensuring strong potential impacts on teachers' practice—the key driver to strong impacts on children's math learning), while protecting the random assignment design against the potential for attrition across such a long study window. Randomly assigning full pre-k sites minimized the likelihood of spillover from one classroom to another while also capitalizing on training and coaching cohorts of teachers within a school.

2.5 Measuring Children's Outcomes Over Time

One reason improving math has seemed promising as an intervention strategy is the set of findings from longitudinal studies suggesting a link between early math skills and children's later outcomes (Duncan et al., 2007; Duncan & Magnuson, 2011). Seminal studies such as Perry Preschool or Abecedarian demonstrate long-term impacts of early cognitive intervention, even in light of early effects "fading out." Some researchers have suggested that the relationship between impacts on early cognitive skills and later reemerging impacts in adulthood may pass through noncognitive or "dark matter" skills that have historically not been measured, such as executive function (Heckman, 2006). However, this theory has not been adequately tested for two primary reasons. First, few intervention studies follow children over time, allowing for an examination of program impacts longitudinally. Additionally, measurement challenges in the field of early childhood make it difficult to parse apart whether a comprehensive set of developmental domains have been assessed at all vs whether they have been assessed well enough to measure program effects.

To test the developmental theory that improving math skills could spill over into children's other outcomes and to examine the role of measurement in the detection of those effects over time, it was imperative to design the study (a) to examine children's outcomes in the short- and long-term and (b) to include a broad, rich set of measures that would allow for a test of math on children's learning in other domains of development and an examination of the role of measurement on previous patterns of program impacts. Such a longitudinal data collection would collect rich data at multiple follow-up time points to fully understand the long-term effects of supporting math skills. The data collected would also include multiple measures within the same domain to parse apart how measurement influences the ability to detect program effects. The MPC study addressed this issue, with a data collection plan designed to collect direct child assessments and school records across a variety of outcome domains from preschool through third grade.

Described in greater detail in the next section, research on early childhood programs shows that effects on some outcomes may fade over time, but long-term impacts on income can sometimes still be observed into adulthood (Chetty et al., 2011). This suggests that certain unmeasured variables may in fact be a pathway through which interventions affect long-term outcomes. An important theoretical reason for this study's focus on math was not only to bolster children's math skills in the short term, but also to assess whether improvements in math capabilities would translate into improvements in other domains of learning and development and across time. Indeed, when examining the effects of an intervention, many studies only assess the targeted domain of interest—in this case, math—limiting our ability to address questions about the cross-domain implications of early math skills. In MPC, we intentionally addressed this gap, by collecting information both on outcomes directly related to the intervention (such as math skills) and other key domains (language, executive function) that are of critical importance to advance the field's understanding of preschool's effects into elementary school, and eventually adolescence and adulthood.

In identifying the constructs to be assessed, two key issues were considered: (1) which constructs were most aligned with the intervention's focus, directly as well as indirectly, and (2) which constructs matter most for children's later outcomes. A careful review of the program's theory of change was used to prioritize constructs aligned with the intervention's focus, while a review of longitudinal research helped identify constructs that matter most for children's outcomes at different time points. For example, while receptive language was an important indicator of a child's understanding of rich language and math talk in the early years, by third grade the project's language focus had shifted to reading comprehension, given the math intervention's focus on children's critical thinking skills.

Measuring children's skills across time presents a number of design challenges. First, children's skills in the preschool-to-third-grade age range change exceptionally fast. As with every longitudinal study, it may not be possible to collect the same measures across multiple time points. For instance, executive function measures of inhibition in preschool (such as a computerized pointing task or a Simon Says-like task) are too easy for third graders and must be adapted for later skills. To address this issue, the MPC study first identified the most promising measures for each construct of interest in preschool. Then, the measures were mapped across time to identify instruments either that could be used across multiple age ranges or that could be mathematically equated across time. For measures or domains where it was not possible to maintain the same instrument across time, the study identified new measures that used similar methodologies and captured the same construct if possible.

In pre-k, the study assessed children's math skills with two direct assessments. The Early Childhood Longitudinal Study-Birth Cohort (ECLS-B; Najarian, Snow, Lennon, & Kinsey, 2010) is a more proximal, IRT-based math measure assessing children's number sense, operations, measurement, geometry, spatial sense, and patterns. The Woodcock-Johnson III Tests of Achievement: Applied Problems (WJ-AP; Woodcock, McGrew, & Mather, 2001) is a more distal, nationally normed math measure of mathematical thinking. Importantly, different measures of the same construct may provide unique information. For example, the study was interested in two aspects in the math domain. A detailed and sensitive measure of children's math ability in multiple subdomains (e.g., numeracy, geometry, patterning)—the ECLS-B—was needed to clearly identify the areas of math where the program had effects. At the same time, a more widely used but less sensitive measure with extensive literature on its predictive power—the Woodcock-Johnson Applied Problems—was needed to better understand what these impacts might mean for future outcomes. Including two differing measures of the same construct, each with differing strengths for assessing program impact and for assessing the long-term implications of such program impacts, strengthened the study substantially.

As indicated earlier, it was also critical to assess other domains of child learning and development in MPC. Receptive language was assessed using the nationally normed Receptive One-Word Picture Vocabulary Test (Martin & Brownell, 2011). Children's executive function (including working memory, inhibition, and cognitive flexibility) was assessed using three measures, each that tapped differing subcomponent skills of executive function. The Pencil Tap (Diamond & Taylor, 1996; Luria, 1966) assesses inhibition, the Spatial Conflict Arrows task (Willoughby, Wirth, Blair, & Family Life Project Investigators, 2012) assesses cognitive flexibility, and Corsi Blocks (Corsi, 1972; Lezak, 1983) assesses working memory.

Collecting information about program impact with differing measures of the same construct, across developmental domains, and across time allows for the most complete assessment of a program's impacts. This may be particularly important for an early childhood math intervention, which is theorized to have not only localized effects on math during the intervention year but also effects that spill over into other domains and across time. The measurement challenges encountered provided an opportunity for new learning into measure sensitivity and overlap, measurement equivalence, and how measurement influences the detection of program impacts.

2.6 Addressing the Fade-Out of Effects

One of the challenges in preschool intervention research is the potential for the "fade-out" of effects (Lee & Loeb, 1995). Frequently, the cognitive and academic benefits from even the most promising preschool interventions have not lasted when the children were assessed later in their elementary school years. As a case in point, MDRC's Foundations of Learning Project found benefits to preschool classrooms and children's outcomes (reduced conflict and increased classroom engagement) from a program targeting teachers' classroom management skills (Morris, Raver, Millenky, Jones, & Lloyd, 2010). However, the effects observed in preschool classrooms were not sustained after children entered elementary school classrooms. Such a phenomenon could be due to either the quality of the elementary schools that children ultimately attend, or to control group "catch up," but it suggests that ongoing support beyond a 1-year preschool intervention may be critical for sustained impacts. A follow-up to one of the Building Blocks trials added on a light-touch follow-through in kindergarten for some children who had received the Building Blocks curriculum in preschool. The follow-through taught teachers about the entering children's skills and some activities that could help support those skills. The follow-up showed that children who received the kindergarten follow-through in addition to preschool Building Blocks had less fade-out of effects in first grade than children who received Building Blocks alone (Sarama, Lange, Clements, & Wolfe, 2012).

In light of this, we wanted to take a multipronged approach to skill development that would include not only a strong approach to changing "pivotal" skills in preschool (i.e., our focus on math for MPC) but also an additional investment in improving children's experiences during kindergarten through ongoing intervention. From our perspective, to do the former without the latter would be insufficient. Moreover, this dual-pronged approach expanded the ability of MPC to answer additional questions about the course of child impacts and the potential fade-out of effects.

In designing this additional investment in kindergarten, we reviewed the literature and had a number of discussions with experts in the field. In doing so, we considered a number of different intervention options for implementation in kindergarten as an add-on to the MPC effort, including changes to the kindergarten curriculum, teacher professional development, in class enrichment, one-on-one tutoring, and pull-out small group enrichment. These options were evaluated against the following six criteria:

1. *Potential to sustain impacts*—the extent to which the program, when offered in kindergarten, had the potential to lead to continued math gains for the students who received the Building Blocks curriculum in preschool.
2. *Evidence of effectiveness*—the degree to which there was evidence to support the effectiveness of the program model.
3. *Logistical complexity*—the ease or difficulty of achieving buy-in from administrators and teachers, scheduling and logistics, and managing tutors/facilitators/coaches/teachers.
4. *Direct program intervention cost*—the degree to which the program had the potential to be cost effective. We considered the costs of developing or modifying an intervention, buying, or developing training materials, employing tutors/coaches/teachers, and providing them with training and supervision. We also considered possible additional implementation costs such as the extent to which technical assistance would be needed.
5. *Long-term scalability*—the likelihood that the program could be scaled in New York City or replicated in other areas.
6. *Maintain integrity of the MPC study design*—the degree to which the design for evaluating an add-on kindergarten intervention would maintain the original design for evaluating the pre-kindergarten math enhancement.

Based on these criteria, pull-out small group enrichment, or what we called a "math club" model, emerged as the most promising approach to support children's ongoing math learning following pre-k. A large number of studies of tutoring for low-achieving students have shown it to be an effective strategy for improving children's skills across domains, although most tutoring programs to date have focused on the literacy domain. In fact, a review found that, across 16 studies, the effect sizes for one-on-one tutoring were positive *in nearly every case* (Wasik & Slavin, 1993). This consistency of findings is quite unusual for intervention research. A meta-analysis conducted by Slavin, Lake, Davis, and Madden (2011) found an overall weighted mean effect size of 0.39 for one-on-one tutoring interventions by teachers, and a similar effect size of 0.31 for small group tutoring. Studies by Brown, Morris, and Fields (2005) and Ehri, Dreyer, Flugman, and Gross (2007) compared certified and paraprofessional tutors and found certified tutors to be much more effective, but both studies still found substantial positive effects for the paraprofessionals. And meta-analyses of tutoring with paraprofessionals shows positive results with struggling readers (weighted mean ES = 0.38; Slavin et al., 2011). The support for the math club option rested heavily on both its feasibility, its lower cost compared to a one-on-one tutoring intervention, and on the consensus that it would maximize the likelihood of large and sustained impacts on children's math skills because math clubs would provide an intensive, direct intervention for the children who had benefited from the *Building Blocks* curriculum.

Previous research demonstrates that high-quality math curricula are effective at raising the achievement of young children (Agodini, Harris, Thomas, Murphy, & Gallagher, 2010; Agodini et al., 2009; Hofer, Lipsey, Dong, & Farran, 2013; Klein, Starkey, Clements, Sarama, & Iyer, 2008; Sarama, Clements, Starkey, Klein, & Wakeley, 2008), and we designed the math club model to incorporate key elements that have been found to be associated with effectiveness including highly engaging activities and differentiated instruction in small groups.

The math club program was developed at the University of Michigan. Called *High 5s*, the program provided a small group enrichment experience for kindergarten students who had received the *Building Blocks* program in kindergarten, using staff who were offered a paraprofessional salary to implement the program. In the *High 5s* program, groups of approximately four kindergarten students met three times a week with a trained facilitator from Bank Street College of Education. Facilitators provided students with targeted instruction as they played fun, engaging math games in a small group setting. The High 5s program was developed with input from the developers of *Building Blocks* and was developed specifically as a follow on to the *Building Blocks* program, building on the "learning trajectories" which form the basis of Building Blocks. It was also designed to be aligned with the Common Core curriculum used by the New York City public schools, with the idea that the activities in the *High 5s* clubs would reinforce concepts learned in the classroom, but also go beyond the content typically covered in kindergarten. Activities were designed to move students along four key mathematical learning trajectories: counting, composition of numbers, early addition and subtraction, and geometry, and also included activities related to measurement and patterning. The program was intended to provide enrichment, not remediation, and to provide students with continued exposure to high-quality math instruction during the kindergarten year.

Once we had identified an approach to this additional boost in math in kindergarten, the next question was whether this should be provided to all children who had experienced enhanced math in pre-k or only a subset of children. In the end, a decision was made to rerandomize the children who had received the preschool math program—providing an embedded evaluation within the larger evaluation—to provide the greatest learning from the overall effort.

More specifically, the design of the kindergarten study was as follows: Students attending the 24 public schools participating in the MPC project who received *Building Blocks* during pre-k were eligible to participate in the High 5s programs. Once parental permission was obtained, eligible students were randomly assigned within schools to either receive the High 5s program or to a "business as usual" control group. A total of 25 facilitators, supported by five supervisors, administered the program to over 300 children in 79 clubs across 24 New York City public schools during the 2015–2016 school year.

This two-stage random assignment design allowed for a wider range of developmental questions to be addressed. The second round of random assignment creates three sets of causal comparisons: (a) the effect of receiving MPC relative to the typical pre-k experience in New York City; (b) the effect of receiving High 5s and MPC compared to MPC alone; and (c) the effect of receiving High 5s and MPC compared to the typical pre-k and kindergarten experiences in New York City. In addition to testing the direct effect of a new, small-group math program (question b), it also addresses questions about the developmental progression of children's skills and how those skills develop in different contexts. The study can examine how sustaining an aligned, enriched math environment over 2 years may affect the course of children's outcomes compared to 1 year of enrichment (or none). Noncausally, High 5s may also let us examine how or if impacts fade or outcomes grow across higher quality and lower quality instructional contexts. Directly speaking to the ongoing debate about the implications of program fade-out in early childhood, the MPC/High 5s two-stage random assignment design will provide much-needed descriptive, correlational, and causal information about the developmental course and sequelae of potential program fade-out.

3. EARLY FINDINGS AND CONCLUSION

Previous research makes it clear that math in the early childhood years could be a potentially meaningful lever for changing children's outcomes in the long term. However, whether a math intervention, implemented *at scale* in a complex and changing *control context*, could effectively leverage these associations between early math and later outcomes to have an impact, *measured across domains* and *sustained across time* for children's outcomes remained to be seen. The MPC and High 5s studies were designed specifically to test these questions. As such, the studies included a number of components to ensure a thorough and comprehensive examination of not just the causal effects of math but also implementation at scale, the role of classroom math context, the developmental progression of fade-out, and the influence of measurement.

This design has already proved fruitful in adding information to the field: Results in the pre-k year demonstrate that while teachers were able to substantially change the amount of math provided to preschoolers—and to a lesser degree, the quality of that math instruction—there were no effects on children's math, language, or executive function outcomes by the end of pre-k (Morris et al., 2016)—although there was evidence of early math impacts in the first half of the pre-k year (that is, based on a fall assessment, that does not appear to be due to a "poorly matched" random assignment process). This in itself is surprising, suggesting that more exploration is needed of the causal relationship between instructional dosage and quality and child outcomes and the role that the counterfactual condition may play in classroom-level and child-level impacts. Future research on early childhood intervention will need to begin to unpack causally which aspects of instructional quality matter, in which contexts, and how, for children's outcomes.

The larger challenge, of course, is that the lack of impacts on math learning at the end of pre-k may limit the potential of this study to address the larger questions about how early math learning might improve outcomes for children, across domains of learning. That said, the jury is still out on these findings. We have yet to fully investigate whether there are some groups of sites, teachers, or children for whom the implementation of Building Blocks produced positive outcomes for children, as well as whether the lack of impacts at the end of pre-k is due to children's "lack" of learning, to the insensitivity of our measurement tools to address children's deeper math learning, or to earlier-than-expected impacts in the *fall* of the pre-k year fading out faster than has previously been documented.

The design considerations laid out in this chapter make this study uniquely suited to shed light on all of these questions. Comprehensive information about the classroom context and student and teacher characteristics will allow for a nuanced exploration of the program's effects for different groups and contexts. The rich data collection reaching into kindergarten and elementary school will provide additional information about how different measures capture children's knowledge and learning and whether more sensitive measurement might detect program effects in a more nuanced manner. Additionally, the longitudinal nature of the data collection will add much-needed information about the trajectories of children's outcomes—and impacts—over time. This will allow for an investigation of whether fall impacts were an anomaly or an example of extremely early impacts and faster-than-expected fade-out by the end of pre-k.

Interestingly, new preliminary results examining the cumulative impacts of MPC and High 5s on children's kindergarten outcomes suggest that there *is* a positive effect of early math intervention on children's math and attitudes toward math in Kindergarten, 1 year later (although it is not yet clear which program contributed to which outcomes and how much; Mattera & Morris, 2017). Moreover, not only do positive impacts on math learning emerge, but executive function skills are also improved as a result of the enhanced math experiences across preschool and kindergarten. Although further analysis is needed to determine whether the effects are due to math learning in preschool vs preschool and kindergarten, the impacts act as a proof of concept to demonstrate that intervening with math in the early years can improve children's math and executive function outcomes, as well as affect how children feel about their math skills.

The unexpected results in pre-k and early kindergarten findings suggest that there is a lot we still do not know about how changes in math instruction affect children's outcomes in the long-term. MPC and High 5s demonstrate that rigorous causal tests, necessary to understand the effects of potential policy levers, can also provide rich descriptive information and a mechanism for investigating additional questions of interest to the field.

Even with considerably more investigation to be done, this and related work is critically important if our interest is in improving the field's knowledge to address the so-called achievement gap—the lower levels of achievement that follow low-income children relative to their more affluent peers as they move through school. If the causal relationship between math and later outcomes exists, then math may be a potentially powerful lever for change if our goal is to dramatically reduce the elementary school achievement gap. At the same time, there is considerable debate about the unique importance of which domain of school readiness matters and is most amenable to curricular intervention at scale, for whom, and in what context (Claessens, Duncan, & Engel, 2009; Duncan et al., 2007; Grimm, Steele, Mashburn, Burchinal, & Pianta, 2010). At the same time, these questions of the focus on effectiveness of early childhood intervention overlap with issues of the longitudinal mechanism of fade-out and measurement specificity and sensitivity. As a field, it is imperative that we begin to build a body of research to address these questions. The design considerations outlined earlier to create a rigorous and richly designed study provide one model to build such a body of research.

REFERENCES

Agodini, R., Harris, B., Atkins-Burnett, S., Heaviside, S., Novak, T., & Murphy, R. (2009). *Achievement effects of four early elementary school math curricula: Findings from first graders in 39 schools. NCEE 2009-4052.* National Center for Education Evaluation and Regional Assistance.

Agodini, R., Harris, B., Thomas, M., Murphy, R., & Gallagher, L. (2010). *Achievement effects of four early elementary school math curricula: Findings for first and second graders.* Princeton, NJ: Mathematica Policy Research.

Bailey, D., Duncan, G. J., Odgers, C. L., & Yu, W. (2017). Persistence and fadeout in the impacts of child and adolescent interventions. *Journal of Research on Educational Effectiveness, 10*(1), 7–39.

Bloom, H. S., Bos, J. M., & Lee, S. W. (1999). Using cluster random assignment to measure program impacts: Statistical implications for the evaluation of education programs. *Evaluation Review, 23*(4), 445–469.

Brown, K. J., Morris, D., & Fields, M. (2005). Intervention after grade 1: Serving increased numbers of struggling readers effectively. *Journal of Literacy Research, 37*(1), 61–94.

Chetty, R., Friedman, J. N., Hilger, N., Saez, E., Schanzenbach, D. W., & Yagan, D. (2011). How does your kindergarten classroom affect your earnings? Evidence from Project STAR. *The Quarterly Journal of Economics, 126*(4), 1593–1660.

Claessens, A., Duncan, G., & Engel, M. (2009). Kindergarten skills and fifth-grade achievement: Evidence from the ECLS-K. *Economics of Education Review, 28*(4), 415–427.

Clements, D. H., & Sarama, J. (2007). *Building blocks: SRA real math, grade PreK.* Columbus, OH: SRA-McGraw Hill.

Clements, D. H., & Sarama, J. (2013). *Building blocks (Vols. 1 and 2).* Columbus: McGraw-Hill Education.

Clements, D. H., Sarama, J., Spitler, M. E., Lange, A. A., & Wolfe, C. B. (2011). Mathematics learned by young children in an intervention based on learning trajectories: A large-scale cluster randomized trial. *Journal for Research in Mathematics Education, 42*(2), 127.

Clifford, R. M., Barbarin, O., Chang, F., Early, D., Bryant, D., Howes, C., & Pianta, R. (2005). What is pre-kindergarten? Characteristics of public pre kindergarten programs. *Applied Developmental Science, 9*(3), 126–143.

Corsi, P. M. (1972). *Human memory and the medial temporal region of the brain.* Unpublished doctoral dissertation Montreal, Canada: McGill University.

Diamond, A. (2013). Executive functions. *Annual Review of Psychology, 64,* 135–168.

Diamond, A., & Taylor, C. (1996). Development of an aspect of executive control: Development of the abilities to remember what I said and to "do as I say, not as I do" *Developmental Psychobiology, 29*(4), 315–334.

Duncan, G. J., Dowsett, C. J., Claessens, A., Magnuson, K., Huston, A. C., Klebanov, P., & Sexton, H. (2007). School readiness and later achievement. *Developmental Psychology, 43*(6), 1428.

Duncan, G. J., & Magnuson, K. (2011). The nature and impact of early achievement skills, attention skills, and behavior problems. In R. J. Murnane & G. J. Duncan (Eds.), *Whither opportunity? Rising inequality, schools, and children's life chances* (pp. 47–70). New York: Russell Sage Foundation.
Ehri, L. C., Dreyer, L. G., Flugman, B., & Gross, A. (2007). Reading rescue: An effective tutoring intervention model for language-minority students who are struggling readers in first grade. *American Educational Research Journal, 44*(2), 414–448.
Gennetian, L. A., Magnuson, K., & Morris, P. A. (2008). From statistical associations to causation: What developmentalists can learn from instrumental variables techniques coupled with experimental data. *Developmental Psychology, 44*(2), 381.
Ginsburg, H. P., Greenes, C., & Balfanz, R. (2003). *Big math for little kids*. Parsippany, NJ: Dale Seymour Publications.
Ginsburg, H. P., Lee, J. S., & Boyd, J. S. (2008). Mathematics education for young children: What it is and how to promote it. In *Social policy report*: *Vol. 22*. Ann Arbor, MI: Society for Research in Child Development. Number 1.
Grimm, K. J., Steele, J. S., Mashburn, A. J., Burchinal, M., & Pianta, R. C. (2010). Early behavioral associations of achievement trajectories. *Developmental Psychology, 46*(5), 976.
Heckman, J. J. (2006). Skill formation and the economics of investing in disadvantaged children. *Science, 312*(5782), 1900–1902.
Hofer, K. G., Lipsey, M. W., Dong, N., & Farran, D. (2013). *Results of the early math project: Scale-up cross-site results*. Nashville, TN: Vanderbilt University. Retrieved from my.vanderbilt.edu/mathfollowup/reports/technicalreports.
Joyce, B. R., & Showers, B. (2002). *Student achievement through staff development*. Alexandria, VA: Association for Supervision and Curriculum Development.
Klein, A., Starkey, P., Clements, D., Sarama, J., & Iyer, R. (2008). Effects of a prekindergarten mathematics intervention: A randomized experiment. *Journal of Research on Educational Effectiveness, 1*(3), 155–178.
Klein, A., Starkey, P., & Ramirez, A. (2002). *Pre-K mathematics curriculum: Early childhood*. Glenview, IL: Scott Foresman.
Lee, V. E., & Loeb, S. (1995). Where do head start attendees end up? One reason why preschool effects fade out. *Educational Evaluation and Policy Analysis, 17*(1), 62–82.
Lezak, M. D. (1983). *Neuropsychological assessments*. New York: Oxford University Press.
Lobman, C., Ryan, S., & McLaughlin, J. (2005). In *Toward a unified system of early childhood teacher education and professional development: Conversations with stakeholders Annual meeting of the American educational research association, Montreal, April* (pp. 11–15).
Luria, A. R. (1966). *Higher cortical functions in man*. New York: Basic Books.
Martin, N., & Brownell, R. (2011). *Receptive one-word picture vocabulary test*. Torrence, CA: Western Psychological Services.
Mattera, S. K., & Morris, P. A. (2017). *Counting on early math skills: Preliminary kindergarten impacts of the making pre-K Count and high 5s programs*. New York: MDRC.
Morris, P. A., Mattera, S. K., & Maier, M. F. (2016). *Making pre-K Count: Improving math instruction in new York City*. New York: MDRC.
Morris, P., Raver, C. C., Millenky, M., Jones, S., & Lloyd, C. M. (2010). *Making preschool more productive: How classroom management training can help teachers*. New York, NY: MDRC.
Najarian, M., Snow, K., Lennon, J., & Kinsey, S. (2010). *Early childhood longitudinal study, birth cohort (ECLS), preschool-kindergarten 2007 psychometric report (NCES 2010–009)*. Washington, DC: National Center for Education Statistics, Institute for Education Sciences, US Dept of Education.
National Research Council. (2009). *Mathematics learning in early childhood: Paths toward excellence and equity*. Washington, DC: National Academy Press.

Sarama, J., Clements, D. H., Starkey, P., Klein, A., & Wakeley, A. (2008). Scaling up the implementation of a pre-kindergarten mathematics curriculum: Teaching for understanding with trajectories and technologies. *Journal of Research on Educational Effectiveness*, *1*(2), 89–119.

Sarama, J., Lange, A. A., Clements, D. H., & Wolfe, C. B. (2012). The impacts of an early mathematics curriculum on oral language and literacy. *Early Childhood Research Quarterly*, *27*(3), 489–502.

Sheridan, S. M., Edwards, C. P., Marvin, C. A., & Knoche, L. L. (2009). Professional development in early childhood programs: Process issues and research needs. *Early Education and Development*, *20*(3), 377–401.

Slavin, R. E., Lake, C., Davis, S., & Madden, N. A. (2011). Effective programs for struggling readers: A best-evidence synthesis. *Educational Research Review*, *6*(1), 1–26.

Wasik, B. A., & Slavin, R. E. (1993). Preventing early reading failure with one-to-one tutoring: A review of five programs. *Reading Research Quarterly*, 179–200.

Weiland, C., & Yoshikawa, H. (2013). Impacts of a prekindergarten program on children's mathematics, language, literacy, executive function, and emotional skills. *Child Development*, *84*(6), 2112–2130.

Willoughby, M. T., Wirth, R. J., Blair, C. B., & Family Life Project Investigators. (2012). Executive function in early childhood: Longitudinal measurement invariance and developmental change. *Psychological Assessment*, *24*(2), 418.

Woodcock, R. W., McGrew, K. S., & Mather, N. (2001). *Woodcock-Johnson III tests of cognitive abilities* (pp. 371–401). Itasca, IL: Riverside Pub.

CHAPTER SEVEN

Interventions for Primary School Children With Difficulties in Mathematics

Ann Dowker[1]
Oxford University, Oxford, United Kingdom
[1]Corresponding author: e-mail address: ann.dowker@psy.ox.ac.uk

Contents

Abstract

Difficulty with arithmetic is a common problem for children and adults, though there has been some work on the topic for a surprisingly long time. This chapter will review some of the research that has been done over the years on interventions with primary

Advances in Child Development and Behavior, Volume 53
ISSN 0065-2407
http://dx.doi.org/10.1016/bs.acdb.2017.04.004

school children. Interventions can be of various levels of intensiveness, ranging from whole-class approaches that take account of individual differences through small-group and limited-time individual interventions to extended-time individual interventions. Interventions discussed here include those involving peer tuition and group collaboration; those involving board and computer games; and those that involve assessing children's strengths and weaknesses in different components of mathematics; and targeting remedial activities to the assessed weaknesses. Most of the interventions discussed in this chapter specifically involve mathematics (usually mainly arithmetic), but there is also some discussion of attempts to improve mathematics by training children in domain-general skills, including Piagetian operations, metacognition, and executive functions.

Difficulty with arithmetic is a common problem for children and adults (Butterworth, Sashank, & Laurillard, 2011). About 22% of the population in the United Kingdom have severe numeracy difficulties that have a serious impact on their occupational and social chances (Bynner & Parsons, 1997; Parsons & Bynner, 2005), whereas only about 5% have similar levels of difficulty in literacy. Less severe, but still significant, mathematical difficulties are still more common. A recent survey indicates that nearly half of working-age adults in the United Kingdom have only the level of numeracy that one would expect at the end of primary school (BIS, 2011). In the United States, over 60% of pupils in both fourth and eighth grade have been found to be achieving below expectations in mathematics (US Department of Education, 2011). There are a number of countries where overall achievement in mathematics is higher, but there are a significant number of children with mathematical difficulties even in the high-achieving countries, such as Finland (Räsänen, Salminen, Wilson, Aunio, & Dehaene, 2009) and the Pacific Rim countries (Chan & Ho, 2010; Lee, Chang, & Lee, 2001).

In recent years, there has been an increasing interest in interventions for children with mathematical difficulties (Butterworth et al., 2011; Chodura, Kuhn, & Holling, 2015; Cohen Kadosh, Dowker, Heine, Kaufmann, & Kucian, 2013; Gervasoni & Sullivan, 2007; Scherer, Beswick, DeBlois, Healy, & Opitz, 2016); though, as will be discussed later in this chapter, there has been some work on the topic for a surprisingly long time. Most intervention studies have been with primary school children, though some have involved preschoolers (Clements & Sarama, 2011; Greenes, Ginsburg, & Balfanz, 2004; Klein, Starkey, Clements, Sarama, & Iyer, 2008).

This chapter will review some of the research that has been done over the years on interventions with primary school children. The age of attending primary school, of course, varies between countries. For the purpose of this review, primary school age will refer to children, who are (a) between the ages of 5 and 11 and (b) are receiving formal school instruction. Interventions aimed primarily at children in nursery school or kindergarten, even if aged 5 or over, are not the focus of the present chapter. They are discussed in the chapter by Clements and Sarama.

It is particularly important to take into account that mathematical difficulties are not unitary. Studies of both atypical and typical mathematical development (Cowan et al., 2011; Desoete, Roeyers, & De Clercq, 2004; Dowker, 2005; Gifford & Rockliffe, 2012; Jordan, Mulhern, & Wylie, 2009; Russell & Ginsburg, 1984) have indicated that arithmetical cognition is made up of multiple components, and that it is quite possible for children and adults to show strong discrepancies, in either direction, between the components. Moreover, although the components often correlate with one another, and some components appear to be easier on average than others, there does not seem to be a clear hierarchy of abilities. People can show weakness in almost any component or combination of components but not in others. Converging evidence for the functional separability of different components of arithmetic comes from studies of adults with acquired dyscalculia (Cappelletti, Butterworth, & Kopelman, 2012; Dehaene, 1997; Delazer, 2003; Demeyere, Lestou, & Humphreys, 2010; Warrington, 1982) and from behavioral and brain imaging studies of adult mathematical cognition (Stanescu-Cosson et al., 2000; Van Eimeren, Grabner, Koschutnik, et al., 2010).

Such findings have suggested the desirability of developing interventions that assess and target children's weaknesses in specific components of numeracy, rather than taking a "one size fits all" approach.

1. INTERVENTIONS OF DIFFERENT LEVELS OF INTENSIVENESS

Interventions in numeracy (and also literacy) have sometimes been classified into three categories of varying degrees of intensiveness, described as "waves" in the United Kingdom and "tiers" in the United States. Wave/Tier 1 involves whole-class teaching designed to be suitable for children of a variety of attainment levels; Wave/Tier 2 involves interventions in small groups with children who are experiencing mild or moderate difficulties in the subject; and Wave/Tier 3, involving more intensive, usually individualized interventions for children with more significant problems.

Tier 1 or 2 interventions may involve allowing for independent individualized or small-group work within a class (El-Naggar, 1996). This may involve progressing through a textbook at one's own pace, the use of individualized worksheets, and/or the individualized use of educational computer software.

Such approaches are far more flexible, and have more potential for taking account of the componential nature of arithmetical ability, than giving all children the same instruction, or streaming or setting. Their potential disadvantages include the risk that, even within one class, some pupils may be labeled as "low attainers" and live down to expectations; and the risk that work will become so individualized that pupils will not benefit from mutual discussion and exploration of ideas.

Lou et al. (1996) carried out a meta-analysis of ability grouping and small group work within classes, in a variety of school subjects. Unlike streaming and setting, which have usually been found to have negative effects on low attainers' performance, this study indicated that within-class grouping had a positive effect on the performance of low achievers; but only if it was accompanied by provision of appropriate materials and activities.

2. EXAMPLES OF SMALL-GROUP INTERVENTIONS

An example of a small-group intervention is that of Askew, Bibby, and Brown (2001) in the United Kingdom. They developed a small-group intervention technique that involved the use of derived fact strategies. Askew et al. (2001) pointed out that there are some children who do not develop such strategies, and often these are also the children who know relatively few arithmetic facts and rely on counting strategies. This may become a vicious circle, whereby their failure to use derived fact strategies may further impede their developing a store of known facts, which in its turn may interfere with the development of derived fact strategies.

In this program, teachers worked with small groups (four per group) of 7- to 8-year-old children, who had performed below average in school achievement at age 7. The children underwent intervention once a week for 20 weeks. The interventions emphasized devising strategies for working out new arithmetical facts on the basis of known facts, e.g., using the fact that $16 + 3 = 19$ to work out 19-3. There were 48 children in the intervention group, who were compared with 48 matched controls. All children were given a diagnostic test of arithmetic problem solving, devised by the researchers, just before intervention, and again shortly after completing the intervention. The children in the program improved significantly more than the controls, both in accuracy and in their use of known and derived facts rather than needing to resort to counting strategies.

One small-group intervention program, which has been used in the United Kingdom, is the Springboard program, used with children in Years 3–7 (7–12 years), developed in the early 2000s. It was mainly used under earlier governments and its description is now archived (e.g., see http://webarchive.nationalarchives.gov.uk/tna/content/20101123073054/ http://www.standards.dfes.gov.uk/primary/publications/mathematics/ springboard_3/ but appears still to be used by some schools. The target group is children with relatively mild arithmetical difficulties. The Springboard program provides additional tuition for small groups of six to eight children as a supplement to the daily mathematics lesson that they undergo with the whole class. Typically, it has provided two 30 min sessions per week, which consolidate the work currently being taught in the daily mathematics lessons. Such interventions are usually delivered by teachers or teaching assistants within the school.

In the United States, there have been several studies of Tier 2 small group interventions for first-grade pupils who were showing signs of mathematical difficulties. Bryant, Bryant, Gersten, Scammacca, and Chavez (2008) delivered 15-min small-group interventions 3–4 days a week for 18 weeks. The interventions dealt with counting, quantity representation, basic facts, and place value concepts. The 26 pupils in the intervention did not differ significantly from controls on a standardized test at the end of the intervention period. However, intervention programs lasting longer than 15 min at a time and/or continuing over a longer period of time have given positive results. Bryant, Bryant, Gersten, Scammacca, Funk, et al. (2008) used similar interventions to the above, but lasting 20 min at a time, 4 days a week, for 23 weeks. The 42 children in this study did perform significantly better than controls on a standardized test.

Fuchs et al. (2005) delivered small-group interventions to first-grade pupils for 40-min sessions 3 days a week for 16 weeks. The interventions involved dealing with number concepts and the operations of addition and subtraction. The intervention was delivered in groups of 2 or 3. Each lesson included both worksheet and manipulative (e.g., base-10 blocks for place value instruction) activities. The 70 children, who underwent the intervention performed significantly better than business-as-usual controls on standardized measures of concepts and applications (effect size 0.67 and story problems (effect size 0.7)), but they performed similarly on addition and subtraction fluency measures and a standardized measure of applied problems.

Fuchs et al. (2006) gave children 50 10-min small-group sessions of on-computer-assisted instruction for 18 weeks, focusing specifically on fluency in addition and subtraction. This intervention briefly presented a number combination on a computer screen. When the number combination disappeared, the computer screen prompted the pupil to retype it from memory. Children were given feedback and received virtual tokens for correct responses. Initially, the number combination was flashed for 1.5 s. The computer adapted to the child's performance by increasing the exposure time by 0.3 s for each incorrect or missed response. After the intervention, the children performed better than controls on tests of addition fluency (effect size 0.82) but not on subtraction fluency or word problems.

Fuchs et al. (2013) gave first-grade children, who were considered at risk of developing mathematical difficulties, tutoring three times a week in number knowledge over a period of 16 weeks. The tutoring sessions lasted for 30 min, including 5 min of practice. The practice was either speeded or nonspeeded. Children who underwent tutoring showed significantly more improvement than a business-as-usual control group in tests of basic arithmetic, two-digit calculations, number knowledge, and word problems. Speeded practice led to more improvement than nonspeeded practice in basic arithmetic and two-digit calculations, while both types of practice led to equal improvement in number knowledge and word problems. However, it is important to be cautious about making any general assumption that speeded practice would be better in all situations, as some studies suggest that timed tests can increase mathematics anxiety and reduce performance (Beilock, 2008; Beilock & Carr, 2005).

Bryant et al. (2011) carried out a study of a Tier 2 intervention program for first-grade pupils with mathematical difficulties. Children were randomly assigned to the intervention program (139 pupils) and a comparison condition (65 pupils). Those in the intervention program were given 25-min sessions 4 days a week in groups of three to five in counting, place value concepts, and addition and subtraction, using visual representations of mathematical concepts and exposure to the teacher modeling "thinking aloud" about tasks. They outperformed controls on whole-number computation (effect size 0.44) but not on word problems.

The conclusions that we may draw are that (1) small-group interventions can have a significant impact on the progress of children with mathematical difficulties, (2) the ways in which such interventions are delivered can affect the level and nature of their effectiveness, and (3) interventions often have selective effects on particular aspects of arithmetic; and this is not solely attributable to children improving just in the aspects on which they have received specific training. This gives important support for the view that arithmetical ability is not unitary, but consists of many components, which may be influenced by different factors.

3. PEER TUITION AND GROUP COLLABORATION

One way in which schools can deal with individual differences is by encouraging children to teach one another. This may be one aspect of the success of some small-group interventions. Peer teaching can involve older children teaching younger children; more able classmates teaching less able classmates; or collaborative learning between peers of similar ability.

All of these techniques have been used extensively in general teaching. Collaborative group projects have been common in schools for many years, and have their adult counterparts in the classes, workshops, and seminars that occur in higher education and professional development. The teaching of younger children by older children has a venerable history, going back at least to the early 19th century "monitorial" system, where older children became teachers as well as learners as a means of coping with the enormous class sizes and high teacher/pupil ratio of Britain's earliest free schools. The present discussion will be confined to the use of such techniques with regard to arithmetic.

The commonest forms of peer teaching involve collaborative group work, where several children cooperate in solving a mathematical problem. This can serve several purposes: increasing motivation; encouraging children to put their mathematical ideas into words, and to reflect on the strategies that they use; and enabling children to transmit mathematical knowledge and ideas to one another. The possible disadvantages of such approaches are that their success may depend on the dynamics of a particular group. There is the risk that a dominant child, or one who is perceived as "clever," may do most of the work, while the others simply accept their decisions. In particular, children who are, or are regarded as being, low achievers in mathematics may be ignored or dismissed in group discussions and decisions. However, some studies have indicated that collaborative learning can be beneficial to children with arithmetical difficulties (e.g., Nastasi & Clements, 1991; Salomon, 1989).

Davenport and Howe (1999) looked at the effect of collaborative learning on children's ability to solve addition and subtraction word problems. Children worked in groups, using problem-solving guidelines that they had been given to solve the problems, and then "taught" their problem to a fellow pupil. In this condition, they first worked on the problem individually, writing down their working; and then shared their answers and calculations with others in the group. If there was disagreement, they discussed them until they got an agreed answer. They were then shown the answer, and, if they were correct, transferred the solution to a group answer sheet, and, if they were incorrect, discussed where they had gone wrong and how to remedy it. Members of the group were paired the next day with children from another group, who had solved a different problem. Each child in a pair took a turn at being the "teacher," guiding the other child through the problem which they had themselves solved in their group. The children in this collaborative condition were compared with children who spent the same amount of time solving the same problems individually. The children in the collaborative condition performed better than those who worked individually, and in particular, children who were below average in arithmetic benefited from being the "learners" who listened to peers.

More intensive forms of peer tutoring involve one-on-one tuition of one child by another: often following some form of explicit training of the child "teacher" (Topping & Bamford, 1998; Vosse, 1999). Usually, older children teach younger children. Studies suggest to show benefits both for the tutors—perhaps due to their need to reflect on the arithmetical concepts and procedures that they teach—and for the tutees (Topping & Bamford, 1998).

Meta-analyses have given somewhat conflicting results as to the benefits of peer tuition. Kroesbergen and Van Luit's (2003) meta-analysis suggest that, while peer tutoring interventions are beneficial for children with arithmetical difficulties, they are less so than some other forms of intervention. Gersten et al. (2009) did not find significant effects of peer tuition in their meta-analysis. However, the meta-analysis by Baker, Gersten, and Lee (2002) did find highly significant effects of peer tuition. This may be due to the samples studied. The children in the studies by Kroesbergen and Van Luit (2003) and Gersten et al. (2009) involved children who were identified as having special needs, either involving general intellectual impairments or very severe specific learning difficulties. Those in Baker et al.'s

(2002) study involved children, who were low attaining in arithmetic but not identified as having special needs. Thus, peer tuition may be a suitable approach for children who have moderate arithmetical weaknesses without severe accompanying problems, while those with more serious or generalized difficulties may depend more on input from trained adults.

4. GAMES AS AN AID TO ARITHMETICAL INTERVENTION

Teachers and parents have often used games as a means of improving children's arithmetical performances; and games form a part of many intervention programs, both group-based and individual.

For generations, many children have enjoyed playing games that include arithmetical activities. Playing cards show numbers both in terms of numerals and quantities, and most card games are based on these numbers. Dominoes involve number recognition and matching. Board games with dice and counters involve counting, addition and often subtraction, and matching between different modes of presentation of numbers (e.g., translating the number of dots on a die into number of moves on a board). Some games, of which the best known is probably Monopoly, involve complex imaginary financial transactions.

The role of numbers is most immediately obvious in games which include such materials as playing cards, dominoes, dice, and boards with numbered squares. However, they are used in many other games. Any game that requires score-keeping will provide further need for arithmetic. Children's playground games often involve numbers as part of the game (e.g., counting up to a certain number while players hide in "Hide and Seek") and/or in scoring; and most of all, perhaps, in some of the elaborate counting-out rituals that are used to assign the roles in a game (One potato, two potato, three potato, four…).

A number of teaching and intervention programs have used games as an opportunity for practicing arithmetical skills (Ainley, 1990; Kirkby, 1992; Rowe, 2001). Some have used existing games; others have devised their own games for the purpose (e.g., Ainley, 1990; Straker, 1996; also see, William Quartey Papafiou's Hoo-Ha games at http://hoohasuper-duper.com). An example is Straker's (1996, p. 68) game "Snake." In this game, a snake made out of eight circles is portrayed on a page. Each player rolls two dice in turn and writes the total number in one of the circles. A number can only be written once, and the numbers in all the circles must be in order. If there is no space for a number, the player misses a turn. The player who has written the most numbers when all the circles are full wins the game.

Most children from middle-class backgrounds will have had some experience of games that incorporate numbers: Snakes/Chutes and Ladders, Ludo, Dominoes, computer games including numbers, etc. Children from more disadvantaged backgrounds may not have this experience, and Ramani and Siegler have suggested that this may inhibit their development of numerical knowledge (Ramani & Siegler, 2008, 2011; Siegler & Ramani, 2008, 2009). They have indeed found that children from poor backgrounds were less good at numerical magnitude recognition and comparison than their middle-class peers; and that there was a significant association between reported experience of playing board games with dice and numerical ability. The researchers gave children in Head Start preschool programs (aimed at children from disadvantaged backgrounds) four sessions, each lasting 15–20 min, of playing a linear numerical board game similar to Chutes and Ladders, and found that their performance on numerical magnitude recognition and comparison improved markedly, to a level indistinguishable from that of middle-class children. Playing a similar board game involving colors rather than numbers did not lead to improvement in number skills, indicating that the important factor was indeed playing a game involving numbers.

In recent years, computer and digital games have of course played an increasing role in children's lives, and are also being increasingly used in interventions for children with mathematical difficulties. A number of computer games have been developed for intervention purposes, though there have been relatively few systematic evaluations of these techniques.

Räsänen et al. (2009) used computer games with 30 Finnish kindergarten children with low numeracy. These games were intended to address not weaknesses in arithmetic itself, but their possible underlying foundations: deficiencies in subitizing and deficiencies in magnitude comparison. The children were randomly assigned to two groups. Fifteen children played a game (Graphogame-Math) which emphasized small sets of exact numerosities by training matching of verbal labels to visual patterns and number symbols. The other 15 played a computer game (The Number Race; designed by Wilson, Revkin, Cohen, Cohen, & Dehaene, 2006), which emphasized numerical comparison and was designed to train approximate magnitude understanding. Both groups participated in a daily intervention session, lasting 10 to 15 min, for 3 weeks. Children's performance in verbal counting, number comparison, object counting, arithmetic, and a control task (rapid serial naming) were measured before and after the intervention. Both interventions improved children's skills in number comparison, compared to a group of 30 typically performing children. However, the children showed no significant improvement in other number skills.

This study was carried out with children who had not yet begun formal primary school instruction, and it may be that results would be somewhat different for children who had already begun such instruction. Moreover, just because they had not yet begun formal teaching, diagnosis of low numeracy may have been less reliable than it would in an older group: some of the children might have improved without intervention. It is, however, interesting in suggesting that computerized interventions can have a rather specific effect on particular components of numeracy, and perhaps that number comparison is particularly susceptible to such intervention.

Kucian et al. (2011) developed Calcularis, a set of computer games mainly for primary school children, designed to improve number representations; and to strengthen links between numbers and spatial processes. This has been tested with both typically developing children and children with a diagnosis of dyscalculia. After 5 weeks of training, involving three or four 20-min sessions per week, both dyscalculic and typically developing children improved in both spatial number representation and arithmetical abilities. The dyscalculic children underwent fMRI scanning before and after training. They showed changes in brain activity, with more activation of parietal areas and less activation of frontal areas in arithmetic tasks: changes of a type that are known to be associated with mathematical improvement.

Butterworth et al. (2011) devised several digital games for use by children with mathematical difficulties. For example, the game Dots2Track was designed to help the learner discern the relationship between the numerosity in a dot pattern and its representation as a digit, and on a number line. The child identifies how many dots there are in a pattern. The program gives the child feedback, and, for an incorrect answer, shows the pattern that the child's input matches, counts the pattern onto a number line, and does the same for the target pattern, showing how the two compare. It then invites the learner to fix their own line by adding or subtracting one dot at a time. Once the regular patterns are known, the next level moves to random patterns, and gradually increases the difficulty. The aim is to move the learner toward recognizing the numerosity, rather having to count it every time. Other similar games (e.g., Dots2Digit) involve matching sets of dots to numerals.

Butterworth et al. (2011) report that teachers have expressed positive views of the games. The children have been found to show more focused attention, spend more time on task, and complete more items within a given period for digital games than for similar games with physical cards.

5. INDIVIDUALIZED INTERVENTIONS: THE HISTORY BEHIND THEM

Individualized interventions, targeted at children's specific strengths and weaknesses, have only come into widespread use relatively recently. However, they have been in existence at least since the 1920s (Brownell, 1929; Tilton, 1947; Williams & Whitaker, 1937).

Weaver (1954) was a particularly strong advocate of differentiated instruction and remediation in arithmetic. He put forward several important points that have since been strongly supported by the evidence, centrally that "arithmetic competence is not a unitary thing but a composite of several types of quantitative ability: e.g., computational ability, problem-solving ability"; "(t)hese abilities overlap to varying degrees, but most are sufficiently independent to warrant separate evaluations"; and "children exhibit considerable variation in their profiles or patterns of ability in the various patterns of arithmetic instruction" (pp. 300–301). He argued (pp. 302–303) that any "effective program of differentiated instruction in arithmetic must include provision for comprehensive evaluation, periodic diagnosis, and appropriate remedial work" and that "[e]xcept for extreme cases of disability, which demand the aid of clinicians and special services, remedial teaching is basically good teaching, differentiated to meet specific instructional needs."

For a long time, some researchers and educators have emphasized the importance of investigating the strategies that individual children use in arithmetic: especially those faulty arithmetical procedures that lead to errors (Buswell & John, 1926; Brownell, 1929; Van Lehn, 1990). Thus, some children might add without carrying (e.g., 34 + 18 = 412); others might add all the digits without any reference to whether they are tens or units (e.g., 34 + 18 = 16); others, when adding a single-digit number to a two-digit number, might add it to both the tens and the units (e.g., 54 + 3 = 87). From the beginning, papers on mathematical difficulties have included lists of such faulty procedures (Tilton, 1947; Williams & Whitaker, 1937). In some more recent studies (Brown & Burton, 1978; Van Lehn, 1990), they are described as "bugs," by analogy with malfunctioning computer programs.

Efforts are made to diagnose such incorrect strategies, so that they can be corrected. In the words of Tilton (1947, pp. 84–85), "...many errors are systematic. In other words, not as many errors are accidental and attributable to carelessness as teachers are inclined to think. ...If the youngsters who have such incorrect rules are to be helped, the teacher should know the child's rule, because the child's need is just as much to unlearn his incorrect rule as it is to learn the correct rule. To work in ignorance of his rules is to give him a feeling of confusion." Brown and Burton (1978) devised a computer program, BUGGY, to diagnose a wide variety of incorrect rules and strategies.

If componential theories of arithmetical ability, and their applications to differentiated instruction and remediation in arithmetic were already being advocated in the early 20th century, why have they had comparatively little impact on theory and practice, at least until recently? Part of the reason is practical: in under-resourced classrooms, it is difficult to provide individualized instruction. In general, class sizes were higher, and teacher–pupil ratios lower, in the past than now.

If differentiated instruction is used at all in overcrowded and understaffed classrooms, it tends to take the form of grouping children according to their perceived overall ability, either in all academic areas, or, at best, in specific school subjects (here, mathematics). Therefore, in some quarters, differentiated instruction may be viewed negatively, due to associations with educational methods which separate those labeled as "less able" from those labeled as "able"; and involve lower expectations and sometimes worse teaching of the former group (see Section 6). In fact, the true aim of differentiated instruction is to identify strengths and weaknesses within the same domain in the same individual, and to use the strengths to overcome or compensate for the weaknesses. Such an aim is quite incompatible with labeling individuals as globally "bad" at a subject, and with regarding their weaknesses as purely innate and not susceptible to intervention (a view which is both incorrect and likely to have a negative effect on the motivation of both pupils and teachers). Nevertheless, some people confuse the appropriate use of differentiated instruction with its misuse.

6. POTENTIAL PROBLEMS WITH INDIVIDUALIZED INSTRUCTION AND REMEDIATION: PAST AND PRESENT

What are the potential problems that can arise with individualized instruction and remediation? Apart from the practical problems arising from inadequate resources, there can be problems with the methods that are used. In particular, there may be gaps in the selection of arithmetical components for remediation. Indeed, in our present state of knowledge, there must be, since we do not yet have a full understanding of the different components of arithmetic, the relationships between them, and the nature and causes of individual differences in them. Nevertheless, there has been sufficient knowledge in the area for quite some time to permit successful work in the area.

Moreover, appropriate individualized instruction depends on appropriate selection of the components of arithmetic to be used in assessment and intervention. This is still an issue for debate and one that requires considerable further research. One of the main potential problems, which was more common in the past than nowadays, is to assume that the components to be addressed must necessarily correspond to specific arithmetical operations; for example, treating addition, subtraction, multiplication, division, etc., as separate components. It is, of course, quite possible for children to have specific problems with a particular arithmetical operation. Nevertheless, it is an oversimplification to assume that these operations are likely to be the primary components of arithmetical processing. Current classifications tend to place greater emphasis on the type of cognitive process; for example, the broad distinctions between factual knowledge (knowing that), procedural knowledge (knowing how), conceptual knowledge (knowing what it all means), and in some theories utilizational knowledge (knowing when to apply it) (see, for example, Greeno, Riley, & Gelman, 1984). A potential danger of overemphasizing the different operations as separate components is that it may encourage children, and perhaps adults, to ignore the relationships between the operations.

Another potential problem—again commoner in the past though still a danger nowadays—is looking at children's difficulties only in terms of procedural errors. Diagnosing such errors is of course important. Children do indeed frequently acquire incorrect strategies, which can become entrenched, especially if the child is given too much of the wrong sort of arithmetical practice. Nonetheless, diagnosing the incorrect strategies is not always the final step. There may be a conceptual reason why the incorrect strategy is acquired and maintained. In the case of the faulty subtraction strategies described earlier, their acquisition could result from an assumption that larger numbers cannot be subtracted from smaller numbers, combined with an inadequate understanding of place value which makes it difficult for children to understand the nature and purpose of borrowing. As Tilton (1947, p. 85) stated many years ago: "Many of the errors made by these… children seemed to be due to an insufficient understanding of the meaning of numbers. It seems as if these children had been asked to learn the rules for the manipulation of numbers in addition, subtraction and multiplication without having learned the meaning of the symbols that they have been asked to manipulate."

7. INDIVIDUALIZED INTERVENTION PROGRAMS WITH YOUNG CHILDREN: RECENT WORK

Until fairly recently, there had been few individualized intervention programs that addressed numeracy in young children, and took componential approaches based on cognitive theories of arithmetic. An increasing number of intervention programs have been developed since around 2005 (Dowker, 2009); though many of these have been small-scale programs, restricted to specific schools and districts, and not subjected to extensive testing and evaluation.

8. "WAVE 3" MATERIALS IN THE UNITED KINGDOM

So-called Wave 3 materials for mathematics were developed in the United Kingdom, under government leadership, from 2003 to 2005 (Gross, 2007). They emphasize individualized diagnosis of the errors and misconceptions shown by primary school children with significant difficulties, specific or nonspecific, with mathematical learning. The materials include number cards, problem cards, cubes, beads, dominoes, number lines, and grids.

The children are given diagnostic interviews to determine conceptual and procedural difficulties in coping with the major areas of the mathematics curriculum. The materials are then used with children to correct the errors and misconceptions that have been spotted. For example, children who have difficulty in comparing two-digit numbers in the early primary years or three-digit numbers in the later primary years may be encouraged to use number lines or squares to order the numbers.

The materials can be used within or outside the daily mathematics lesson, and are used individually by the child with the class teacher, a teaching assistant, or a special needs teacher. According to the original government suggestions, the child typically undergoes one 20-min individual session per week, and one 5-min "spotlight" sessions on each of the following days, in addition to following the mathematics curriculum with or without specialist sup support. One difficulty that teachers have found in using the materials is that the diagnostic criteria and methods of testing were not at first very clearly specified. Some schools and local authorities since developed their own diagnostic tests to enable more targeted use of these intervention materials (Dowker, 2009). The above describes the materials and their proposed use in the period from 2005 to 2010. Since then, there have been changes of government, and ongoing curricular changes, and although schools still have and use the Wave 3 materials, the original suggestions for their use are archived.

9. INTENSIVE INTERVENTION PROGRAMS FOR CHILDREN WITH SEVERE DIFFICULTIES

Children with severe mathematical difficulties (e.g., those scoring in the bottom 5% on mathematics-related tests) are likely to need intensive intervention. Two such programs are the Mathematics Recovery program mentioned earlier, and the more recently developed Numbers Count program. They will be described later.

10. MATHEMATICS RECOVERY

The Mathematics Recovery program was designed in Australia by Wright and his colleagues (Wright, Martland, & Stafford, 2006). In this program, teachers provide intensive individualized intervention to low-attaining 6- and 7-year olds. Children in the program undergo 30 min of individualized instruction per day over a period of 12–14 weeks.

The choice of topics within the program is based on the Learning Framework in Number, devised by the Steffe (1992). This divides the learning of arithmetic into five broad stages: emergent (some simple counting, but few numerical skills), perceptual (can count objects and sometimes add small sets of objects that are present), figurative (can count well and use "counting-all" strategies to add), counting-on (can add by "counting on from the larger number" and subtract by counting down; can read numerals up to 100 but have little understanding of place value), and facile (know some number facts; are able to use some derived fact strategies; can multiply and divide by strategies based on repeated addition; may have difficulty with carrying and borrowing).

Children are assessed, before and after intervention, in a number of key topics. They undergo interventions based on their initial performance in each of the key topics. The key topics that are selected vary with the child's overall stage. For example, the key topics at the Emergent stage are (i) number word sequences from 1 to 20, (ii) numerals from 1 to 10, (iii) counting visible items (objects), (iv) spatial patterns (e.g., counting and recognizing dots arranged in domino patterns and in random arrays), (v) finger patterns (recognizing and demonstrating quantities up to five shown by number of fingers), and (vi) temporal patterns (counting sounds or movements that take place in a sequence). The key topics at the next, perceptual, stage are (i) number word sequences from 1 to 30, (ii) numerals from 1 to 20, (iii) figurative counting (counting on and counting back, where some objects are visible but others are screened), (iv) spatial patterns (more sophisticated use of domino patterns; grouping sets of dots into "lots of 2"; "lots of 4"; etc.), (v) finger patterns (recognizing, demonstrating, and manipulating patterns up to 10 shown by numbers of fingers), and (vi) equal groups and sharing (identifying equal groups and partitioning sets into equal groups). The key topics at later stages place greater emphasis on arithmetic and less on counting. Despite the overall division into stages, the program acknowledges and adapts to the fact that some children can be at a later stage for some topics than for others.

There are many activities that are used for different topics and stages within the Mathematics Recovery program. For example, activities dealing with temporal patterns at the Emergent stage include children counting the number of chopping movements made with the adult's hands, producing a requested number of chopping movements with their own hands, counting the number of times they hear the adult clap, and clapping their own hands a requested number of times. Activities dealing with number word sequences in fives at the Counting-On stage include children being presented with sets of 5-dot cards; counting the dots as each new card is presented; counting to 30 in fives without counting the dots; counting to 30 in fives without the cards; counting to 50 in fives without the cards; and counting backward in fives from 30, first with and then without the cards.

Early evidence indicated that children in the program improved significantly on the topics that form the focus of the problem, often reaching age-appropriate levels in these topics. The teachers who worked on the program found the experience very useful; they felt that it helped them to gain a better understanding of children's mathematical development, and used ideas and techniques from the program in their subsequent classroom teaching. Several quantitative empirical studies have shown that test scores of children in the program improve significantly in comparison with controls (e.g., Smith, Cobb, Farran, Cordray, & Munter, 2013; Willey, Holliday, & Martland, 2007; Wright, Martland, & Stafford, 2006); though it is not clear how long the effects last. Smith et al. (2013) found strong effects at the end of first grade on a measure devised by Mathematics Recovery, and moderate effects at the end of first grade for an independent evaluation, but the effects were not significant by the time the children reached second grade. It should, however, be noted that we do not know what effects might emerge at later stages of education, and that most interventions have not been subjected to extremely long-term evaluation.

11. NUMBERS COUNT

Every Child Counts was a partnership initiative in the United Kingdom between the Every Child a Chance charity (a coalition of business partners and charitable trusts) and government (Dunn, Matthews, & Dowrick, 2010). The aim was to enable the lowest attaining children to make greater progress toward expected levels of attainment in mathematics. The original intention was to provide intensive support in mathematics to 30,000 Year 2 (6- to 7-year old) children annually, though that has been significantly reduced due to the economic crash of 2008 and later government spending cuts.

The Numbers Count program was developed for the purpose of early intensive intervention (Dowrick, 2015; Dunn et al., 2010). This program draws on aspects of three existing interventions: Mathematics Recovery, MultiSensory Mathematics (developed in Leeds using Numicon materials), and Numeracy Recovery (a program developed in Hackney, London). It involves careful assessment of individual children's strengths and weaknesses, followed by individualized intervention targeted to addressing specific weaknesses, and emphasizes the development of number concepts through multisensory teaching.

In the initial stages of the project, 2621 Year 2 children, across 27 English local authorities, took part in Numbers Count. They received an average of 40 half-hour individualized Numbers Count lessons in a term, delivered by teachers who had received Masters-level training. The participating children were given the Sandwell Test, a standardized arithmetic test, before and after entering the program, and were retested 3 and 6 months later.

Torgerson et al. (2011) carried out an independent evaluation of the program. Before entering the program, their Number Age was on average 11 months below their Chronological Age. On average, they gained 14 months in Number Age in one term, a ratio gain of over 4 (months gained in Mathematical Age divided by mean duration of intervention in months). They were performing at chronological age level by the time they exited the program. The gains were maintained over 6-month follow-up. A long-term evaluation is currently being carried out to investigate whether the effects of the intervention persists to the end of primary school and into secondary school.

12. LIGHTER-TOUCH INTERVENTIONS

These results are impressive; however, only a minority of children will need such intensive intervention. Very intensive interventions with children with less severe difficulties may not be cost-effective or even practically possible to implement on a large scale; and also risk seriously reducing the children's time spent in other beneficial school activities. Nevertheless, there are many children outside the very lowest achieving group who need intervention with arithmetic, and may suffer serious long-term consequences if such intervention is not available. Although intensive interventions are aimed at approximately the bottom 5% of mathematics learners, we know that at least 20% of people have persistent numeracy difficulties in adulthood which seriously impair their social, practical, and economic lifestyles (Bynner & Parsons, 1997; Parsons & Bynner, 2005). There are clearly many people outside the bottom 5% who have serious need of intervention. Lighter-touch, less intensive interventions are needed for such individuals.

A somewhat less intensive intervention developed by Edge Hill University in association with Numbers Count is Ist Class @ Number (see https://everychildcounts.edgehill.ac.uk/mathematics/1stclassnumber/). In this program, a specially trained teaching assistant delivers 30 half-hour sessions to a group of up to four children, for 12–15 weeks. The lessons focus on number and calculation, developing children's mathematical understanding, communication, and reasoning skills. It is used for both primary and secondary pupils. So far, it has been used with over 45,000 children. Evaluations so far by Edge Hill University have shown an average Number Age gain of over three times the expected amount of progress. The intervention is now receiving an independent evaluation organized by the Education Endowment Foundation.

13. CATCH UP NUMERACY

A less intensive intervention, in terms of time spent, is Catch Up™ Numeracy, which was developed through a collaboration between the author and Catch-Up™, a not-for-profit British educational charity (Dowker & Sigley, 2010; Holmes & Dowker, 2013). The main target pupils for the Catch Up™ Numeracy intervention are pupils in Years 2–6 who have numeracy difficulties. Over 45,000 children in England and Wales are estimated to have taken part in this intervention since its development in 2007.

Catch Up Numeracy is based on an analysis of the particular subskills that children bring to arithmetical tasks, with remediation of the specific areas where children show problems. The components addressed here are not to be regarded as an all-inclusive list of components of arithmetic, either from a mathematical or educational point of view. Rather, the components were selected because earlier research studies and discussions with teachers have indicated them to be important in early arithmetical development, and because research has shown them to vary considerably between individual children in the early school years.

The 10 components are:

(1) Counting verbally (subcomponents: counting verbally from 0 or 1, counting on from a given number, counting back from a given number).

(2) Counting objects (subcomponents: counting objects, order irrelevance, repeated addition of objects, repeated subtraction of objects).

(3) Reading and writing numbers (subcomponents: reading numerals, reading number words, writing numerals).

(4) Hundreds, tens and units (subcomponents: number comparison, adding tens and units, subtracting tens and units).

(5) Ordinal numbers. The ordinal task involves showing children pictures of bead strings and telling them that the bead with the dot in the middle is the first (or in some pictures 10th) bead. One bead in each picture is shaded, and the children are asked to say which bead it is (e.g., second, fourth, and ninth).

(6) Word problems (solving arithmetical problems presented in story format).

(7) Translation between different formats (subcomponents: translating from objects to numerals, translating from numerals to objects, translating from number words to numerals, translating from number words to objects).

(8) Derived fact strategies (using arithmetical principles to derive a new fact from a known fact; e.g., using knowledge of commutativity of addition to work out that if $44 + 23 = 67.23 + 44$ must also be 67).

(9) Estimation of quantities and of answers to arithmetic problems (guessing approximate answers when not required to be exact).

(10) Remembered number facts (retrieving number bonds from memory).

The teachers and teaching assistants involved in the program receive three half days of training. Children in this project receive individual interventions from trained teachers or teaching assistants during two 15-min sessions per week. Initially, each child is assessed individually by a trained teacher/teaching assistant. This creates a personalized profile to determine the entry level for each of 10 Catch Up Numeracy components. Children are given mathematical games and activities targeted to their specific levels in specific components. Teachers and teaching assistants are able to choose from a list of such games and activities. Where possible, these games and activities involve the use of materials that are commonly available in schools, rather than specially created apparatus (Dowker & Sigley, 2010). Such materials include, for example, cards and sheets showing numerals and number operation symbols; number lines and number tracks; pictures illustrating quantities and operations; tape measures, rulers, and other measuring devices; beads and strings; cubes; spinners; and dice and dominoes.

Each 15-min teaching session includes a review and introduction to outline the focus of the current session; a numeracy activity; and a linked recording activity where the child records the results of the activity, in oral, written, and/or concrete fashion. Children stay in the program for one term (approximately 12 weeks); though teachers may adjust the length of intervention in the light of the pupils' progress. A first evaluation looked at 440 primary school children (Holmes & Dowker, 2013). Of these, 348 received the Catch Up intervention program. Fifty received the same amount of time for individualized mathematics work that was *not* specifically targeted to preassessed individual strengths and weaknesses;

but usually involved reviewing work done in the school mathematics classes. Forty-two children received no intervention, except for the usual school instruction.

The children were given the Hodder Number Screening Test (Gillham & Hesse, 2001) before and after the intervention to determine their Number Ages. The initial and final tests were given approximately 5 months apart.

Although the group assignments at this stage were not randomized, analyses of variance confirmed there were no significant differences in either chronological age or number age at the start of intervention. The 348 children who received Catch Up intervention showed very significant improvement. Indeed, they improved *more than twice* as much as would be expected by the passage of time. Their mean ratio gain was 2.51 (standard deviation of 1.9). The other children also improved, but not nearly as much. The 50 children who received matched time intervention showed a mean ratio gain of 1.47 (standard deviation 1.78). The 42 children who received no intervention showed a mean ratio gain of 1.25 (standard deviation of 1.8).

An ANOVA, with Group as the factor and Ratio Gain as the dependent variable, showed a highly significant effect of group. Post hoc tests confirmed that the Catch Up intervention group made significantly higher ratio gains than either of the other groups, which did not differ significantly from one another.

Funding by the Education Endowment Fund has since made it possible to carry out randomized controlled trials, where children are randomly assigned by an independent body to intervention and no-intervention groups. A study with 334 participants showed again that children who underwent Catch Up intervention improved more than twice as much as would be expected by the passage of time alone, and that their improvement and their test scores at the end of intervention were significantly better than those who only received the usual school instruction. Current research is using similar randomized controlled trials to further investigate the differences between those who undergo Catch Up and matched-time intervention. We are also planning fine-grained analyses of the longitudinal relationships between different components at the beginning and end of intervention, and in particular the extent to which interventions in particular components improve performance in other components, and whether there are closer relationships between some components than others.

14. SOME CONCLUSIONS ABOUT TARGETED INTERVENTIONS

In summary, the findings so far suggest that any extra help in arithmetic is likely to give some benefit. But interventions that focus on the particular components with which an individual child has difficulty are likely to be much more effective. Moreover, the amount of time given to such individualized work does not, in many cases, need to be very large to be effective. Catch Up Numeracy resulted in very significant gains despite only involving half an hour's intervention per week. We still do not know, however, what is the lower limit of the amount of time that can be spent on an intervention while still seeing effects.

More broadly, it appears that individualized, component-based approaches to intervention can be highly effective. It would be desirable to investigate whether different approaches to such intervention (e.g., age when intervention starts, degree of intensiveness) may be differentially appropriate to different groups of children. It would also be desirable to investigate the extent to which such methods may be adapted to younger or older children than those typically included in such interventions. Already, Catch Up Numeracy is being applied in some secondary schools for pupils who have gaps in numeracy that are affecting their subsequent progress. At the other end of the scale, it would be of great interest to see whether very early targeted interventions in specific areas of weaknesses in preschoolers' or school beginners' early numerical abilities might prevent at least some of them from falling behind in the first place. It would also be desirable to investigate the potential for similar types of intervention in areas of mathematics other than numeracy (e.g., geometry and measurement).

15. HOW IMPORTANT IS AGE TO INTERVENTION IN MATHEMATICS?

Findings suggest that, at least within the primary school age range, age at intervention has little effect on level of improvement (Chodura et al., 2015; Dowker, 2016). This suggests that there is no "critical period" or rigid timescale for learning, consistent with findings that age of starting formal education has little impact on the final outcome (Mullis, Martin, Foy, & Arora, 2012). It is still desirable to intervene relatively early to prevent mathematical difficulties from impeding progress in other school subjects, or contributing to the development of mathematics anxiety; but at least in the case of primary school children, intervention can be effective at any age.

16. TRAINING IN DOMAIN-GENERAL COGNITIVE SKILLS

So far, this chapter has been focusing on interventions that directly address aspects of arithmetic and number concepts. Some intervention techniques, rather than focusing directly on number and arithmetic, seek to teach domain-general cognitive skills, which are seen as crucial to mathematical skills. Some years ago, the focus tended to be on Piagetian operations (e.g., Adey & Shayer, 1994; Ginsburg, 1977). Recently, it has been more on executive functions and working memory (e.g., Diamond & Lee, 2011; Klingberg, 2010).

17. TRAINING IN PIAGETIAN OPERATIONS

At certain times in the past, training in Piagetian operations was considered an important form of intervention to improve children's arithmetic. The basis for this approach was that understanding the underlying logical concepts is a necessary prerequisite to arithmetic. This approach was susceptible to criticism by some Piagetians, because Piaget and some of his followers have considered that such operations are developed through maturation and experience rather than direct training, and because it may seem unnecessary to train children in concepts that develop universally (Ginsburg, 1977). More recently, it has become less popular because Piaget's theory in its pure form is not fully accepted by many researchers. Some approaches that have emphasized training in Piagetian concrete operations have been relatively unsuccessful in improving arithmetical performance (Snorre Ostad, personal communication). On the other hand, some approaches that have combined such training with other forms of intervention have led to significant improvement. An example is Van de Rijt and Van Luit's (1998). Additional Early Mathematics program, which combined, a counting intervention for 4- to 7-year olds in The Netherlands, who were struggling with mathematics, was combined with practice in seriation, correspondence, and classification. The children were given small-group interventions (two half-hour sessions per week in groups of 4 or 5).

The acquisition of *formal* operations (involving the manipulation of symbols and abstractions) in adolescence may be more susceptible to, and dependent, on training than the acquisition of concrete operations. There are considerable individual and cross-cultural differences as regards whether and when individuals succeed in tasks designed to measure formal operations, and they seem to depend highly on schooling. There has been some research involving training older children in formal operations. Typically, this has been aimed at groups of children of all abilities, rather than specific interventions for children with difficulties. Research by Adey and Shayer (1994) indicate that training in formal operations does indeed appear to have a positive impact on the mathematical development of older children and adolescents. This research was predominantly aimed at improving performance in science; improvement in mathematics was a by-product. Subsequently, Adhami, Johnson, and Shayer (1998) have used training in formal operations more specifically to enhance mathematical learning: Cognitive Acceleration in Mathematics Education or CAME.

18. TRAINING IN METACOGNITION

Several intervention projects have involved training children, with and without arithmetical difficulties to reflect on their knowledge and plan and monitor their arithmetical strategies (Adey & Shayer, 1994; Desoete et al., 2004; Verschaffel et al., 1999) and have generally produced positive results.

It is not always absolutely clear which aspects of such programs are important to their success. "Metacognitive training" can mean somewhat different things in different studies. Some studies may emphasize a form of reflection that is really an aspect of Piagetian formal operations, and thus comes within the previous section; some emphasize the ability to discuss and compare arithmetical strategies; others may emphasize a planning ability more akin to aspects of working memory; still others may emphasize awareness of one's own knowledge state, including "knowing when one doesn't know." More research is needed on exactly which aspects of metacognition are important here.

19. TRAINING IN WORKING MEMORY AND EXECUTIVE FUNCTION

A number of studies suggest that working memory, and other executive functions, such as shifting and inhibition, are correlated with arithmetic (Ashkenazi, Rosenberg-Lee, Metcalfe, Swigart, & Menon, 2013;

Bull & Scerif, 2001; Passolunghi & Cornoldi, 2008; Szucs, Devine, Soltesz, Nobes, & Gabriel, 2014). This has led to attempts to train children in working memory and/or other executive functions, with the hope both of improving their cognitive abilities in general and their mathematical abilities in particular.

There are two questions: is it possible to train children in executive functions, and, if it is, does it improve their mathematical development? This topic is discussed by Day-Hess and Clements in this volume. There is much controversy about both issues (Clements, Sarama, & Germeroth, 2016). Bodrova and Leong (2007) developed the Tools of the Mind program to train pre-K and kindergarten children in executive function in a play context. Some studies have suggested that it does lead to improved executive function (Barnett et al., 2008; Diamond, Barnett, Thomas, & Munro, 2007). However, several other studies reported no effect (Clements, Sarama, Layzer, Unlu, & Germeroth, 2012).

Computerized executive function training programs have been developed for somewhat older primary school children (Klingberg, 2010; Thorell, Lindqvist, Bergman Nutley, Bohlin, & Klingberg, 2009), with conflicting results. Diamond and Lee (2011) found overall evidence of effectiveness, as did Klingberg (2010) specifically for working memory training. However, Melby-Lervåg and Hulme (2013) carried out a meta-analysis suggesting relatively little effect of such training. Bergman Nutley et al. (2011) found that training in working memory improved performance on similar working memory tasks, but not on a nonverbal reasoning task.

There is also controversy as to the effect of such training on academic skills, including mathematics. Redick, Shipstead, Wiemers, Melby-Lervag, and Hulme (2015) found that working memory training had some effects on performance on untrained working and short-term memory task, but not on mathematical or other academic performance. By contrast, Dunning, Holmes, and Gathercole (2013) found that computerized working memory training in 10-year olds did not lead to many far transfer effects *except* for mathematical reasoning. Holmes, Gathercole, and Dunning (2009) found effects of such training on untrained working memory tasks *and* on mathematical performance. Thus, training in working memory and other executive functions is a potentially promising approach to improving mathematical skills; but more research is needed on whether it is effective, and, if so, under what conditions.

20. LARGER-GROUP INTERVENTIONS

There are many other interventions in use. While so far, I have focused on individualized or very small-group interventions, schools use a variety of interventions of varying degrees of intensiveness, ranging in intensiveness from individualized to whole-class intervention.

On the whole, whole-class approaches have not been regarded as interventions, but in the context of curriculum development. However, as several countries have become increasingly concerned with improving numeracy standards, there is an increasing interest in investigating the possible role of certain new whole-class programs in reducing the incidence of numeracy difficulties. One program which is attracting current interest from this point of view in the United Kingdom is Mathematics Mastery, a program inspired by some aspects of Singapore mathematics education. Compared to traditional curricula, fewer topics are covered in more depth and greater emphasis is placed on problem solving and on encouraging mathematical thinking. A current evaluation by the Education Endowment Fund has so far indicated that use of the program in Year 1 results in an average increased gain in mathematics age of 2 months in the first year, and use of the program in Year 7 results in an average increased gain in mathematics age of 1 month. Further investigation is desirable to see whether these gains are maintained or extended over time.

21. THE RESPONSE TO INTERVENTION APPROACH

In the United States, the Response to Intervention approach is one approach to providing interventions for struggling learners in both reading and mathematics (Forbringer & Fuchs, 2014; Fuchs, Fuchs, & Prentice, 2004; Fuchs et al., 2008). It aims to identify children at risk of educational failure early enough to prevent it from occurring. The aim is to monitor all pupils' performance two or three times a year to identify those who may need intervention. They are then at first given relatively nonintensive interventions and, according to their response to these interventions, may receive interventions of increasing "tiers" of intensiveness, from appropriate whole-class instruction, through small-group intervention, to individualized intervention and in severe cases assignment to special education programs.

22. CONCLUSIONS

Mathematics interventions for primary school children have been in existence for many years, but only since the 1990s have they been in relatively widespread use, and they are still not nearly as widely used as interventions for language or literacy. Research has indicated that such interventions, especially those targeted to individuals' specific strengths and weaknesses, are likely to have a positive impact at least in the short term. More research is still needed to evaluate their long-term impact. A particularly intriguing form of intervention involves addressing not the mathematical difficulties themselves, but weaknesses in underlying domain-general cognitive abilities. If these interventions could be proven to be successful in addressing both the underlying cognitive deficits and mathematical skills themselves, then they would have wide-reaching impact; but the evidence is still conflicting.

Most interventions have received only limited long-term follow-up and it is important to learn more about the extent to which they influence educational success in the long term; and whether there are any ways of increasing such effects, either by modifications of the original interventions or by the introduction of later "boosters."

As well as the important practical implications of intervention studies, they have implications for theories about mathematical development. There are important bidirectional relationships between research and intervention. We can integrate the implementation and evaluation of intervention schemes with the investigation of individual differences in arithmetic or in selected components of arithmetic, and on how different numerical abilities may relate to one another and/or to certain domain-general abilities. Thus, theories about number and arithmetic may inspire and inform intervention projects, which may in their turn serve to test these theories.

REFERENCES

Adey, P., & Shayer, M. (1994). *Really raising standards*. London: Routledge.

Adhami, M., Johnson, D., & Shayer, M. (1998). *Thinking maths: Accelerated learning in mathematics*. Oxford: Heinemann.

Ainley, J. (1990). Playing games and learning mathematics. In L. Steffe & T. Wood (Eds.), *Transforming children's mathematics education* (pp. 84–91). Hillsdale, NJ: Erlbaum.

Ashkenazi, S., Rosenberg-Lee, M., Metcalfe, A., Swigart, A., & Menon, V. (2013). Visuo-spatial working memory is an important source of domain-general vulnerability in the development of arithmetic cognition. *Neuropsychologia*, *51*, 2305–2317.

Askew, M., Bibby, T., & Brown, M. (2001). *Raising attainment in primary number sense: From counting to strategy.* London: BEAM Education.

Baker, S., Gersten, R., & Lee, D. (2002). A synthesis of empirical research on teaching mathematics to low-achieving students. *Elementary School Journal, 103*, 51–73.

Barnett, W. S., Jung, K., Yarosz, D. J., Thomas, J., Hornbeck, A., Stechuk, R., et al. (2008). Educational effects of the tools of the mind curriculum: A randomized trial. *Early Childhood Research Quarterly, 23*, 299–313.

Beilock, S. L. (2008). Math performance in stressful situations. *Current Directions in Psychological Science, 17*, 339–343.

Beilock, S. L., & Carr, T. H. (2005). When high-powered people fail: Working memory and "choking under pressure" in math. *Psychological Science, 16*, 101–105.

Bergman Nutley, S., Söderqvist, S., Bryde, S., Thorell, L. B., Humphreys, K., & Klingberg, T. (2011). Gains in fluid intelligence after training non-verbal reasoning in 4-year-old children: A controlled, randomized study. *Developmental Science, 14*, 591–601.

BIS. (2011). *BIS research paper 57: Skills for life survey headline findings.* London: Department of Business, Innovation and Skills.

Bodrova, E., & Leong, D. J. (2007). *Tools of the mind: The Vygotskian approach to early childhood education.* New York, NY: Merrill/Prentice-Hall.

Brown, J. S., & Burton, R. (1978). Diagnostic models for procedural bugs in basic mathematical skills. *Cognitive Science, 2*, 155–192.

Brownell, W. (1929). Remedial cases in arithmetic. *Peabody Journal of Education*, 7, 100–107.

Bryant, D. P., Bryant, B. R., Gersten, R., Scammacca, N., & Chavez, M. (2008a). Mathematics intervention program for first-and second-grade students with mathematics difficulties: The effects of tier 2 intervention delivered as booster lessons. *Remedial and Special Education, 29*, 20–32.

Bryant, D. P., Bryant, B. R., Gersten, R., Scammacca, N., Funk, C., Winter, A., et al. (2008b). The effects of tier 2 intervention on first-grade students who are at risk for mathematics difficulties. *Learning Disability Quarterly, 31*, 47–63.

Bryant, D. P., Bryant, B. R., Roberts, G., Vaughn, S., Pfannenstiel, K. H., Porterfield, J., et al. (2011). Early numeracy intervention program for first-grade students with mathematics difficulties. *Exceptional Children, 78*, 7–23.

Bull, R., & Scerif, G. (2001). Executive functioning as a predictor of children's mathematics ability: Inhibition, switching, and working memory. *Developmental Neuropsychology, 19*, 273–293.

Buswell, G. T., & John, L. (1926). Diagnostic studies in arithmetic. *Elementary School Journal, 27*, 232–234.

Butterworth, B., Sashank, V., & Laurillard, D. (2011). Dyscalculia: From brain to education. *Science, 332*, 1049–1053.

Bynner, J., & Parsons, S. (1997). *Does numeracy matter?* London: Basic Skills Agency.

Cappelletti, M., Butterworth, B., & Kopelman, M. (2012). Numeracy skills in patients with degenerative disorders and focal brain lesions: A neuropsychological investigation. *Neuropsychology, 26*, 1–19.

Chan, B. M. Y., & Ho, C. S. H. (2010). The cognitive profile of Chinese children with mathematics difficulties. *Journal of Experimental Child Psychology, 107*, 260–279.

Chodura, S., Kuhn, J.-T., & Holling, H. (2015). Interventions for children with mathematical difficulties: A meta-analysis. *Zeitschrift für Psychologie, 223*, 129–144.

Clements, D. H., & Sarama, J. (2011). Early childhood mathematics intervention. *Science, 333*, 968–970.

Clements, D. H., Sarama, J., & Germeroth, C. (2016). Learning executive function and early mathematics: Directions of causal relations. *Early Childhood Research Quarterly, 36*, 79–90.

Clements, D. H., Sarama, J., Layzer, C., Unlu, F., & Germeroth, C. (2012). In *The efficacy of an intervention synthesizing scaffolding designed to promote self-regulation with an early mathematics curriculum: Effects on executive function Paper presented at the society for research on educational effectiveness.*

Cohen Kadosh, R., Dowker, A., Heine, A., Kaufmann, L., & Kucian, K. (2013). Interventions for improving numerical abilities: Present and future. *Trends in Neuroscience and Education, 2*, 85–93.

Cowan, R., Donlan, C., Shepherd, D. L., Cole-Fletcher, R., Saxton, M., & Hurry, J. (2011). Basic calculation proficiency and mathematics achievement in elementary school children. *Journal of Educational Psychology, 103*, 786–803.

Davenport, P., & Howe, C. (1999). Conceptual gain and successful problem-solving in primary school mathematics. *Educational Studies, 25*, 55–78.

Dehaene, S. (1997). *The number sense.* London: Macmillan.

Delazer, M. (2003). Neuropsychological findings on conceptual knowledge of arithmetic. In A. Baroody & A. Dowker (Eds.), *The development of arithmetical concepts and skills* (pp. 385–407). Mahwah, NJ: Erlbaum.

Demeyere, N., Lestou, V., & Humphreys, G. W. (2010). Neuropsychological evidence for a dissociation in counting and subitizing. *Neurocase, 16*, 137–219.

Desoete, A., Roeyers, H., & De Clercq, A. (2004). Children with mathematics learning disabilities in Belgium. *Journal of Learning Disabilities, 37*, 32–41.

Diamond, A., Barnett, W. S., Thomas, J., & Munro, S. (2007). Preschool program improves cognitive control. *Science, 318*, 1387–1388.

Diamond, A., & Lee, K. (2011). Interventions shown to aid executive function development in children 4 to 12 years old. *Science, 333*(6045), 959–964.

Dowker, A. D. (2005). *Individual differences in arithmetic: Implications for psychology, neuroscience and education.* Hove: Psychology Press. Chapter 10.

Dowker, A. D. (2009). *What works for children with mathematical difficulties? The effectiveness of intervention schemes.* London: DCSF.

Dowker, A. D. (2016). Factors that influence improvement in numeracy, reading and comprehension in the context of a numeracy intervention. *Frontiers in Psychology*, 7, 1929.

Dowker, A., & Sigley, G. (2010). Targeted interventions for children with arithmetical difficulties. *British Journal of Educational Psychology II*, 7, 65–81.

Dowrick, N. (2015). Numbers count: A large-scale intervention for young children who struggle with mathematics. In R. Cohen Kadosh & A. Dowker (Eds.), *The Oxford handbook of numerical cognition* (pp. 1099–1117). Oxford: Oxford University Press.

Dunn, S., Matthews, L., & Dowrick, N. (2010). Numbers count: Developing a national approach to intervention. In I. Thompson (Ed.), *Issues in teaching numeracy in primary schools* (pp. 224–234). Maidenhead: Open University Press.

Dunning, D. L., Holmes, J., & Gathercole, S. E. (2013). Does working memory training lead to generalized improvements in children with low working memory? A randomized controlled trial. *Developmental Science, 16*(6), 915–925.

El-Naggar, O. (1996). *Specific learning difficulties in mathematics: A classroom approach.* London: NASEN.

Forbringer, L., & Fuchs, W. (2014). *Response to intervention in math: Evidence-based interventions for struggling students.* London: Routledge.

Fuchs, L. S., Compton, D. L., Fuchs, D., Paulsen, K., Bryant, J. D., & Hamlett, C. L. (2005). The prevention, identification, and cognitive determinants of math difficulty. *Journal of Educational Psychology, 87*, 155–173.

Fuchs, L. S., Fuchs, D., Hamlett, C. L., Powell, S. R., Capizzi, A. M., & Seethaler, P. M. (2006). The effects of computer-assisted instruction on number combination skill in at-risk first-graders. *Journal of Learning Disabilities, 39*, 467–475.

Fuchs, L. S., Fuchs, D., & Prentice, K. (2004). Responsiveness to mathematical problem-solving instruction: Comparing students at risk of mathematics disability with and without risk of reading disability. *Journal of Learning Disabilities, 37*, 293–306.

Fuchs, L. S., Geary, D. C., Compton, D. L., Fuchs, D., Schatschneider, C., Hamlett, C. L., et al. (2013). Effects of first-grade number knowledge tutoring with contrasting forms of practice. *Journal of Educational Psychology, 105*, 58–77.

Fuchs, L. S., Seethaler, P. M., Powell, S. R., Fuchs, D., Hamlett, C. L., & Fletcher, J. M. (2008). Effects of preventative tutoring on the mathematical problem solving of third-grade students with math and reading difficulties. *Exceptional Children, 74*, 155–173.

Gersten, R., Chard, D., Jayanthi, M., Baker, S., Morphy, P., & Flojo, J. (2009). Mathematics instruction for students with learning disabilities: A meta-analysis of instructional components. *Review of Educational Research, 79*, 1202–1242.

Gervasoni, A., & Sullivan, P. (2007). Assessing and teaching children who have difficulty learning arithmetic. *Educational & Child Psychology, 24*(2), 40–53.

Gifford, S., & Rockliffe, F. (2012). Mathematics difficulties: Does one approach fit all? *Research in Mathematics Education, 14*, 1–16.

Gillham, B., & Hesse, K. (2001). *Basic number screening test: National numeracy strategy edition: Forms A & B, for ages 7 to 12 years* (3rd ed.). London: Hodder Education.

Ginsburg, H.P. (1977). Children's arithmetic: How they learn it and how you teach it. New York, NY: Teachers' College Press.

Greenes, C., Ginsburg, H. P., & Balfanz, R. (2004). Big math for little kids. *Early Childhood Research Quarterly, 19*, 159–166.

Greeno, T., Riley, M., & Gelman, R. (1984). Young children's counting and understanding of principles. *Cognitive Psychology, 16*, 94–143.

Gross, J. (2007). Supporting children with gaps in their mathematical understanding. *Educational and Child Psychology, 24*, 146–156.

Holmes, W., & Dowker, A. D. (2013). Catch up numeracy: A targeted intervention for children who are low attaining in mathematics. *Research in Mathematics Education, 15*, 249–265.

Holmes, J., Gathercole, S. E., & Dunning, D. L. (2009). Adaptive training leads to sustained enhancement of poor working memory in children. *Developmental Science, 12*, 9–15.

Jordan, J. A., Mulhern, G., & Wylie, J. (2009). Individual differences in trajectories of arithmetical development in typically achieving 5-7-year-olds. *Journal of Experimental Child Psychology, 103*, 455–468.

Kirkby, D. (1992). *Games in the teaching of mathematics*. Cambridge: Cambridge University Press.

Klein, A., Starkey, P., Clements, D., Sarama, J., & Iyer, R. (2008). Effects of a pre-kindergarten mathematics intervention: A randomized experiment. *Journal of Research on Educational Effectiveness, 1*, 155–178.

Klingberg, T. (2010). Training and plasticity of working memory. *Trends in Cognitive Sciences, 14*(7), 317–324.

Kroesbergen, E., & Van Luit, J. (2003). Mathematics interventions for children with special educational needs: A meta-analysis. *Remedial and Special Education, 24*, 97–114.

Kucian, K., Grond, U., Rotzer, S., Henzi, B., Schönmann, C., Plannger, F., et al. (2011). Mental number line training in children with dyscalculia. *NeuroImage, 57*, 782–795.

Lee, N. H., Chang, S. C. A., & Lee, P. Y. (2001). The role of metacognition in the learning of mathematics among the low-achieving students. *Teaching and Learning, 22*, 18–30.

Lou, Y., Abrami, P. C., Spence, J. C., Poulsen, C., Chambers, B., & D'Appolinia, S. (1996). Within-class grouping: A meta-analysis. *Review of Educational Research, 66*, 423–458.

Melby-Lervåg, M., & Hulme, C. (2013). Is working memory training effective? A meta-analytic review. *Developmental Psychology, 49*, 270–291.

Mullis, I. V. S., Martin, M. O., Foy, P., & Arora, A. (2012). *TIMSS international results in mathematics*. Chestnut Hill, MA: TIMSS & PIRLS International Study Center, Boston College.

Nastasi, B. K., & Clements, D. H. (1991). Research on cooperative learning: Implications for practice. *School Psychology Review, 20*, 110–131.

Parsons, S., & Bynner, J. (2005). *Does numeracy matter more?* London: NRDC.
Passolunghi, M. C., & Cornoldi, C. (2008). Working memory failures in children with arithmetical difficulties. *Child Neuropsychology, 14*, 387–400.
Ramani, G. B., & Siegler, R. S. (2008). Promoting broad and stable improvements in low-income children's numerical knowledge through playing number board games. *Child Development, 79*, 375–394.
Ramani, G. B., & Siegler, R. S. (2011). Reducing the gap in numerical knowledge between low- and middle-income preschoolers. *Journal of Applied Developmental Psychology, 32*, 146–159.
Räsänen, O., Salminen, J., Wilson, A., Aunio, P., & Dehaene, S. (2009). Computer-assisted intervention for children with low numeracy skills. *Cognitive Development, 24*, 450–472.
Redick, T. S., Shipstead, Z., Wiemers, E. A., Melby-Lervag, M., & Hulme, C. (2015). What's working in working memory training? An educational perspective. *Educational Psychology Review, 27*, 617–633.
Rowe, J. C. (2001). An experiment in the use of games in the teaching of mental arithmetic. *Philosophy of Mathematics Education, 14*, 1–23.
Russell, R., & Ginsburg, H. P. (1984). Cognitive analysis of children's mathematical difficulties. *Cognition and Instruction, 1*, 217–244.
Salomon, G. (1989). When teams do not function the way they ought. *International Journal of Educational Research, 13*, 89–99.
Scherer, P., Beswick, K., DeBlois, L., Healy, L., & Opitz, E. M. (2016). Assistance of students with mathematical learning difficulties: How can research support practice? *ZDM Mathematics Education, 48*, 633.
Siegler, R. S., & Ramani, G. B. (2008). Promoting broad and stable improvements in low-income children's numerical knowledge through playing number board games. *Developmental Science, 11*, 655–661.
Siegler, R. S., & Ramani, G. B. (2009). Playing linear board games—But not circular ones—Improves low-income preschoolers' numerical understanding. *Journal of Educational Psychology, 101*, 545–560.
Smith, T. M., Cobb, P., Farran, D. C., Cordray, D. S., & Munter, C. (2013). Evaluating math recovery: Assessing the causal impact of a diagnostic tutoring program on student achievement. *American Educational Research Journal, 50*, 397–428.
Stanescu-Cosson, R., Pinel, P., Van de Moortele, P. F., Le Bihan, D., Cohen, L., & Dehaene, S. (2000). Understanding dissociations in dyscalculia: A brain-imaging study of the impact of number size on the cerebral networks for exact and approximate calculation. *Brain, 123*, 2240–2255.
Steffe, L. P. (1992). Learning stages in the construction of the number sequence. In J. Bideaud, C. Meljac, & J. Fischer (Eds.), *Pathways to number: Children's developing numerical abilities* (pp. 83–88). Hillsdale, NJ: Lawrence Erlbaum.
Straker, A. (1996). *Mental maths for ages 5 to 7: Teachers' book*. Cambridge: Cambridge University Press.
Szucs, D., Devine, A., Soltesz, F., Nobes, A., & Gabriel, F. (2014). Cognitive components of a mathematical processing network in 9-year-old children. *Developmental Science, 17*, 506–524.
Thorell, L. B., Lindqvist, S., Bergman Nutley, S., Bohlin, G., & Klingberg, T. (2009). Training and transfer effects of executive functions in preschool children. *Developmental Science, 12*(1), 106–113.
Tilton, J. W. (1947). Individualized and meaningful instruction in arithmetic. *Journal of Educational Psychology, 38*, 83–88.
Topping, K. J., & Bamford, J. (1998). *Parental involvement and peer tutoring in mathematics and science*. London: Fulton.

Torgerson, C. J., Wiggins, A., Torgerson, D. J., Ainsworth, H., Barmby, P., Hewitt, C., et al. (2011). *Every child counts: The independent evaluation executive summary*. London: Department for Education (DfE).
US Department of Education. (2011). *The condition of education, 2011*. Washington, DC: National Center for Education Statistics.
Van de Rijt, B., & Van Luit, J. (1998). Effectiveness of the additional early mathematics program for teaching children early mathematics. *Instructional Science, 26*, 337–358.
Van Eimeren, L., Grabner, R. H., Koschutnik, K., et al. (2010). Structure-function relationships underlying calculation: A combined diffusion tensor imaging and fMRI study. *NeuroImage, 52*, 358–363.
Van Lehn, K. (1990). *Mind bugs: The origin of procedural misconceptions*. Cambridge, MA: MIT Press.
Verschaffel, L., DeCorte, E., Lasure, S., Van-Vaerenburgh, G., Bogaerts, H., & Ratinckxx, E. (1999). Learning to solve mathematical application problems: A design experiment with fifth graders. *Mathematical Thinking and Learning, 1*, 195–229.
Vosse, A. J. M. (1999). Effects and implementation of a peer tutoring program for children at risk. *Pedagogische Studien, 76*, 201–210.
Warrington, E. K. (1982). The fractionation of arithmetical skills: A single case study. *Quarterly Journal of Experimental Psychology, 34A*, 31–51.
Weaver, J. (1954). Differentiated instruction in arithmetic: An overview and a promising trend. *Education, 74*, 300–305.
Willey, R., Holliday, A., & Martland, J. (2007). Achieving new heights in Cumbria: Raising standards in early numeracy through mathematics recovery. *Educational and Child Psychology, 24*, 108–118.
Williams, C., & Whitaker, R. L. (1937). Diagnosis of arithmetical difficulties. *Elementary School Journal, 37*, 592–600.
Wilson, A. J., Revkin, S. K., Cohen, D., Cohen, L., & Dehaene, S. (2006). An open trial assessment of "The Number Race", an adaptive computer game for remediation of dyscalculia. *Behavioural and Brain Functioning, 2*, 20.
Wright, R. J., Martland, J., & Stafford, A. (2006). *Early numeracy: Assessment for teaching and intervention* (2nd ed.). London: Paul Chapman.

CHAPTER EIGHT

Early Executive Function and Mathematics Relations: Correlation Does Not Ensure Concordance

Michèle M.M. Mazzocco*[,1], Jenny Yun-Chen Chan*, Allison M. Bock†
*Institute of Child Development, University of Minnesota, Minneapolis, MN, United States
†Center for Early Education and Development, University of Minnesota, Minneapolis, MN, United States
[1]Corresponding author: e-mail address: mazzocco@umn.edu

Contents

Abstract

In this chapter, we address one potentially overlooked component of the relation between executive function (EF) skills and early mathematics, a relation for which there is widespread empirical support. Evidence for this relation has, thus far, been largely correlational. Here we emphasize that because positive correlations do not guarantee concordance among all members of a sample or population, a small but meaningful number of children may either fare well in mathematics despite poor EF skills, or may have strong EF skills despite weak mathematics skills. We propose that attention to different profiles of discordance for EF and mathematics may help identify individualized learning needs for students at risk for mathematics difficulties and disabilities.

There is little doubt that executive function (EF) skills are related to mathematics achievement. EF skills refer to deliberate, goal-oriented cognitive processes that guide behavior (Diamond, 2013; Miyake et al., 2000)—and

Advances in Child Development and Behavior, Volume 53
ISSN 0065-2407
http://dx.doi.org/10.1016/bs.acdb.2017.05.001

that seem to have an intuitive connection to mathematics. *Cognitive flexibility* allows us to consider multiple approaches to problem solving and to switch between operations during arithmetic computations when necessary (at least those we execute by hand), such as adding when regrouping during multidigit multiplication or subtraction, or subtracting when regrouping during addition. *Inhibitory control* supports selective attention to relevant components of problems when solving word problems or learning new mathematics principles that counter well-established mathematical knowledge, such as when information about negative numbers (e.g., "−2" is larger than "−4") or fractions (e.g., ½ is larger than ¼) conflicts with whole-number knowledge (e.g., that 4 is larger than 2). *Working memory*, which involves the active mental storage, manipulation, or updating of information, supports the mental processes we engage to keep track of multistep problem solving (e.g., multidigit multiplication and long division) or even to track items we have already counted (and those we need to count) when determining "how many" items are in a set. EF skills also support the mental reflection that guides self-evaluation of generated solutions, an important metacognitive support for successful mathematics (Garrett, Mazzocco, & Baker, 2006; Prager, under review). In this chapter, we do not review the full extent of the EF and mathematics relation, nor different theoretical perspectives on this relation, as many others have already done (e.g., Clements, Sarama, & Germeroth, 2016). Instead, we proffer a potentially overlooked and critical view that individual differences in the nature of this relation should be considered when studying the role of EF skills in early mathematics.

The notion that EF and mathematics are related is not based solely on intuition and examples like those described earlier. Confidence in the existence of this relation stems from the large and rapidly growing body of studies demonstrating that EF skills predict concurrent and later mathematics achievement in school-age children (e.g., Bull & Scerif, 2001; Geary, Hoard, & Nugent, 2012; Mazzocco & Kover, 2007), adolescents (Best, Miller, & Naglieri, 2011), and adults (Hubber, Gilmore, & Cragg, 2014). Predictive relations between EF and mathematics are also evident in early childhood and primary school years (Blair, McKinnon, & The Family Life Project Investigators, 2016; Clark, Sheffield, Wiebe, & Espy, 2013; Fuhs, Nesbitt, Farran, & Dong, 2014; Gashaj, Uehlinger, & Roebers, 2016; Hassinger-Das, Jordan, Glutting, Irwin, & Dyson, 2014; Hornung, Schiltz, Brunner, & Martin, 2014; Prager, Sera, & Carlson, 2016; Purpura, Schmitt, & Ganley, 2017), although the mechanism is unclear. There is emerging evidence that EF mediates the relation between early numerical knowledge and later mathematics achievement from kindergarten to Grade 1 (Hassinger-Das et al., 2014), but also evidence that number decomposition skills mediate the relation between EF and mathematics achievement (Fuhs, Hornburg, & McNeil, 2016). Emerging evidence also raises the possibility that early mathematics skills predict and may support later EF development (e.g., Clements et al., 2016; Welsh, Nix, Blair, Bierman, & Nelson, 2010). At issue is whether and how the EF and mathematics relation can best be measured to help identify, understand, and support children at risk for poor mathematics achievement outcomes.

Evidence of an early EF/mathematics relation, and the challenges in understanding this relation, have implications for early mathematics education. As discussed elsewhere in this volume (chapter "Interventions for Primary School Children With Difficulties in Mathematics" by Dowker; chapter "Notice, Explore, and Talk About Mathematics: Making a Positive Difference for Preschool Children, Families, and Educators in Australian Communities That Experience Multiple Disadvantages" by Gervasoni and Perry; chapter "Designing Studies to Test Causal Questions About Early Math: The Development of Making Pre-K Count" by Mattera et al.; chapter "Interventions in Early Mathematics: Avoiding Pollution and Dilution" by Sarama and Clements) early mathematics skills provide a foundation for successful mathematics achievement throughout the school-age years (e.g., Geary, Hoard, Nugent, & Bailey, 2013). This early foundation is not trivial, given that school entry number system knowledge is more predictive of functional numeracy in adolescents than is growth in number system knowledge in primary school (Geary et al., 2013). Therefore, an understanding of if and how EF is related to early mathematics may facilitate efforts to strengthen that foundation. Here we argue that, in addition to understanding the general nature of EF/mathematics relations, it is essential to understand how the relation varies *across children* if we seek to strengthen that foundation for all children, including those at risk for poor mathematics outcomes. We offer three considerations that may guide our understanding of this variability, each of which speaks to why we should be cautious about generalizing findings of EF/mathematics relations to decisions about individuals.

The first consideration is that measures of EF and of specific mathematical skills differ across studies, thereby leading to inconsistent conclusions about their relation with one another. For instance, Fuhs and McNeil (2013) reported that inhibitory control accounts for variation in the acuity of preschoolers' approximate number comparison, whereas Bull and Scerif (2001) showed that inhibitory control and cognitive flexibility (i.e., shifting) were *less* predictive of primary school-age students' mathematics performance than was working memory. Geary (2011) found that each of the working memory components plays a unique role in mathematical thinking, even in early childhood. For example, visuospatial working memory supports learning number line concepts, whereas verbal working memory supports learning the verbal number sequence and, later, fact retrieval. Numerical verbal working memory (as measured by backward digit span) is correlated with early math fluency and word problem solving in 5–7 year olds (Martin, Cirino, Sharp, & Barnes, 2014). Performance on nonnumerical verbal working memory tasks also predicts number skills performance and growth. For instance, preschoolers' listening recall is correlated with their quantity comparison and combination skills (Purpura et al., 2017). Apparent inconsistencies in the EF/math relation may reflect differences in which EF and mathematics skills were evaluated across studies, which may influence evidence for the existence and strength of EF/mathematics relations. This consideration cautions us about generalizing conclusions across findings.

Another consideration is that the relation between EF and mathematics is dynamic, such that it varies throughout early childhood, given the extremely rapid growth in EF skills between 3 and 5 years of age (Carlson, Zelazo, & Faja, 2013) and the qualitative changes in mathematical thinking that also emerge during this time period (Chu, vanMarle, & Geary, 2016; Clements & Sarama, 2014; Sarama & Clements, 2009). This means that relations between EF and mathematics are subject to developmental differences within individuals (and, as we discuss later, to individual differences across persons). The EF/mathematics relation may be further complicated by the evidence that EF first emerges as a unitary construct (as reviewed by Bull & Lee, 2014), such that EF subcomponents are not reliably differentiated until late in school age or even adolescence (Willoughby, Wirth, & Blair, 2012). This has implications for whether EF (or math) should be measured in terms of an absolute level of a specific skill, or as an age-referenced relative score, both at a point in time, or in terms of the rate of growth over a specified time period (Willoughby, Magnus, Vernon-Feagans, Blair, & The Family Life Project Investigators, 2016). The latter may help determine sensitive periods during which performance or growth is most indicative of optimal (or suboptimal) development.

Finally, the directionality of this relation is unclear, and causal mechanisms have not been determined. Many researchers consider EF as a set of foundational skills supporting mathematical development, while others raise the possibility of bidirectional and synergistic relation in EF and mathematics development (e.g., Clements et al., 2016). A few researchers consider the EF and mathematics correlations as spurious (e.g., Willoughby, Kupersmidt, & Voegler-Lee, 2012), potentially linked by other factors that impact the development of both EF and mathematics, including literacy skills like letter recognition and word reading (Chu et al., 2016), IQ (Chu et al., 2016), and even teacher–student relationships (Blair et al., 2016). Despite these caveats, the strength of EF/mathematics correlations is approximately 0.31 (measured by r_p) regardless of participants' age, the measures used in studies, the EF subcomponent evaluated in a study, and whether the correlations tested are concurrent or predictive (based on a meta-analysis by Jacob & Parkinson, 2015). The contribution of factors related to these associations justifies the need to pursue more causal studies of the EF/mathematics relation (Willoughby, Kupersmidt, et al., 2012).

1. BINARY CORRELATIONS AND CONCORDANT PERFORMANCE LEVELS

It is worth noting that researchers emphasize *concordance* in discussing studies of an EF and mathematics relation, at least implicitly, by focusing on the direction of the overall correlation—such as drawing inferences that higher EF scores implicate higher (or lower) mathematics scores, that low EF is associated with poor performance in mathematics (e.g., Morgan et al., 2016, and vice versa), and that poor EF or working memory is a primary cognitive signature of mathematics learning difficulties or disabilities (e.g., Alloway & Passolunghi, 2011; Bull, Johnston, & Roy, 1999). Of course, only a perfect correlation *guarantees* concordance, but researchers typically do not evaluate the prevalence of discordance in their studies of EF and mathematics when correlations fall well below 1.0 (and correlations around 0.30, like those reported for EF/mathematics, are considered strong for behavioral studies). Perhaps the possibility of discordance is dismissed given the large literature base demonstrating positive correlations between EF and mathematics. When one author of this chapter referred to children with a profile of high mathematics and low EF skills, or low mathematics and high EF skills, to a group of early mathematics researchers, some questioned whether these combinations are even possible!

Given that correlations do not guarantee *dependence* (an argument as old as the correlation itself; Aldrich, 1995), here we explore two examples of discordance as a potential window on individual differences in early math learning profiles. That is, perhaps in addition to the well-known warning that correlation does not by itself indicate causation, it may be prudent to be reminded that correlation does not guarantee concordance among all members of a sample or population.

2. EXPLORING DISCORDANCE FOR EF AND MATHEMATICS: A GLIMPSE OF TWO STUDIES

2.1 Prekindergarten Children From a Low SES Sample

In an ongoing study of EF and mathematics skills among students in public school-based prekindergarten classrooms (as discussed by Day-Hess and Clements in chapter "The DREME Network: Research and Interventions in Early Childhood Mathematics"), participant samples were drawn from urban schools in high-poverty communities, in order to oversample children at risk for poor mathematics outcomes. We included children who met criteria for special education services, except for two nonverbal students who were excluded. From two study sites, we collected data from 15 classrooms within three such schools across two urban centers (one in the Midwest and one in a Rocky Mountain state). The school-wide rate of free and reduced meal program enrollment was >91% in the schools at the Midwest site and 49% at the Rocky Mount site. At each site, 40 or 45 participants enrolled. Our final sample was 85 children ranging in age from 4 years 0 months to 5 years 3 months (M=4 years 9 months, SD=3.74 months).

At each site, we administered two baseline measures during the fall of prekindergarten. The first measure was the Minnesota Executive Function Scale (MEFS; Carlson & Zelazo, 2014), which is an adaptive, iPad-administered, virtual dimensional card sort task that increases in difficulty based on up to seven levels of EF demands. We used the highest level passed on the MEFS as the EF variable in our analyses. This score can be converted to an age-referenced percentile score, but we used highest level passed in the descriptive summary of discordance that follows. The second measure was a 20-item research-based early mathematics assessment (REMA-Brief, Clements, Sarama & Wolfe, 2008/2016). The REMA-Brief is an interactive assessment of early mathematics skills including counting, set size matching, number comparison, cardinality, and shape identification. We used total percent accuracy as a performance measure on this task.

The MEFS and REMA-Brief scores were comparable across sites, so data were combined. Both measures showed a wide range of performance levels (Table 1). The REMA-Brief was normally distributed; although there were no outliers among the MEFS scores, the distribution was slightly right-skewed, so a Spearman *Rho* was computed for the correlation, using highest level passed. As anticipated, these two scores were correlated with each other, $r_s(83) = 0.273$, $P = 0.011$. Age was not correlated with either measure, Ps > 0.08.

Table 1 MEFS and REMA-Brief Scores Among PreK Students ($N = 85$)

	Mean	*SD*	Range
MEFS—highest level passed	3.07	1.34	0–5
MEFS—national percentile score	31.98	22.97	<1–82
REMA-Brief % correct	35	13	10–67

This correlation is not surprising. It replicates correlations of similar magnitude, as reported earlier (e.g., Jacob & Parkinson, 2015) and replicates findings of concurrent correlations in preschool children specifically. But unlike other studies, here we evaluate discordance between EF and mathematics performance. Toward this goal, we first divided our group of 85 participants into "low" vs "high" performers on the MEFS and REMA, separately, using a group-based median split for each set of scores. The rate of discordance was comparable between sites, so the data were combined. The uneven distribution across cells supports the notion that, because the data are positively correlated, the frequency distribution across cells was not random, X^2 (1, $N = 85$) = 4.289, $P < 0.05$. Still, 33 of the 85 participants had a discordant profile (shown in shaded cells), suggesting that the correlation may not be meaningful for 39% of individuals (Table 2).

A conclusion of a high discordance rate in this case is, however, premature, because some "discordant" cases in Table 2 represent individuals with average scores that hover around the midpoint. Thus, an inflated discordance rate is a likely artifact of relying on a median split to test discordance.

As a more conservative alternative, we divided our sample to capture the bottom and top quartiles, and a large average group, for each measure separately, representative of a normal distribution. The range of MEFS scores support the appropriateness of "low" and "high EF" groups designations in this sample, because the national percentiles for these scores ranged from <1% to the 80th percentile; the mean percentile score was, however, well below 50% (Table 1), consistent with our deliberate oversampling of at-risk students. REMA-Brief scores were not nationally normed, but the range of scores was wide and normally distributed. Adopting this approach and more stringent criteria for discordance (depicted by shaded cells), 10 of the 85 participants (11.7%) were discordant for EF and math performance levels (Table 3). On the one hand, this discordance level is low, and some may dismiss it as inconsequential. On the other hand, the rate is comparable with the reported incidence of mathematics learning disabilities, which is approximately 6%–13% of all school-age children (e.g., Barbaresi, Katusic,

Table 2 Number of Children Per Each of Four Cells Based on Sample Median ($N = 85$)

	REMA-Brief	
MEFS Level	**Low**	**High**
Low	25	18
High	15	27

Table 3 Number of Children Per Each of Nine Cells Based on Sample Quartiles ($N = 85$)

	MEFS Level		
REMA-Brief	**Low**	**Middle**	**High**
Low	6	7	8
Middle	8	15	19
High	2	2	18

Colligan, Weaver, & Jacobsen, 2005; Shalev, 2007). The rate is also consistent with the 9% rate of students identified by Willoughby and colleagues as at risk for poor academic achievement based on inadequate growth in EF during preschool (2016). Moreover, this "discordant" group is one-third the size of the strictly *concordant* pairs (low/low and high/high), and so cannot be said to represent statistical outliers. In this sample, it was not unusual for some children to have high EF and low mathematics scores, suggesting that EF is beneficial but not sufficient for early mathematics, for some children.

It is unclear what the implications are for children who fall in one discordant group or the other, whether a classification like this is stable over time, whether discordance stability is even relevant to early childhood mathematics education goals, or whether other criteria for discordance are more sound. (We return to these questions in our conclusion.)

2.2 Primary School Children From an Average SES Sample

It is also unclear whether this discordance pattern would emerge in other samples, so we explored a second sample drawn from an existing data set from a previously reported longitudinal study (Mazzocco & Kover, 2007). This second sample differed from the first sample in two important ways. First, it was *not* derived from a population at high risk for poor mathematics achievement; instead, the sample was drawn from seven suburban schools that had relatively low indices of annual mobility (7%–20%) and free and reduced meal eligibility (2%–29%). Second, the sample was older but nevertheless fell within the early childhood period (Grades 1 and 3). Specifically, the sample was comprised of 190 first graders ranging in age from 6.05 to 7.43 years (*M* age = 6 years 9 months, *SD* = 3.7 months), who were also seen 2 years later, in Grade 3 (*M* age = 8 years 8 months, *SD* = 3.6 months). This afforded an opportunity to evaluate discordance rates across time (in this sample) and across two samples of different ages. However, we are cautious in our inferences between these two samples of convenience, because they also differed in the measures used to assess EF and math.

At both grades, EF skills were measured via the Contingency Naming Test (CNT), an experimental Stroop-like task that requires naming colored geometric shapes according to four increasingly challenging rules (Anderson, Anderson, Northam, & Taylor, 2000). For example, the third rule, referred to as the "one-attribute switching rule," imposes EF demands requiring that children name the color *or* shape of geometric stimuli that contain an inner shape, such as ✪, depending on this instruction: "if the inner and outer shapes match, name the color; if the inner and outer shapes do not match, name the outer shape." Moreover, the naming trial is explicitly timed, and children are asked to answer as quickly as possible "trying not to make mistakes." (Children complete several learning and practice trials before the experimental trials.) The "two-attribute switching rule" builds on the former rule, by adding, "but only follow this rule if there is no backwards arrow over a design. If there is a backward arrow over the design, then you do the rule the backwards way…" by reversing the one-attribute instructions. As exemplified, these rules involve cognitive shifting and increased working memory loads. In Grade 1, the one-attribute rule poses a developmentally appropriate challenge, whereas most first graders cannot complete trials that require applying the the two-attribute rule. By Grade 3, most children can complete the two-attribute trial, albeit with difficulty (Mazzocco & Kover, 2007). These CNT trials each yield an efficiency score reflecting speed and accuracy, with higher scores indicating more efficient performance.

In Grade 1, we obtained mathematics scores from the Test of Early Mathematics Ability—2nd Edition (Ginsburg & Baroody, 1990), a standardized measure normed for ages 3–8 years. The TEMA-2 consists of a wide range of primary school level counting, arithmetic, and number line-related items and yields an age-referenced standardized composite score. At Grade 3, the TEMA-2 is subject to ceiling effects, so we used the Woodcock Johnson Calculation subtest instead. The number of correct items solved on this paper-and-pencil computation task is converted to an age-referenced standard score.

Table 4 TEMA-2 and CNT Scores Among Students at Grades 1 and 3 ($N = 190$)

	Mean	*SD*	Range
Gr 1 CNT one-attribute switch trial—efficiency	0.76	0.45	0.14–1.96
Gr 1 TEMA-2 composite score	107.78	12.85	80–140
Gr 3 CNT two-attribute switch trial—efficiency	0.82	0.42	0.19–2.17
Gr 3 WJR—calculation standard score	109.41	17.48	62–144

Table 5 Number of Children Per Cell Based on Sample Quartiles ($N = 190$)

(A) Grade 1				(B) Grade 3			
	CNT 1 Attribute				CNT 2 Attribute		
TEMA-2	**Low**	**Middle**	**High**	**WJR Calc**	**Low**	**Middle**	**High**
Low	17	24	6	Low	20	27	3
Middle	23	52	24	Middle	22	41	30
High	7	19	18	High	6	26	15

Like the prekindergarten data reported earlier, first and third graders in this study (one group of children seen longitudinally) showed a wide range of performance levels on their EF and math scores (Table 4). As anticipated, their CNT scores were correlated with TEMA-2 scores at Grade 1, $r_s(188) = 0.332$, $P < 0.001$ and with WJR calculation scores at Grade 3, $r_s(188) = 0.315$, $P < 0.001$. (Age was correlated with TEMA-2 scores at Grade 1, r_s (188) $= -0.208$, $P = 0.004$, and with WJR calculation scores at Grade 3, r_s (188) $= -0.153$, $P = 0.035$, but not with CNT scores at Grade 1 or 3, $Ps > 0.08$.)

Using the same discordance criteria reflected in Table 3, we found that the discordant rate in this sample was 6.8% in Grade 1. This is more than one-third the rate of strict concordant cells (Table 5A; Fig. 1) and cannot be considered a result of outliers. The cells were roughly equally divided across those with low EF/high math and high EF/low math. In Grade 3, the discordance rate was only 4.7%, but it would be irresponsible to speculate whether these discordance rates actually reflect any decline with age, given the very different study samples and different measures of EF and mathematics used across our examples.

Our main position is simply that EF/mathematics discordance should be examined more frequently in empirical studies, even if just descriptively, as a possible indicator of specific at-risk status for children with special learning needs. In an informal review of over 40 recent publications on the EF and

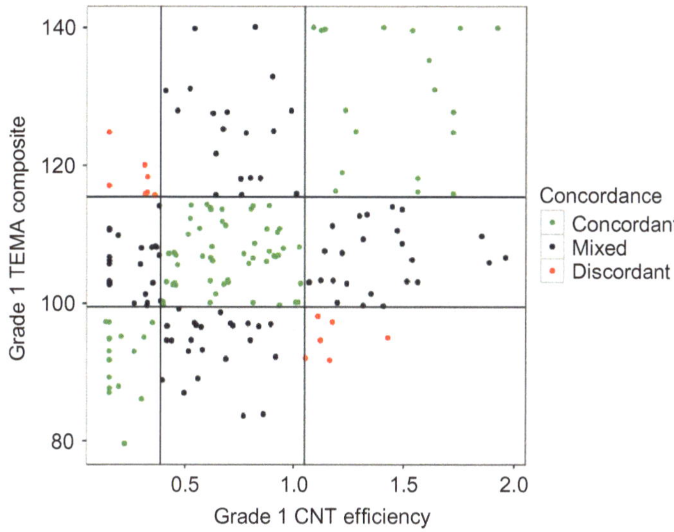

Fig. 1 Scatterplot depicting data for strict discordance criteria (and lenient concordance criteria) reported in Table 5A.

mathematics relation, we found little if any direct discussion of discordance on measures of EF and mathematics, with a few exceptions. In a recent longitudinal study of early math ability, Blair et al. (2016) noted that EF skills at 4 years moderated growth in math ability from preschool to kindergarten, such that preschoolers with low math ability in preschool fared better than expected in kindergarten if their EF at preschool was higher vs lower. This discordant combination improved the odds of children at risk based on low mathematics performance. (Teacher–child relationships also moderated the relation between early and later mathematics.) Next we consider this discordance as one of the three hypothetical variations in EF/mathematics relations that may reflect distinct profiles of individual differences in mathematics achievement, including various profiles of mathematics learning difficulties.

2.3 Three Models of Mathematics Difficulty Based on EF and Mathematics Relations

In Fig. 2, we illustrate how EF and mathematics skills may broadly conform to one of the four alternative combinations depicted in four separate panels, three of which may be associated with mathematics difficulties. Across all panels, the terms "EF" and "mathematics ability" are used here to generally

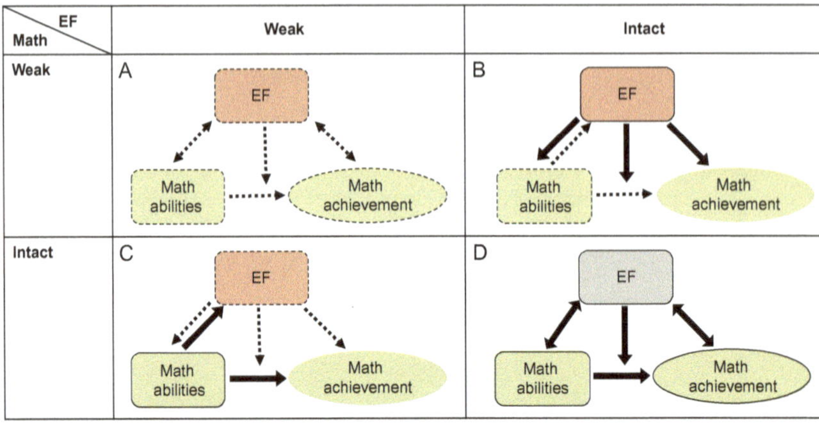

Fig. 2 A conceptual model of the relations among executive function (EF) skills (e.g., cognitive flexibility, inhibitory control, or working memory, or a composite thereof); mathematics abilities (a range of specific abilities such as number processing or spatial rotation); and formal school mathematics achievement. The panels collectively depict four variations of potential outcomes resulting from weak vs intact EF and math abilities.

refer to a range of possible specific measures or constructs of EF, specific mathematics abilities such as symbolic number knowledge, as contributions to formal school mathematics achievement typically measured via global standardized testing. This means that the terms EF and math can theoretically be substituted, with more specific terms, depending on which aspect of EF or math are related to each other, in Fig. 2. Within each panel of Fig. 2, shapes depict these constructs of EF and mathematics abilities, and formal mathematics achievement. (Again, each model may theoretically apply to either specific EF and mathematics abilities or to a composite within either domain.) Solid outlines indicate *intact* abilities and cognitive processes, whereas dashed outlines and arrows indicate *weak* abilities and cognitive processes. The arrows depict relations between these three constructs.

2.3.1 Concordant Profiles

One combination, Panel (D), conforms to the general view that high levels of EF and mathematics do cooccur. It may represent that intact EF and mathematics skills synergistically support each other, leading to age appropriate or even above-average mathematics achievement (e.g., Hoard, Geary, Byrd-Craven, & Nugent, 2008); that one of these skills supports the other; or that additional nondepicted factors mediate the relations implicated by the arrows. These scenarios are supported (but not specified) by data on positive correlations between EF and mathematics, and the outcome depicted in Panel (D) represents the most desirable of the four depicted variations from the perspective of individuals' mathematics achievement. It also represents the expected concordance implied by a significant correlation statistic, but it does not rule out that the concordance is influenced by other cooccurring factors.

Another combination, Panel (A), also represents concordance, wherein both EF and mathematics abilities are weak. It is consistent with evidence that students with numerical difficulties in kindergarten do more poorly if their EF skills are also weak (e.g., Blair et al., 2016; Hassinger-Das et al., 2014), and it may represent a relatively common manifestation of mathematical learning difficulties. It bears reminding that Panels (A) and (D) each represent concordance on a continuum such that, when combined, spans the full range of "deficient" to "highly capable" EF and mathematics abilities.

2.3.2 Discordant Profiles

Whereas Panel (A) may represent the "expected" profile of a young student struggling with mathematics—and it may indeed account for many children who are at risk for poor mathematics outcomes, as the research suggests (Alloway & Passolunghi, 2011; Geary, 2011)—the discordant profiles depicted in Panels (B) and (C) also represent different constellations of student abilities and learning needs that other students with mathematics difficulties may experience. In these panels, the absence of an outline (solid or dashed) for the construct *math achievement* serves to depict the potential variability in mathematics achievement outcomes. Unlike in Panels (A) and (D), here it is unclear whether that outcome will be positive (intact math achievement) or negative (poor achievement). This uncertainty stems from the discordance between intact EF and weak math abilities (Panel B) or the reverse of this (Panel C), along respective continua; and the absence of an arrow leading from *mathematics achievement* to EF in the figure results from this unknown directionality.

Specifically, Panel (B) illustrates that intact EF skills may support weak or aberrant mathematics abilities in a compensatory manner (such as the "low-achieving" group described by Geary, 2011, in contrast to children with math learning disability, or the children in Blair et al.'s 2016 study, reported earlier), such as intact attention skills facilitating mathematics learning (Geary, 2013). This profile may describe students with special needs who are identified as having a specific learning disability in mathematics (or, for younger students, at-risk status) based on slow, effortful problem solving and below average to average mathematics scores, but it may also result in unidentified learning needs that are masked by age-appropriate achievement resulting from compensatory strategies and successful effort. The latter scenario is empirically supported in studies of children with neurodevelopmental profiles whose numerical processing deficits appear to be overcome by intact verbal memory skills, at least during select phases of development (such as when overlearning math facts; e.g., Murphy & Mazzocco, 2008), at least for some children. An alternative outcome corresponding to Fig. 2B is that weak mathematics skills (such as symbolic number automaticity) lead to an increase in the EF demands of mathematics tasks, such as by overloading working memory, such that intact working memory skills are insufficient for supporting mathematics learning or performance.

A fourth scenario, depicted by Panel (C), is that EF limitations interfere with otherwise intact mathematics abilities, especially for tasks that inherently require greater EF demands. In early childhood, such tasks may involve counting large or nonlinear item sets; in older children or adults, they may involve exact vs approximate arithmetic (Kalaman & Lefevre, 2007), or long division vs basic fact fluency. Although *manifestation* for Panels (B) and (C) may be effortful, inefficient, and error prone mathematics performance, the underlying explanations for this outcome differs. That is, weak EF skills *or* weak basic mathematics abilities are the primary explanation for poor mathematics outcomes across the two situations. Thus, these scenarios are aligned with evidence of either *domain general* underpinnings of poor mathematics outcomes (Panel C), such as mathematics learning disabilities observed in children with neurofibromatosis Type 1 or fragile X syndrome that are secondary to nonmathematical cognitive or neurologic sequela; or weak abilities more specific to mathematics (Panel B), such as numerical or spatial processing deficits or delays that characterize Williams and Turner syndromes (e.g., reviewed by Dennis, Berch, & Mazzocco, 2009). There is growing acceptance that domain-general skills play essential roles in mathematics abilities and disabilities (Hohol, Cipora, Willmes, & Nuerk, 2017) and that their *relative* contribution to mathematics difficulties is subject to developmental and individual differences (e.g., Geary, 2011). Importantly, the pathways to mathematics difficulties observed in neurodevelopmental disorders are models that may apply to other children whose special learning needs are not linked to a known neurodevelopmental condition (Dennis et al., 2009).

2.4 Caveats Concerning Figs. 1 and 2

Three major issues that we do not address here concern the criteria for classifying discordance and the factors that influence its likelihood. The first issue was illustrated in Fig. 1, wherein some data points within cells designated as "mixed" may represent individuals whose profile may be more aligned with discordant vs concordant performance levels. Cut points are prone to such problematic classification outcomes. The second issue pertains to all four profiles depicted in Fig. 2; it concerns the likelihood that the existence, direction, and strength of EF/mathematics relations will vary depending on what precise mathematics or EF skills are measured (and if each is measured as a static performance level or indicator of growth), the developmental period(s) in which the skills are measured, and other factors discussed previously in this chapter (and elsewhere in this volume) that we do not address in the model. Finally, the model in Fig. 2 does not address that the relation between mathematics and EF may be direct or indirect, moderated or mediated by these and other factors, although each of these types of relations between EF and mathematics has been reported (e.g., Cragg, Keeble, Richardson, Roome, & Gilmore, 2017). All three of these issues should be addressed when determining the nature of individuals' discordant performance levels on EF and mathematics tasks and its significance.

3. CONCLUSION

Our emphasis on discordance does not negate the existing evidence for a correlation between EF and mathematics, nor its importance. As Fig. 1 and dozens of studies have shown, the correlations are repeatedly replicated because this predominantly linear relation *does* emerge among many samples, with many relevant measures. One open question concerns what causal mechanisms explain it. Another question, which we address informally—and which may pertain to a small proportion but meaningful number of individuals—is what to make of cases that do not conform to the correlation. We must be mindful that nonconforming profiles may include students with special learning needs whose mathematics achievement is at risk. The implications of these profiles must be considered when designing instructions or selecting interventions for individuals. It would be unfortunate, and a bit ironic, if interventions based primarily on central tendencies are intended for children whose individual profile conforms to another tendency altogether.

ACKNOWLEDGMENTS

All authors are either members (M.M.M.M.) or affiliates (J.Y.-C.C., A.M.B.) of the Development and Research on Early Mathematics Education (DREME) Network funded by Heising-Simons. The authors thank DREME affiliates Taylor Praus and Crystal Day-Hess for their project management roles for the study on which we base the first illustrated case study in this chapter. This work was partially supported by the Heising-Simons Foundation.

REFERENCES

Aldrich, J. (1995). Correlations genuine and spurious in Pearson and Yule. *Statistical Science, 10*(4), 364–376.

Alloway, T. P., & Passolunghi, M. C. (2011). The relationship between working memory, IQ, and mathematical skills in children. *Learning and Individual Differences, 21*(1), 133–137. http://doi.org/10.1016/j.lindif.2010.09.013.

Anderson, P., Anderson, V., Northam, E., & Taylor, H. G. (2000). Standardization of the contingency naming test (CNT) for school-aged children: A measure of reactive flexibility. *Clinical Neuropsychological Assessment, 1*, 247–273.

Barbaresi, W. J., Katusic, S. K., Colligan, R. C., Weaver, A. L., & Jacobsen, S. J. (2005). Math learning disorder: Incidence in population-based birth cohort, 1976-82, Rochester, Minn. *Ambulatory Pediatrics, 5*(5), 281–289.

Best, J. R., Miller, P. H., & Naglieri, J. A. (2011). Relations between executive function and academic achievement from ages 5 to 17 in a large, representative national sample. *Learning and Individual Differences, 21*(4), 327–336.

Blair, C., McKinnon, R., & The Family Life Project Investigators. (2016). Moderating effects of executive functions and the teacher-child relationship on the development of mathematics ability in kindergarten. *Learning and Instruction, 41*, 85–93. 10.1016/j.learninstruc.2015.10.001.

Bull, R., Johnston, R. S., & Roy, J. A. (1999). Exploring the roles of the visual-spatial sketch pad and central executive in children's arithmetical skills: Views from cognition and developmental neuropsychology. *Developmental Neuropsychology, 15*(3), 421–442. http://doi.org/10.1080/87565649909540759.

Bull, R., & Lee, K. (2014). Executive functioning and mathematics achievement. *Child Development Perspectives, 8*(1), 36–41. http://doi.org/10.1111/cdep.12059.

Bull, R., & Scerif, G. (2001). Executive functioning as a predictor of children's mathematics ability: Inhibition, switching, and working memory. *Developmental Neuropsychology, 19*(3), 273–293. http://doi.org/10.1207/S15326942DN1903.

Carlson, S. M., & Zelazo, P. D. (2014). *Minnesota executive function scale: Test manual*. St. Paul, MN: Reflection Sciences, Inc.

Carlson, S. M., Zelazo, P. D., & Faja, S. (2013). Executive function. In P. D. Zelazo (Ed.), *Body and mind: Vol. 1. The Oxford handbook of developmental psychology* (pp. 706–743). New York: Oxford University Press. http://doi.org/10.1093/oxfordhb/9780199958450.013.0025.

Chu, F. W., vanMarle, K., & Geary, D. C. (2016). Predicting children's reading and mathematics achievement from early quantitative knowledge and domain-general cognitive abilities. *Frontiers in Psychology*, 7, 1–14. http://doi.org/10.3389/fpsyg.2016.00775.

Clark, C. A., Sheffield, T. D., Wiebe, S. A., & Espy, K. A. (2013). Longitudinal associations between executive control and developing mathematical competence in preschool boys and girls. *Child Development, 84*(2), 662–677.

Clements, D. H., & Sarama, J. (2014). *Learning and teaching early math: The learning trajectories approach* (2nd ed.). New York, NY: Routledge.

Clements, D. H., Sarama, J., & Germeroth, C. (2016). Learning executive function and early mathematics: Directions of causal relations. *Early Childhood Research Quarterly*, *36*, 79–90. http://doi.org/10.1016/j.ecresq.2015.12.009.

Clements, D. H., Sarama, J., & Wolfe, C. B. (2008/2016). *REMA—Research-based early mathematics assessment*. Denver, CO: Kennedy Institute, University of Denver.

Cragg, L., Keeble, S., Richardson, S., Roome, H. E., & Gilmore, C. (2017). Direct and indirect influences of executive functions on mathematics achievement. *Cognition*, *162*, 12–26. http://dx.doi.org/10.1016/j.cognition.2017.01.014.

Dennis, M., Berch, D. B., & Mazzocco, M. M. (2009). Mathematical learning disabilities in special populations: Phenotypic variation and cross-disorder comparisons. *Developmental Disabilities Research Reviews*, *15*(1), 80–89.

Diamond, A. (2013). Executive functions. *Annual Review of Psychology*, *64*, 135–168.

Fuhs, M. W., Hornburg, C. B., & McNeil, N. M. (2016). Specific early number skills mediate the association between executive functioning skills and mathematics achievement. *Developmental Psychology*, *52*, 1217–1235. http://dx.doi.org/10.1037/dev0000145.

Fuhs, M. W., & McNeil, N. M. (2013). ANS acuity and mathematics ability in preschoolers from low-income homes: Contributions of inhibitory control. *Developmental Science*, *16*, 136–148.

Fuhs, M. W., Nesbitt, K. T., Farran, D. C., & Dong, N. (2014). Longitudinal associations between executive functioning and academic skills across content areas. *Developmental Psychology*, *50*, 1698–1709. http://dx.doi.org/10.1037/a0036633.

Garrett, A. J., Mazzocco, M. M. M., & Baker, L. (2006). Development of the metacognitive skills of prediction and evaluation in children with or without math disability. *Learning Disabilities Research and Practice*, *21*(2), 77–88.

Gashaj, V., Uehlinger, Y., & Roebers, C. M. (2016). Numerical magnitude skills in 6-years-old children: Exploring specific associations with components of executive function. *Journal of Educational and Developmental Psychology*, *6*(1), 157–172. http://doi.org/10.5539/jedp.v6n1p157.

Geary, D. C. (2011). Consequences, characteristics, and causes of mathematical learning disabilities and persistent low achievement in mathematics. *Journal of Developmental & Behavioral Pediatrics*, *32*(3), 250–263. http://doi.org/10.1097/DBP.0b013e318209edef.

Geary, D. C. (2013). Early foundations for mathematics learning and their relations to learning disabilities. *Current Directions in Psychological Science*, *22*(1), 23–27. http://doi.org/10.1177/0963721412469398.

Geary, D. C., Hoard, M. K., & Nugent, L. (2012). Independent contributions of the central executive, intelligence, and in-class attentive behavior to developmental change in the strategies used to solve addition problems. *Journal of Experimental Child Psychology*, *113*, 49–65. http://doi.org/10.1016/j.jecp.2012.03.003.

Geary, D. C., Hoard, M. K., Nugent, L., & Bailey, D. H. (2013). Adolescents' functional numeracy is predicted by their school entry number system knowledge. *PLoS One*, *8*(1), e54651.

Ginsburg, H. P., & Baroody, A. J. (1990). *Test of early mathematics ability* (2nd ed.). Austin, TX: Pro-Ed.

Hassinger-Das, B., Jordan, N. C., Glutting, J., Irwin, C., & Dyson, N. (2014). Domain-general mediators of the relation between kindergarten number sense and first-grade mathematics achievement. *Journal of Experimental Child Psychology*, *118*(1), 78–92. http://doi.org/10.1016/j.jecp.2013.09.008.

Hoard, M. K., Geary, D. C., Byrd-Craven, J., & Nugent, L. (2008). Mathematical cognition in intellectually precocious first graders. *Developmental Neuropsychology*, *33*(3), 251–276. http://doi.org/10.1080/87565640801982338.

Hohol, M., Cipora, K., Willmes, K., & Nuerk, H.-C. (2017). Bringing back the balance: Domain-general processes are also important in numerical cognition. Frontiers in Psychology, *8*, 499. http://dx.doi.org/10.3389/fpsyg.2017.00499.

Hornung, C., Schiltz, C., Brunner, M., & Martin, R. (2014). Predicting first-grade mathematics achievement: The contributions of domain-general cognitive abilities, nonverbal number sense, and early number competence. *Frontiers in Psychology*, *5*, 272. http://doi.org/10.3389/fpsyg.2014.00272.

Hubber, P. J., Gilmore, C., & Cragg, L. (2014). The roles of the central executive and visuospatial storage in mental arithmetic: A comparison across strategies. *The Quarterly Journal of Experimental Psychology*, *67*(5), 936–954.

Jacob, R., & Parkinson, J. (2015). The potential for school-based interventions that target executive function to improve academic achievement: A review. *Review of Educational Research*, *41*, 41. http://doi.org/10.3102/0034654314561338.

Kalaman, D. A., & Lefevre, J.-A. (2007). Working memory demands of exact and approximate addition. *European Journal of Cognitive Psychology*, *19*(2), 187–212. http://doi.org/10.1080/09541440600713445.

Martin, R. B., Cirino, P. T., Sharp, C., & Barnes, M. (2014). Number and counting skills in kindergarten as predictors of grade 1 mathematical skills. *Learning and Individual Differences*, *34*, 12–23.

Mazzocco, M. M. M., & Kover, S. T. (2007). A longitudinal assessment of executive function skills and their association with math performance. *Child Neuropsychology*, *13*, 18–45. http://doi.org/10.1080/09297040600611346.

Miyake, A., Friedman, N. P., Emerson, M. J., Witzki, A. H., Howerter, A., & Wager, T. D. (2000). The unity and diversity of executive functions and their contributions to complex "frontal lobe" tasks: A latent variable analysis. *Cognitive Psychology*, *41*, 49–100.

Morgan, P. L., Li, H., Farkas, G., Cook, M., Pun, W. H., & Hillemeier, M. M. (2016). Executive functioning deficits increase kindergarten children's risk for reading and mathematics difficulties in first grade. *Contemporary Educational Psychology*. (in press). http://doi.org/10.1016/j.cedpsych.2016.01.004.

Murphy, M. M., & Mazzocco, M. M. M. (2008). Rote numeric skills may mask underlying mathematical disabilities in girls with fragile x syndrome. *Developmental Neuropsychology*, *33*(3), 345–364. http://doi.org/10.1080/87565640801982429.

Prager E. O., Executive function and early numeracy in preschoolers: Can training help? Under review.

Prager, E. O., Sera, M. D., & Carlson, S. M. (2016). Executive function and magnitude skills in preschool children. *Journal of Experimental Child Psychology*, *147*, 126–139. http://doi.org/10.1016/j.jecp.2016.01.002.

Purpura, D. J., Schmitt, S. A., & Ganley, C. M. (2017). Foundations of mathematics and literacy: The role of executive functioning components. *Journal of Experimental Child Psychology*, *153*, 15–34. http://doi.org/10.1016/j.jecp.2016.08.010.

Sarama, J., & Clements, D. H. (2009). *Early childhood mathematics education research: Learning trajectories for young children*. New York, NY: Routledge.

Shalev, R. S. (2007). Prevalence of developmental dyscalculia. In D. B. Berch & M. M. M. Mazzocco (Eds.), *Why is math so hard for some children? The nature and origins of mathematical learning difficulties and disabilities* (pp. 49–60). Baltimore, MD: Paul H Brookes Publishing.

Welsh, J. A., Nix, R. L., Blair, C., Bierman, K. L., & Nelson, K. E. (2010). The development of cognitive skills and gains in academic school readiness for children from low-income families. *Journal of Educational Psychology*, *102*(1), 43–53. http://doi.org/10.1037/a0016738.

Willoughby, M. T., Kupersmidt, J. B., & Voegler-Lee, M. E. (2012a). Is preschool executive function causally related to academic achievement? *Child Neuropsychology*, *18*(1), 79–91. http://doi.org/10.1080/09297049.2011.578572.

Willoughby, M. T., Magnus, B., Vernon-Feagans, L., Blair, C. R., & The Family Life Project Investigators. (2016). Developmental delays in executive function from 3 to 5 years of age predict kindergarten academic readiness. *Journal of Learning Disabilities*, 1–14. https://doi.org/10.1177/0022219415619754.
Willoughby, M. T., Wirth, R. J., & Blair, C. B. (2012b). Executive function in early childhood: Longitudinal measurement invariance and developmental change. *Psychological Assessment*, *24*, 418–431.

Papier | Fördert gute Waldnutzung
FSC® C083411

Kolibri 360 GmbH
Ferdinand-Jühlke-Straße 7
99095 Erfurt, Deutschland
produktsicherheit@kolibri360.de